OPTICAL FIBER COMMUNICATIONS

McGraw-Hill Series in Electrical Engineering

Consulting Editor
Stephen W. Director, Carnegie-Mellon University

Networks and Systems
Communications and Information Theory
Control Theory
Electronics and Electronic Circuits
Power and Energy
Electromagnetics
Computer Engineering
Introductory and Survey
Radio, Television, Radar, and Antennas

Previous Consulting Editors

Communications and Information Theory

OPTICAL
FIBER
COMMUNICATIONS

Gerd Keiser

GTE Communications
Systems Division

McGraw-Hill Book Company

New York   St. Louis   San Francisco   Auckland   Bogotá   Hamburg
Johannesburg   London   Madrid   Mexico   Montreal   New Delhi
Panama   Paris   São Paulo   Singapore   Sydney   Tokyo   Toronto

This book was set in Times Roman by Interactive Composition Corporation.
The editors were T. Michael Slaughter and Madelaine Eichberg;
the production supervisor was Leroy A. Young.
The drawings were done by Fine Line Illustrations, Inc.
The cover was designed by Myrna Sharpe.
Halliday Lithograph Corporation was printer and binder.

OPTICAL FIBER COMMUNICATIONS

1234567890HALHAL89876543

ISBN 0-07-033467-6

Library of Congress Cataloging in Publication Data

Keiser, Gerd.
    Optical fiber communications.

    (McGraw-Hill series in electrical engineering.
Communications and information theory)
    Includes bibliographical references and index.
    1. Optical communications.   2. Fiber optics.
I. Title.   II. Series.
TK5103.59.K44   1983   621.381'0414   82-18655
ISBN 0-07-033467-6

CONTENTS

PREFACE

This book has been written specifically as a textbook with the purpose of providing the basic material for an introductory senior or first-year graduate course in the theory and application of optical fiber communication technology. It will also serve well as a working reference for practicing engineers dealing with optical fiber communication system designs. To aid students in learning the material and in applying it to practical designs, a collection of 150 homework problems is included. These problem sets are an important and integral part of this book. They are intended not only to help readers test their comprehension of the material covered but also to extend and elucidate the text. To assist the instructor the problem sets are conveniently grouped at the end of each chapter. A solution manual to all the problems is available to the instructor from the publisher.

The background required to study the book is only that of typical senior-level engineering students. Specifically it is assumed that the student has been introduced to electromagnetic theory, calculus and elementary differential equations, basic concepts of optics as presented in a freshman physics course, and the basic concepts of electronics. Courses in communication theory and solid-state physics are not essential but would be helpful for gaining a full understanding of the material in this book. Realizing that students tend to forget much of this material, mainly because of a lack of opportunity to apply it in later courses, I have included introductory reviews of several of the major topics. Some of these, such as the principles of optics, are included in the main body of the text, while others, such as semiconductor physics and communication theory, exist as appendixes at the end of the book.

The book is organized to give a clear and logical sequence of topics. It progresses systematically from descriptions of the individual elements of an optical communication system to an analysis of system designs, and ends with discussions of measurement techniques for evaluating components and systems. The introductory chapter gives a brief historical background of communication systems and shows how the operating region of optical fibers fits into the electromagnetic spectrum. It also includes an overview of the fundamental components of an optical fiber link and the types of networks that can be constructed with these devices. Chapter 1 also notes the advantages of optical fibers relative to other transmission media, and points to application areas showing rapidly growing interest in optical fiber technology.

Chapter 2 begins with a review of the basic laws and definitions of optics that are relevant to optical fibers. Following a description of optical fiber structure, two methods are used to show how a fiber guides light. The first approach uses the geometrical or ray optics concepts of light reflection and refraction to provide an intuitive picture of the light propagation mechanism. In the second approach light is treated as an electromagnetic wave that propagates along the optical fiber waveguide. This is analyzed exactly for a step-index fiber by solving Maxwell's equations subject to the cylindrical boundary conditions of the fiber. For a graded-index fiber an approximation based on the WKB method is used.

Signal degradation arising from attenuation and distortion mechanisms in fibers is addressed in Chap. 3. We first describe signal attenuation causes, such as absorption, scattering, and radiation of optical power. We then analyze the major mechanisms that cause a signal to become distorted as it travels along a fiber.

To transmit data through a fiber waveguide, an optical source compatible with the fiber size and transmission properties is required. This is the subject of Chap. 4 where the structures, materials, and operating characteristics of various types of semiconductor light sources are discussed. The question of how to couple the light emitted from these sources into an optical fiber is dealt with in Chap. 5. To have this launched optical power propagate over long distances, several lengths of fiber cable must be joined together to form a long, continuous transmission chain. The effects on optical power loss resulting from mechanical misalignments at fiber-to-fiber joints and from mismatched fiber geometries and structures are also addressed in Chap. 5.

When the optical power emerges from the end of the optical fiber it must be converted into an electric signal. This function is performed by a photodetector which is described in Chap. 6. Here a discussion of the photodetection principles is followed by an examination of specific detector performance parameters for *pin* and avalanche photodiodes, which are the two types of detectors used almost exclusively in fiber optics systems. Once the optical signal has been reconverted to an electric signal, the receiver restores the fidelity to the signal which has become attenuated and distorted as it traversed the transmission channel. The factors associated with this function are addressed in extensive detail in Chap. 7.

The combination of all the pieces discussed above results in an optical fiber communication system. The possible link configurations (including data buses and WDM architectures), performance tradeoffs, component selections, and line-coding schemes needed for implementing a total system are described in Chap. 8. To verify that the components used meet certain performance specifications and to check the proper operation of the installed system, various measurements must be made. Measurement techniques are therefore examined in Chap. 9.

As a final topic, Chap. 10 describes the various materials used to make fibers, shows how fibers are manufactured, analyzes the strengths of optical fibers, and illustrates how optical fibers are incorporated into cable structures.

The inception of this book is attributable to Dr. Tri Ha, GTE International Systems Corporation, who urged me to write it. His encouragement and assistance at various stages of this work were most helpful. I am also indebted to my colleague Mr. Don Rice for numerous technical discussions and for a critical proofreading of part of the manuscript. Special thanks are extended to two anonymous reviewers whose suggestions enhanced the content and organization of the book. Since the development of optical fiber communications is the result of the work of many people from a wide variety of disciplines, this book would not have been possible without the many contributions that have appeared in the open literature which served as the basic source material for this work. I am therefore very grateful to the scientists and engineers who have taken the time and energy to share their results through publication in books and recognized technical journals and conferences. My sincerest thanks also go to Mrs. Maureen Thomas for typing the entire manuscript and its revisions in an exceptionally expert fashion. As a final personal note, I am grateful to my wife Ching-yun and my daughter Nishla for their patience and encouragement during the time I devoted to writing this book.

Gerd Keiser

CHAPTER ONE

OVERVIEW OF OPTICAL FIBER COMMUNICATIONS

Ever since ancient times, one of the principal interests of human beings has been to devise communication systems for sending messages from one distant place to another. The fundamental elements of any such communication system are shown in Fig. 1-1. These elements include at one end an information source which inputs a message to a transmitter. The transmitter couples the message onto a transmission channel in the form of a signal which matches the transfer properties of the channel. The channel is the medium bridging the distance between the transmitter and the receiver. This can be either a guided transmission line such as a wire or waveguide, or it can be an unguided atmospheric or space channel. As the signal traverses the channel, it may be progressively both attenuated and distorted with increasing distance. For example, electric power is lost through heat generation as an electric signal flows along a wire, and optical power is attenuated through scattering and absorption by molecules in an atmospheric channel. The function of the receiver is to extract the weakened and distorted signal from the channel, amplify it, and restore it to its original form before passing it on to the message destination.

1-1 FORMS OF COMMUNICATION SYSTEMS

Many forms of communication systems have appeared over the years. The principal motivations behind each new one were either to improve the transmission fidelity, to increase the data rate so that more information could be sent, or to increase the transmission distance between relay stations. Prior to the nineteenth century all communication systems were of a very low data rate type and basically involved

Figure 1-1 Fundamental elements of a communication system.

only optical or acoustical means, such as signal lamps or horns. One of the earliest known optical transmission links,[1] for example, was the use of a fire signal by the Greeks in the eighth century B.C. for sending alarms, calls for help, or announcements of certain events. Only one type of signal was used, its meaning being prearranged between the sender and the receiver. In the fourth century B.C. the transmission distance was extended through the use of relay stations, and by approximately 150 B.C. these optical signals were encoded in relation to the alphabet so that any message could then be sent. Improvements of these systems were not actively pursued because of technology limitations. For example, the speed of the communication link was limited since the human eye was used as a receiver, line-of-sight transmission paths were required, and atmospheric effects such as fog and rain made the transmission path unreliable. Thus it generally turned out to be faster and more efficient to send messages by a courier over the road network.

The discovery of the telegraph by Samuel F. B. Morse in 1838 ushered in a new epoch in communications—the era of electrical communications.[2] The first commercial telegraph service using wire cables was implemented in 1844 and further installations increased steadily throughout the world in the following years. The use of wire cables for information transmission expanded with the installation of the first telephone exchange in New Haven, Connecticut, in 1878. Wire cable was the only medium for electrical communication until the discovery of long-wavelength electromagnetic radiation by Heinrich Hertz in 1887. The first implementation of this was the radio demonstration by Guglielmo Marconi in 1895.

In the ensuing years an increasingly larger portion of the electromagnetic spectrum was utilized for conveying information from one place to another. The reason for this is that, in electrical systems, data is usually transferred over the communication channel by superimposing the information signal onto a sinusoidally varying electromagnetic wave which is known as the *carrier*. At the destination the information is removed from the carrier wave and processed as desired. Since the amount of information that can be transmitted is directly related to the frequency range over which the carrier wave operates, increasing the carrier frequency theoretically increases the available transmission bandwidth and, consequently, provides a larger information capacity. Thus the trend in electrical communication system developments was to employ progressively higher frequencies (shorter wavelengths), which offered corresponding increases in bandwidth and, hence, an increased information capacity. This activity led, in turn, to the birth of television, radar, and microwave links.

The portion of the electromagnetic spectrum that is used for electrical communications[3] is shown in Fig. 1-2. The transmission media used in this spectrum include millimeter and microwave waveguides, metallic wires, and radio waves. Among the communication systems using these media are the familiar telephone, AM and FM radio, television, CB (citizen's band radio), radar, and satellite links all of which have become a part of our everyday lives. The frequencies of these applications range from about 300 Hz in the audioband to about 90 GHz for the millimeter band.

Another important portion of the electromagnetic spectrum encompasses the

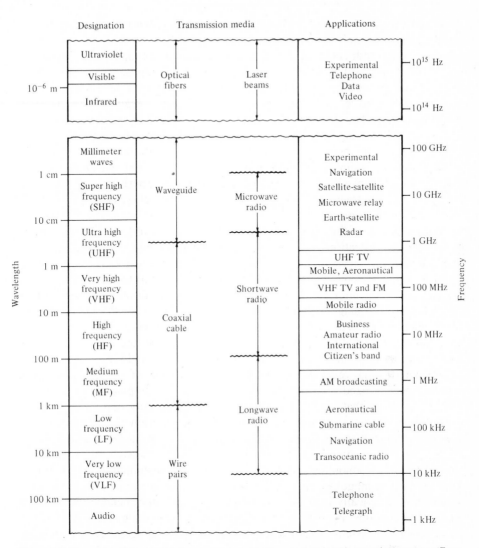

Figure 1-2 Examples of communication systems applications in the electromagnetic spectrum. *(From Carlson,[3] Copyright 1975. Used with the permission of McGraw-Hill Book Company.)*

optical region shown in Fig. 1-2. In this region it is customary to specify the band of interest in terms of wavelength instead of frequency as is used in the radio region. The optical spectrum ranges from about 50 nm (ultraviolet) to approximately 100 μm (far infrared), with the visible spectrum being the 400- to 700-nm region. Similar to the radio-frequency spectrum two transmission media can be used: an atmospheric channel or a guided-wave channel.

A great interest in communication at these optical frequencies was created in 1960 with the advent of the laser[4] which made available a coherent optical source. Since optical frequencies are on the order of 5×10^{14} Hz, the laser has a theoretical information capacity exceeding that of microwave systems by a factor of 10^5, which is approximately equal to 10 million TV channels.

With the potential of such wideband transmission capabilities in mind, a number of experiments[5,6] using atmospheric optical channels were carried out in the early 1960s. These experiments showed the feasibility of modulating a coherent optical carrier wave at very high frequencies. However the high installation expense that would be required, the tremendous costs that would be incurred to develop all the necessary components, and the limitations imposed on the atmospheric channel by rain, fog, snow, and dust make such extremely high-speed systems economically unattractive in view of present demands for communication channel capacity. Nevertheless, numerous developments[7-9] of free-space optical channel systems operating at baseband frequencies are in progress for earth-to-space communications.

Concurrent with the work on atmospheric optical channels were the investigations of optical fibers, since they can provide a much more reliable and versatile optical channel than the atmosphere. Initially, the extremely large losses of more than 1000 dB/km observed in the best optical fibers made them appear impractical. This changed in 1966 when Kao and Hockman[10] and A. Werts[11] almost simultaneously speculated that these high losses were a result of impurities in the fiber material, and that the losses could be reduced to the point where optical waveguides would be a viable transmission medium. This was realized in 1970 when Kapron, Keck, and Maurer[12] of the Corning Glass Works fabricated a fiber having a 20-dB/km attenuation. A whole new era of optical fiber communications was thus launched.

The ensuing development of optical fiber transmission systems grew from the combination of semiconductor technology, which provided the necessary light sources and photodetectors, and optical waveguide technology upon which the optical fiber is based. The result was a transmission link that had certain inherent advantages over conventional copper systems in telecommunications applications. For example, optical fibers have lower transmission losses and wider bandwidths as compared to copper wires. This means that with optical fiber cable systems more data can be sent over longer distances, thereby decreasing the number of wires and reducing the number of repeaters needed over these distances. In addition the low weight and the small hair-sized dimensions of fibers offer a distinct advantage over heavy, bulky wire cables in crowded underground city ducts. This is also of importance in aircraft where small light-weight cables are advantageous and in tactical military applications where large amounts of cable must be unreeled and retrieved rapidly. An especially important feature of optical fibers relates to their dielectric nature. This provides optical waveguides with immunity to electromagnetic interference, such as inductive pickup from signal carrying wires and lightning, and freedom from electromagnetic pulse (EMP) effects, the latter being of particular interest to military applications. Furthermore there is no need to worry about ground loops, fiber-to-fiber cross talk is very low, and a high degree of data security is afforded since the optical signal is well confined within the waveguide (with any emanations being absorbed by an opaque jacketing around the fiber). Of additional importance is the advantage that silica is the principal material of which optical fibers are made. This material is abundant and inexpensive since the main source of silica is sand.

Early recognition of all these advantages in the early 1970s created a flurry of activity in all areas related to fiber optic transmission systems. This resulted in significant technological advances in optical sources, fibers, photodetectors, and fiber cable connectors. By 1980 this activity had led to the development and world-wide installation of practical and economically feasible optical fiber communication systems that carry live telephone, cable TV, and other types of telecommunications traffic. These installations all operate as baseband systems in which the data are sent by simply turning the transmitter on and off. Despite their apparent simplicity such systems have already offered very good solutions to some vexing problems in conventional applications.

1-2 ELEMENTS OF AN OPTICAL FIBER TRANSMISSION LINK

An optical fiber transmission link comprises the elements shown in Fig. 1-3. The key sections are a transmitter consisting of a light source and its associated drive circuitry, a cable offering mechanical and environmental protection to the optical fibers contained inside, and a receiver consisting of a photodetector plus amplification and signal-restoring circuitry. The cabled optical fiber is one of the most important elements in an optical fiber link, as we shall see in Chaps. 2 and 3. In addition to protecting the glass fibers during installation and service, the cable may contain copper wires for powering repeaters which are needed for periodically amplifying and reshaping the signal when the link spans long distances. The cable generally contains several cylindrical hair-thin glass fibers, each of which is an independent communication channel.

Analogous to copper cables, the installation of optical fiber cables can be either aerial, in ducts, undersea, or buried directly in the ground. As a result of installation and/or manufacturing limitations, individual cable lengths will range from several hundred meters to several kilometers for long-distance applications. The actual length of a single cable section is determined by practical considerations such as reel size and cable weight. The shorter lengths tend to be used when the cables are pulled through ducts. Longer cable lengths are used in aerial or direct-burial applications. The complete long-distance transmission line is formed by splicing together these individual cable sections.

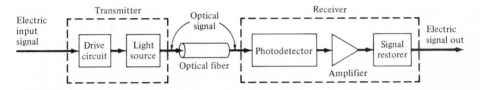

Figure 1-3 Basic elements of an optical fiber transmission link.

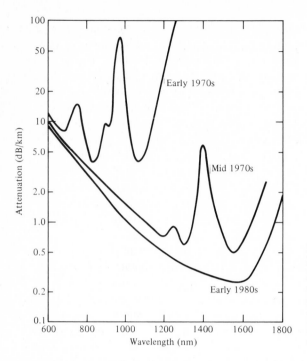

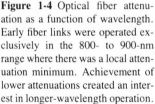

Figure 1-4 Optical fiber attenuation as a function of wavelength. Early fiber links were operated exclusively in the 800- to 900-nm range where there was a local attenuation minimum. Achievement of lower attenuations created an interest in longer-wavelength operation.

One of the principal optical fiber characteristics is its attenuation as a function of wavelength, as shown in Fig. 1-4. Early technology made exclusive use of the 800- to 900-nm wavelength band, since in this region the fibers made at that time exhibited a local minimum in the attenuation curve, and optical sources and photodetectors operating at these wavelengths were available. By reducing the concentrations of hydroxyl ions and metallic ion impurities in the fiber material, fiber manufacturers were eventually able to fabricate optical waveguides with very low losses in the 1100- to 1600-nm region. This spectral bandwidth is usually referred to as the *long-wavelength region*. Increased interest thus developed at the 1300-nm wavelength since this is the region of minimum signal distortion in pure silica fibers. Bell Laboratories, for example, began the first U.S. field trial of a 1300-nm LED-based telephone system in Sacramento, California, on June 26, 1980, and other U.S., European, and Japanese telecommunications companies are actively experimenting with advanced long-wavelength trial installations.

Research has also commenced[13] on new types of fiber materials for use in the 3- to 5-μm wavelength band. The initial interest centered on the polycrystalline metal halides such as zinc chloride ($ZnCl_2$), thallium bromide (TlBr), and thallium bromoiodide (known as KRS-5). Theoretical predictions expect these fibers to have attenuations of less than 0.01 dB/km. The successful fabrication of fibers having these types of losses can have a profound effect on long-distance communications.

Once the cable is installed a light source which is dimensionally compatible with the fiber core is used to launch optical power into the fiber. Semiconductor light-emitting diodes (LEDs) and laser diodes are suitable transmitter sources for this purpose since their light output can be modulated rapidly by simply varying the bias current. The electric input signals to the transmitter can be either of an analog or of a digital form. The transmitter circuitry converts these electric signals to an optical signal by varying the current flow through the light source. An optical source is a *square-law device,* which means that a linear variation in drive current results in a corresponding linear change in the optical output power. In the 800- to 900-nm region the light sources are generally alloys of GaAlAs. At the longer wavelengths (1100 to 1600 nm), an InGaAsP alloy is the principal optical source material. These optical sources are discussed in detail in Chap. 4.

After an optical signal has been launched into the fiber, it will become progressively attenuated and distorted with increasing distance because of scattering, absorption, and dispersion mechanisms in the waveguide. At the receiver the attenuated and distorted modulated optical power emerging from the fiber end will be detected by a photodiode. Analogous to the light source, the photodetector is also a square-law device since it converts the received optical power directly into an electric current output (photocurrent). Semiconductor *pin* and avalanche photodiodes (APDs) are the two principal photodetectors used in a fiber optic link. Both device types exhibit high efficiency and response speed. For applications in which a low-power optical signal is received, an avalanche photodiode is normally used since it has a greater sensitivity owing to an inherent internal gain mechanism (avalanche effect). Silicon photodetectors are used in the 800- to 900-nm region. A variety of optical detectors are potentially available at the longer wavelengths. The prime material candidate in the 1100- to 1600-nm region is an InGaAs alloy. We address these photodetectors in Chap. 6.

The design of the receiver is inherently more complex than that of the transmitter since it has to both amplify and reshape the degraded signal received by the photodetector. The principal figure of merit for a receiver is the minimum optical power necessary at the desired data rate to attain either a given error probability for digital systems or a specified signal-to-noise ratio for an analog system. As we shall see in Chap. 7, the ability of a receiver to achieve a certain performance level depends on the photodetector type, the effects of noise in the system, and the characteristics of the successive amplification stages in the receiver.

When an optical signal has traveled a certain distance along a fiber, the signal has become attenuated and distorted to such a degree that a repeater is needed in the transmission line to amplify and reshape the signal. An optical repeater consists of a receiver and a transmitter placed back to back. The receiver section detects the optical signal, converts it to an electric signal which is amplified, reshaped, and sent to the electric input of the transmitter section. The transmitter section converts this electric signal back to an optical signal and sends it on down the optical fiber waveguide. The conditions that specify when repeaters are required in an optical link are presented in Chap. 8, along with discussions of wavelength division multiplexing systems, optical fiber data buses, and line-coding schemes.

Measurement techniques are an important and integral part of the design, fabrication, and installation of any optical fiber communication system. These are necessary to verify the characteristics of individual components of the link and to ensure proper operation of the installed system. Various measurement methods used to characterize optical fibers, electro-optic components, and installed systems are presented in Chap. 9.

1-3 OPTICAL FIBER SYSTEMS

Initial applications of optical fiber technology have centered on point-to-point digital telecommunications links. These first-generation links have principally been designed for bit rates from 2 to 50 Mb/s. The electro-optical components used in these systems operate in the 810- to 890-nm wavelength range, where the nominal fiber cable transmission loss is between 4 and 6 dB/km after installation and splicing. This allows repeaterless spans from 5 to 10 km, which eliminates the need for manhole-based repeaters in most inner-city areas and reduces the number of repeaters in other longer metropolitan telecommunications networks. Second-generation systems operating at 1.3 μm, where transmission losses of optical fiber cables tend to be about 1 dB/km, can substantially increase the repeaterless transmission span.

In addition to telecommunications applications, optical fiber links have been installed in electric power plants.[14] These links are used to transmit information for system protection, supervision, and control, which are extremely important in large and complex modern power plants. Such information is conventionally transmitted by microwave communication systems in conjunction with various power line carriers and communication cables. However, in large cities interference from the installation of 275-kV power cable networks makes it difficult to maintain a reliable, high-quality microwave communication system. Also high-rise buildings often block the microwave signal path. The immunity of optical fibers to inductive interference and their high data transmission capacity thus offer an excellent alternative communication system.

Another growing application of optical fiber technology is in the subscriber loop plant.[15, 16] This is the portion of the telecommunications network that connects the subscriber's terminal set to the nearest switching center. The traditional transmission medium in this network is a twisted pair of copper wires. These wires have proved to be adequate, reliable, and cost effective for many decades to transmit voice signals. However, they are inadequate for the increasing demand for new broadband multiservices to the home or office, such as pay or educational TV, library and information retrieval, video word processing, and electronic mail, banking, and shopping. This has led to a great interest in applying fiber optic technology in the subscriber loop from the local switching point to the home or office. In contrast with twisted wire pairs, optical fibers offer low attenuation, large information-carrying capacity, immunity to lightning and electromagnetic interference, freedom from cross talk between fibers, and a virtual independence from signal frequency, which eliminates the need for equalization circuitry. By making use of these features, fiber optic subscriber loop plants could thus be designed to meet both present and future broadband multiservice demands.

The widespread interest in such broadband multiservices is evidenced by a number of field trials that are underway or planned in various parts of the world. Some of the major activities include the Highly Interactive Optical Visual Information System (Hi-OVIS) field trial in Japan,[17] the Yorkville (Bell Canada) and Elie Rural (Manitoba Telephone) Systems in Canada,[15] the ambitious Biarritz project[18] in France sponsored by the French PTT, and the Multiservice System for Business Applications[19] developed by GTE.

1-4 USE AND EXTENSION OF THE BOOK

The following chapters are intended as an introduction to the field of optical fiber communication systems. The sequence of topics systematically takes the reader from a description of the major building blocks to an analysis of a complete optical fiber link. Although the material is introductory, it provides a broad and firm basis with which to analyze and design optical fiber communication systems.

Numerous references are provided at the end of each chapter as a start for delving deeper into any given topic. Since optical fiber communications brings together research and development efforts from many different scientific and engineering disciplines, there are hundreds of articles in the literature relating to the material covered in each chapter. Even though not all of these articles were cited in the references, the selections represent some of the major contributions to the fiber optics field and can be considered as a good introduction to the literature. All references were chosen on the basis of easy accessibility and should be available in a good technical library.

Additional references which, in general, are not readily accessible are various conference proceedings. These proceedings can normally be obtained from the organizers of the conference or through a major technical library. Some of these proceedings are devoted exclusively to fiber optics components and systems. Examples are given in Refs. 20 through 24. A number of review articles and books of reprints are also available for a rapid introduction to optical fiber communications. Among these are the ones cited in Refs. 25 through 30. In addition, supplementary material can be found in the books listed in Refs. 31 through 40.

To help the reader understand and use the material in the book, an overview of units, listings of mathematical formulas, and discussions of semiconductor physics and communications theory are given in Apps. A through F.

As is the case for any active scientific and engineering discipline, optical fiber technology is constantly undergoing changes and improvements. New concepts are presently being pursued in optical multiplexing, integrated optics, and device configurations; new materials are being introduced for fibers, sources, and detectors; and measurement techniques are improving. These changes should not have a major impact on the material presented in this book since it is primarily based on enduring fundamental concepts. The understanding of these concepts will allow a rapid comprehension of any new technological developments that will undoubtedly arise.

REFERENCES

1. V. Aschoff, "Optische Nachrichtenübertragung im klassischen Altertum," *Nachrichtentechn. Z. (NTZ),* **30,** 23–28, Jan. 1977.
2. H. Busignies, "Communication channels," *Sci. Amer.,* **227,** 99–113, Sept. 1972.
3. A. B. Carlson, *Communication Systems,* McGraw-Hill, New York, 1975.
4. T. H. Maiman, "Stimulated optical radiation in ruby," *Nature,* **187,** 493–494, Aug. 1960.
5. D. W. Berreman, "A lens or light guide using convectively distorted thermal gradients in gases," *Bell Sys. Tech. J.,* **43,** 1469–1475, July 1964.
6. R. Kompfner, "Optical communications," *Science,* **150,** 149–155, Oct. 1965.
7. A. R. Kraemer, "Free-space optical communications," *Signal,* **32,** 26–32, Oct. 1977.
8. J. H. McElroy, N. McAvoy, E. H. Johnson, J. J. Degnan, F. E. Goodwin, D. M. Henderson, T. A. Nussmeier, L. S. Stokes, B. J. Peyton, and T. Flattau, "CO_2 laser communication systems for near-earth space applications," *Proc. IEEE,* **65,** 221–251, Feb. 1977.
9. L. F. Staff, "Blue-green laser links to subs," *Laser Focus,* **16,** 14–18, Apr. 1980.
10. K. C. Kao and G. A. Hockman, "Dielectric fiber surface waveguides for optical frequencies," *Proc. IEE,* **133,** 1151–1158, July 1966.
11. A. Werts, "Propagation de la lumière cohérente dans les fibres optiques," *L'Onde Électrique,* **46,** 967–980, 1966.
12. F. P. Kapron, D. B. Keck, and R. D. Maurer, "Radiation losses in glass optical waveguides," *Appl. Phys. Lett.,* **17,** 423–425, Nov. 1970.
13. (*a*) D. A. Pinnow, A. L. Gentile, A. G. Standlee, A. J. Timper, and L. M. Hobrock, "Polycrystalline fiber optical waveguides for infrared transmission," *Appl. Phys. Lett.,* **33,** 28–29, July 1978.
    (*b*) L. G. Van Uitert and S. H. Wemple, "$ZnCl_2$ glass: A potential ultralow loss optical fiber material," *ibid.,* 57–59.
14. F. Aoki and H. Nabeshima, "Optical fiber communications for electrical power companies in Japan," *Proc. IEEE,* **68,** 1280–1285, Oct. 1980.
15. K. Y. Chang, "Fiberguide systems in the subscriber loop," *Proc. IEEE,* **68,** 1291–1299, Oct. 1980.
16. J. E. Midwinter, "Potential broad-band services," *Proc. IEEE,* **68,** 1321–1327, Oct. 1980.
17. T. Nakahara, H. Kumamaru, and S. Takeuchi, "An optical fiber video system," *IEEE Trans. Commun.,* **COM-26,** 955–961, July 1978.
18. M. Niguil, "A major fiber optic program in France: The wired city of Biarritz," *FOC-81 East Proc.,* 119–121, Mar. 1981 (published by Information Gatekeepers, Inc., 1981).
19. D. Gray, "A multiservice optical fiber system for business applications," *Natl. Telecomm. Conf.,* Paper 34.4, Dec. 1980, Houston, Tex.
20. *Topical Meetings on Optical Fiber Transmission* (OSA/IEEE), Williamsburg, Va., 7–9 Jan. 1975 and 22–24 Feb. 1977; Washington, D.C., 6–8 Mar. 1979; Phoenix, Ariz., Apr. 1982.
21. *Fiber Optics & Communications (FOC) Proceedings* (sponsored by Information Gatekeepers), Chicago, Sept. 1978 and 1979; San Francisco, Sept. 1980–1982.
22. *European Conf. on Optical Fibre Communications:* 1st London, Sept. 1975; 2d Paris, Sept. 1976; 3d Munich, Sept. 1977; 4th Geneva, Sept. 1978; 5th Amsterdam, Sept. 1979; 6th York (England), Sept. 1980; 7th Copenhagen, Sept. 1981; 8th Cannes (France), Sept. 1982.
23. *Proceedings of the International Conf. on Communications* (ICC) and the *National Telecommunications Conf.* (NTC), both sponsored by IEEE, generally have several sessions devoted to the latest developments in fiber communications.
24. Other conference proceedings with varying emphasis and coverage of optical fiber components, equipment, and systems are the following:
    (*a*) *OSA/IEEE Conf. on Lasers & Electro-Optics* (CLEO)
    (*b*) *Soc. of Photo-Optical Instrumentation Engineers* (SPIE)
    (*c*) *OSA/IEEE Int. Conf. Integrated Opt. & Opt. Fiber Comm*
    (*d*) *Electro-Optics/Laser (Indus. & Sci. Conf. Management)*

25. S. E. Miller, E. A. J. Marcatili, and T. Li, "Research toward optical fiber transmission systems," *Proc. IEEE,* **61,** 1703–1751, Dec. 1973.
26. J. Conradi, F. P. Kapron, and J. C. Dyment, "Fiber optical transmission between 0.8 and 1.4 µm," *IEEE Trans. Electron. Devices,* **ED-25,** 180–193, Feb. 1978.
27. T. G. Giallorenzi, "Optical communications research and technology: Fiber optics, " *Proc. IEEE,* **66,** 744–780, July 1978.
28. D. Botez and G. J. Herskowitz, "Components for optical communications systems: A review," *Proc. IEEE,* **68,** 698–731, June 1980.
29. D. Gloge, *Optical Fiber Technology,* IEEE Press, New York, 1976. This is a book of reprints of some of the fundamental research articles appearing in the literature from 1969 to 1975.
30. C. K. Kao, *Optical Fiber Technology, II,* IEEE Press, New York, 1980. This is a continuation of Ref. 29. It contains reprints of articles appearing in the literature from 1976 to 1979.
31. M. K. Barnoski (Ed.), *Fundamentals of Optical Fiber Communications,* Academic, New York, 1976; 2d ed., 1982.
32. J. E. Midwinter, *Optical Fibers for Transmission,* Wiley, New York, 1979.
33. S. E. Miller and A. G. Chynoweth, *Optical Fiber Telecommunications,* Academic, New York, 1979.
34. C. P. Sandbank (Ed.), *Optical Fibre Communication Systems,* Wiley, New York, 1980.
35. M. J. Howes and D. V. Morgan, *Optical Fibre Communications,* Wiley, New York, 1980.
36. Technical Staff of CSELT, *Optical Fibre Communication,* McGraw-Hill, New York, 1981.
37. H. Kressel (Ed.), *Semiconductor Devices for Optical Communications,* Springer, New York, 1980.
38. S. D. Personick, *Optical Fiber Transmission Systems,* Plenum, New York, 1981.
39. A. B. Sharma, S. J. Halme, and M. M. Butusov, *Optical Fiber Systems and Their Components,* Springer, New York, 1981.
40. C. K. Kao, *Optical Fiber Systems,* McGraw-Hill, New York, 1982.

TWO

OPTICAL FIBERS: STRUCTURES AND WAVEGUIDING FUNDAMENTALS

One of the most important components in any optical fiber system is the optical fiber itself, since its transmission characteristics play a major role in determining the performance of the entire system. Some of the questions that arise concerning optical fibers are:

1. What is the structure of an optical fiber?
2. How does light propagate along a fiber?
3. What is the signal loss or attenuation mechanism in a fiber?
4. Why and to what degree does a signal get distorted as it travels along a fiber?
5. Of what materials are fibers made?
6. How is the fiber fabricated?
7. How are fibers incorporated into cable structures?

The purpose of this chapter is to present some of the fundamental answers to the first two questions in order to attain a good understanding of the physical structure and waveguiding properties of optical fibers. Questions 3 and 4 are answered in Chap. 3, and the remaining three questions are addressed in Chap 10.

Since fiber optics technology involves the emission, transmission, and detection of light, we begin our discussion by first considering the nature of light and then we shall review a few basic laws and definitions of optics.[1] Following a description of the structure of optical fibers, two methods are used to describe how an optical fiber guides light. The first approach uses the geometrical or ray optics concepts of light reflection and refraction to provide an intuitive picture of the propagation mechanisms. In the second approach light is treated as an electromagnetic wave which propagates along the optical fiber waveguide. This involves solving Maxwell's equations subject to the cylindrical boundary conditions of the fiber.

2-1 THE NATURE OF LIGHT

The concepts concerning the nature of light have undergone several variations during the history of physics. Until the early seventeenth century it was generally believed that light consisted of a stream of minute particles that were emitted by luminous sources. These particles were pictured as traveling in straight lines, and it was assumed that they could penetrate transparent materials but were reflected from opaque ones. This theory adequately described certain large-scale optical effects such as reflection and refraction, but failed to explain finer-scale phenomena such as interference and diffraction.

The correct explanation of diffraction was given by Fresnel in 1815. Fresnel showed that the approximately rectilinear propagation character of light could be interpreted on the assumption that light is a wave motion, and that the diffraction fringes could thus be accounted for in detail. Later the work of Maxwell in 1864 theorized that light waves must be electromagnetic in nature. Furthermore observation of polarization effects indicated that light waves are transverse (that is, the wave motion is perpendicular to the direction in which the wave travels). In this *wave* or *physical optics viewpoint* the electromagnetic waves radiated by a small optical source can be represented by a train of spherical wave fronts with the source at the center as shown in Fig. 2-1. A *wave front* is defined as the locus of all points in the wave train which have the same phase. Generally one draws wave fronts passing either through the maxima or the minima of the wave, such as the peak or trough of a sine wave, for example. Thus the wave fronts (also called *phase fronts*) are separated by one wavelength.

When the wavelength of the light is much smaller than the object (or opening) which it encounters, the wave fronts appear as straight lines to this object or opening. In this case the light wave can be represented as a *plane wave,* and its direction of travel can be indicated by a *light ray* which is drawn perpendicular to the phase front. Thus large-scale optical effects such as reflection and refraction can be analyzed by the simple geometrical process of *ray tracing.* This view of optics is referred to as *ray* or *geometrical optics.* The concept of light rays is very useful because the rays show the direction of energy flow in the light beam.

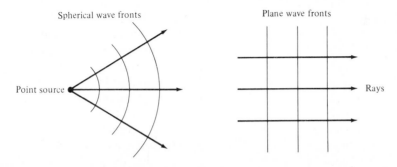

Figure 2-1 Representations of spherical and plane wavefronts and their associated rays.

A train of *plane* or *linearly polarized waves* traveling in a direction **k** can be represented in the general form

$$\mathbf{A}(\mathbf{x}, t) = \mathbf{e}_i A_0 \exp\left[j(\omega t - \mathbf{k} \cdot \mathbf{x})\right] \tag{2-1}$$

with $\mathbf{x} = x\mathbf{e}_x + y\mathbf{e}_y + z\mathbf{e}_z$ representing a general position vector and $\mathbf{k} = k_x\mathbf{e}_x + k_y\mathbf{e}_y + k_z\mathbf{e}_z$ representing the wave propagation vector.

Here A_0 is the maximum amplitude of the wave, $\omega = 2\pi\nu$, where ν is the frequency of the light, the magnitude of the wavevector **k** is $k = 2\pi/\lambda$, which is known as the *wave propagation constant* with λ being the wavelength of the light, and $\mathbf{e}_j$ is a unit vector lying parallel to an axis designated by j.

Note that the components of the actual (measurable) electromagnetic field represented by Eq. (2-1) are obtained by taking the real part of this equation. For example, if $\mathbf{k} = k\mathbf{e}_z$ and if **A** denotes the electric field **E** with the coordinate axes chosen such that $\mathbf{e}_j = \mathbf{e}_x$, then the real measurable electric field varies harmonically in the x direction and is given by

$$\mathbf{E}_x(z, t) = \text{Re}(\mathbf{E}) = \mathbf{e}_x E_{0x} \cos(\omega t - kz) \tag{2-2}$$

which represents a plane wave traveling in the z direction. The reason for using the exponential form is that it is more easily handled mathematically than equivalent expressions given in terms of sine and cosine. In addition, the rationale for using harmonic functions is that any waveform can be expressed in terms of sinusoidal waves using Fourier techniques.

The electric and magnetic field distributions in a train of plane electromagnetic waves at a given instant in time are shown in Fig. 2-2. The waves are moving in the direction indicated by the vector **k**. Based on Maxwell's equations it is easily shown[2] that **E** and **H** are both perpendicular to the direction of propagation. Such a wave is called a *transverse wave*. Furthermore **E** and **H** are mutually perpendicular, so that **E, H,** and **k** form a set of orthogonal vectors.

The plane wave example given by Eq. (2-2) has its electric field vector always pointing in the $\mathbf{e}_x$ direction. Such a wave is *linearly polarized* with polarization vector $\mathbf{e}_x$. A general state of polarization is described by considering another

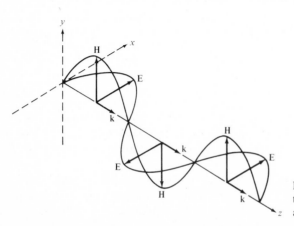

Figure 2-2 Field distributions in a train of plane electromagnetic waves at a given instant in time.

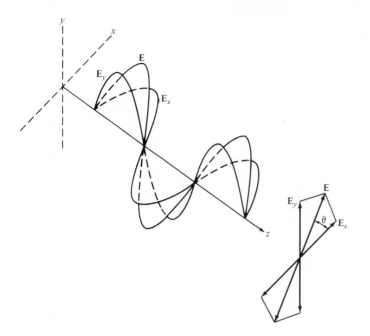

Figure 2-3 Addition of two linearly polarized waves having zero relative phase between them.

linearly polarized wave which is independent of the first wave and orthogonal to it. Let this wave be

$$\mathbf{E}_y(z, t) = \mathbf{e}_y E_{0y} \cos{(\omega t - kz + \delta)} \tag{2-3}$$

where δ is the relative phase difference between the waves. The resultant wave is then simply

$$\mathbf{E}(z, t) = \mathbf{E}_x(z, t) + \mathbf{E}_y(z, t) \tag{2-4}$$

If δ is zero or an integer multiple of 2π, the waves are in phase. Equation (2-4) is then also a linearly polarized wave with a polarization vector making an angle

$$\theta = \arctan{\frac{E_{0y}}{E_{0x}}}$$

with respect to $\mathbf{e}_x$ and having a magnitude

$$E = (E_{0x}^2 + E_{0y}^2)^{1/2}$$

This case is shown schematically in Fig. 2-3. Conversely, just as any two orthogonal plane waves can be combined into a linearly polarized wave, an arbitrary linearly polarized wave can be resolved into two independent orthogonal plane waves that are in phase.

For general values of δ the wave given by Eq. (2-4) is elliptically polarized. The resultant field vector $\mathbf{E}$ will both rotate and change its magnitude as a function of the angular frequency ω. As shown in Fig. 2-4 the endpoint of $\mathbf{E}$ will trace out

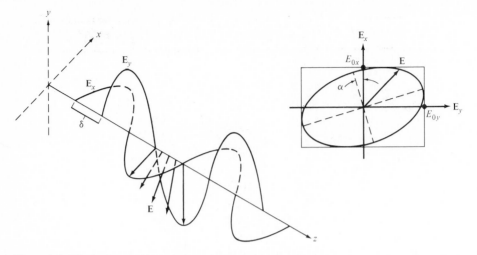

Figure 2-4 Elliptically polarized light resulting from the addition of two linearly polarized waves of unequal amplitude having a nonzero phase difference between them.

an ellipse at a given point in space with the axis of the ellipse related to the x axis by an angle α given by (see Prob. 2-4)

$$\tan 2\alpha = \frac{2E_{0x}E_{0y} \cos \delta}{E_{0x}^2 - E_{0y}^2} \tag{2-5}$$

When $E_{0x} = E_{0y}$ the resultant wave is circularly polarized.

The wave theory of light adequately accounts for all phenomena involving the transmission of light. However, in dealing with the interaction of light and matter, such as occurs in dispersion and in the emission and absorption of light, neither the particle theory nor the wave theory of light is appropriate. Instead we must turn to quantum theory which indicates that optical radiation has particle as well as wave properties. The particle nature arises from the observation that light energy is always emitted or absorbed in discrete units called *quanta* or *photons*. In all experiments used to show the existence of photons, the photon energy is found to depend only on the frequency ν. This frequency, in turn, must be measured by observing a wave property of light.

The relationship between the energy E and the frequency ν of a photon is given by

$$E = h\nu \tag{2-6}$$

where $h = 6.625 \times 10^{-34}$ J·s is Planck's constant. When light is incident on an atom, a photon can transfer its energy to an electron within this atom, thereby exciting it to a higher energy level. In this process either all or none of the photon energy is imparted to the electron. The energy absorbed by the electron must be exactly equal to that required to excite the electron to a higher energy level.

Conversely, an electron in an excited state can drop to a lower state separated from it by an energy $h\nu$ by emitting a photon of exactly this energy.

2-2 BASIC OPTICAL LAWS AND DEFINITIONS

We shall next review some of the basic optical laws and definitions relevant to optical fibers. A fundamental optical parameter of a material is the *refractive index* (or *index of refraction*). In free space a light wave travels at a speed $c = 3 \times 10^8$ m/s. The speed of light is related to the frequency ν and the wavelength λ by $c = \nu\lambda$. Upon entering a dielectric or nonconducting medium the wave now travels at a speed v, which is characteristic of the material and less than c. The ratio of the speed of light in a vacuum to that in matter is the index of refraction n of the material and is given by

$$n = \frac{c}{v} \tag{2-7}$$

Typical values of n are 1.00 for air, 1.33 for water, 1.50 for glass, and 2.42 for diamond.

The concepts of reflection and refraction can be interpreted most easily by considering the behavior of light rays associated with plane waves traveling in a dielectric material. When a light ray encounters a boundary separating two different media, part of the ray is reflected back into the first medium and the remainder is bent (or refracted) as it enters the second material. This is shown in Fig. 2-5 where $n_2 < n_1$. The bending or refraction of the light ray at the interface is a result of the difference in the speed of light in two materials having different refractive indices. The relationship at the interface is known as Snell's law and is given by

$$n_1 \sin \phi_1 = n_2 \sin \phi_2 \tag{2-8}$$

or equivalently as

$$n_1 \cos \theta_1 = n_2 \cos \theta_2 \tag{2-9}$$

where the angles are defined in Fig. 2-5.

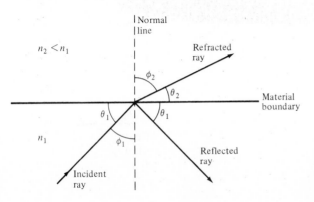

Figure 2-5 Refraction and reflection of a light ray at a material boundary.

According to the law of reflection the angle θ_1 at which the incident ray strikes the interface is exactly equal to the angle the reflected ray makes with the same interface. In addition, the incident ray, the normal to the interface, and the reflected ray all lie in the same plane which is perpendicular to the interface plane between the two materials. When light traveling in a certain medium is reflected off an optically denser material (one with a higher refractive index), the process is referred to as *external reflection*. Conversely, the reflection of light off of less optically dense material (such as light traveling in glass being reflected at a glass-to-air interface) is called *internal reflection*.

As the angle of incidence θ_1 in an optically denser material (higher refractive index) becomes smaller, the refracted angle θ_2 approaches zero. Beyond this point no refraction is possible and the light rays become *totally internally reflected*. The conditions required for total internal reflection can be determined by using Snell's law [Eq. (2-9)]. Consider Fig. 2-6, which shows a glass surface in air. A light ray gets bent toward the glass surface as it leaves the glass in accordance with Snell's law. If the angle of incidence θ_1 is decreased, a point will eventually be reached where the light ray in air is parallel to the glass surface. This point is known as the *critical angle of incidence θ_c*. When the incident angle θ_1 is less than the critical angle, the condition for total internal reflection is satisfied; that is, the light is totally reflected back into the glass with no light escaping from the glass surface. (This is an idealized situation. In practice there is always some tunneling of optical energy through the interface. This can be explained in terms of the electromagnetic wave theory of light which is presented in Sec. 2-4.)

As an example consider the glass-air interface shown in Fig. 2-6. When the light ray in air is parallel to the glass surface then $\theta_2 = 0$ so that $\cos\theta_2 = 1$. The critical angle in the glass is thus

$$\theta_c = \text{arc cos } \frac{n_2}{n_1} \qquad (2\text{-}10)$$

Using $n_1 = 1.50$ for glass and $n_2 = 1.00$ for air, θ_c is about 48°. Any light in the glass incident on the interface at an angle θ_1 less than 48° is totally reflected back into the glass.

In addition, when light is totally internally reflected, a phase change δ occurs in the reflected wave. This phase change depends on the angle $\theta_1 < \theta_c$ according to the relationships[1]

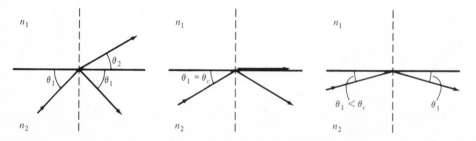

Figure 2-6 Representation of the critical angle and total internal reflection at a glass-air interface.

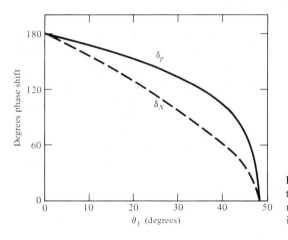

Figure 2-7 Phase shifts occurring from the reflection of wave components normal (δ_N) and parallel (δ_p) to the reflecting surface.

$$\tan \frac{\delta_N}{2} = \frac{\sqrt{n^2 \cos^2 \theta_1 - 1}}{n \sin \theta_1} \qquad (2\text{-}11a)$$

$$\tan \frac{\delta_p}{2} = \frac{\sqrt{n^2 \cos^2 \theta_1 - 1}}{\sin \theta_1} \qquad (2\text{-}11b)$$

Here δ_N and δ_p are the phase shifts of the wave components normal and parallel to the reflecting surface, respectively, and $n = n_1/n_2$. A derivation of Eqs. (2-11a) and (2-11b) can be found in Klein.[1] These phase shifts are shown in Fig. 2-7 for a glass-air interface ($n = 1.5$ and $\theta_c = 48°$). The values range from zero immediately at the critical angle to π when $\theta_1 = 0°$.

These basic optical principles will now be used to illustrate how optical power is transmitted along a fiber.

2-3 OPTICAL FIBER MODES AND CONFIGURATIONS

Before going into details on optical fiber characteristics in Sec. 2-3-3, we first present a brief overview of the underlying concepts of optical fiber modes and optical fiber configurations.

2-3-1 Fiber Types

An optical fiber is a dielectric waveguide that operates at optical frequencies. This fiber waveguide is normally cylindrical in form. It confines electromagnetic energy in the form of light to within its surfaces and guides the light in a direction parallel to its axis. The transmission properties of an optical waveguide are dictated by its structural characteristics, which have a major effect in determining how an optical signal is affected as it propagates along the fiber. The structure basically establishes the information-carrying capacity of the fiber and also influences the response of the waveguide to environmental perturbations.

The propagation of light along a waveguide can be described in terms of a set of guided electromagnetic waves called the *modes* of the waveguide. These guided modes are referred to as the *bound* or *trapped* modes of the waveguide. Each guided mode is a pattern of electric and magnetic field lines that is repeated along the fiber at intervals equal to the wavelength. Only a certain discrete number of modes are capable of propagating along the guide. As will be seen in Sec. 2-4, these modes are those electromagnetic waves that satisfy the homogeneous wave equation in the fiber and the boundary condition at the waveguide surfaces.

Although many different configurations of the optical waveguide have been discussed in the literature,[3] the most widely accepted structure is the single solid dielectric cylinder of radius a and index of refraction n_1 shown in Fig. 2-8. This cylinder is known as the *core* of the fiber. The core is surrounded by a solid dielectric *cladding* having a refractive index n_2 that is less than n_1. Although, in principle, a cladding is not necessary for light to propagate along the core of the fiber, it serves several purposes. The cladding reduces scattering loss resulting from dielectric discontinuities at the core surface, it adds mechanical strength to the fiber, and it protects the core from absorbing surface contaminants with which it could come in contact.

In low- and medium-loss fibers the core material is generally glass which is surrounded by either a glass or a plastic cladding. Higher-loss plastic core fibers with plastic claddings are also widely in use. In addition, most fibers are encapsulated in an elastic, abrasion-resistant plastic material. This material adds further strength to the fiber and mechanically isolates or buffers the fibers from small geometrical irregularities, distortions, or roughnesses of adjacent surfaces. These perturbations could otherwise cause scattering losses induced by random microscopic bends that can arise when the fibers are incorporated into cables or supported by other structures.

Variations in the material composition of the core give rise to the two commonly used fiber types shown in Fig. 2-9. In the first case the refractive index of the core is uniform throughout and undergoes an abrupt change (or step) at the cladding boundary. This is called a *step-index fiber*. In the second case the core refractive index is made to vary as a function of the radial distance from the center of the fiber. This type is a *graded-index fiber*.

Both the step- and the graded-index fibers can be further divided into single-mode and multimode classes. As the name implies, a single-mode fiber sustains only one mode of propagation, whereas multimode fibers contain many hundreds of modes. A few typical sizes of single- and multimode fibers are given in Fig. 2-9 to provide an idea of the dimensional scale. Multimode fibers offer several advantages compared to single-mode fibers. As we shall see in Chap. 5, the larger core

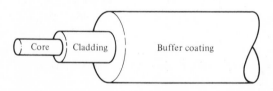

Figure 2-8 Schematic of a single-fiber structure. A circular solid core of refractive index n_1 is surrounded by a cladding having a refractive index $n_2 < n_1$. An elastic plastic buffer encapsulates the fiber.

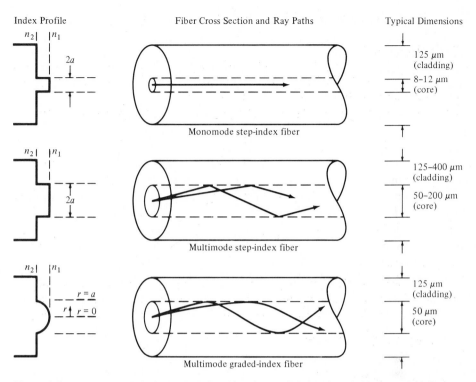

Figure 2-9 Comparison of single-mode and multimode step-index and graded-index optical fibers.

radii of multimode fibers make it easier to launch optical power into the fiber and facilitate the connecting together of similar fibers. Another advantage is that light can be launched into a multimode fiber using a light-emitting-diode (LED) source, whereas single-mode fibers must be excited with laser diodes. Although LEDs have less optical output power than laser diodes (as we shall discuss in Chap. 4), they are easier to make, are less expensive, require less complex circuitry, and have longer lifetimes than laser diodes, thus making them more desirable in many applications.

A disadvantage of multimode fibers is that they suffer from intermodal dispersion. We shall describe this effect in detail in Sec. 3-4. Briefly, intermodal dispersion can be described as follows. When an optical pulse is launched into a fiber, the optical power in the pulse is distributed over all (or most) of the modes of the fiber. Each of the modes that can propagate in a multimode fiber travels at a slightly different velocity. This means that the modes in a given optical pulse arrive at the fiber end at slightly different times, thus causing the pulse to spread out in time as it travels along the fiber. This effect, which is known as *intermodal dispersion,* can be reduced by using a graded-index profile in the fiber core. This allows graded-index fibers to have much larger bandwidths (data rate transmission capabilities) than step-index fibers. Even higher bandwidths are possible in single-mode fibers where intermodal dispersion effects are not present.

2-3-2 Rays and Modes

As we mentioned in Sec. 2-3-1, the electromagnetic light field that is guided along an optical fiber can be represented by a superposition of bound or trapped modes. Each of these guided modes is composed of a set of simple electromagnetic field configurations which form a standing-wave pattern in the transverse direction (that is, transverse to the waveguide axis). Thus as we shall see in Sec. 2-4, for monochromatic light fields of radian frequency ω, a mode traveling in the positive z direction (that is, along the fiber axis) has a time and z dependence given by

$$e^{j(\omega t - \beta z)}$$

The factor β is the z component of the wave propagation constant $k = 2\pi/\lambda$ and is the main parameter of interest in describing fiber modes. For guided modes β can only assume certain discrete values, which are determined from the requirement that the mode field must satisfy Maxwell's equations and the electric and magnetic field boundary conditions at the core-cladding interface. This is described in detail in Sec. 2-4.

Another method for theoretically studying the propagation characteristics of light in an optical fiber is the geometrical optics or ray-tracing approach. This method provides a good approximation to the light acceptance and guiding properties of optical fibers when the ratio of the fiber radius to the wavelength is large. This is known as the *small wavelength limit*. Although the ray approach is strictly valid only in the zero wavelength limit, it is still relatively accurate and extremely valuable for nonzero wavelengths when the number of guided modes is large, that is, for multimode fibers. The advantage of the ray approach is that, compared to the exact electromagnetic wave (modal) analysis, it gives a more direct physical interpretation of the light propagation characteristics in an optical fiber.

Since the concept of a light ray is very different from that of a mode, let us see qualitatively what the relationship is between them. (The mathematical details of this relationship are beyond the scope of this book but can be found in the literature.[4,5]) A guided mode traveling in the z direction (along the fiber axis) can be decomposed into a family of superimposed plane waves that collectively form a standing-wave pattern in the direction transverse to the fiber axis.[4] Since with any plane wave we can associate a light ray that is perpendicular to the phase front of the wave, the family of plane waves corresponding to a particular mode forms a set of rays called a *ray congruence*. Each ray of this particular set travels in the fiber at the same angle relative to the fiber axis.[5] We note here that, since only a certain number M of discrete guided modes exist in a fiber, the possible angles of the ray congruences corresponding to these modes are also limited to the same number M. Although a simple ray picture appears to allow rays at any angle less than the critical angle to propagate in a fiber, the allowable quantized propagation angles result when the phase condition for standing waves is introduced into the ray picture. This is discussed further in Sec. 2-3-5.

Despite the usefulness of the approximate geometrical optics method, a number of limitations and discrepancies exist between it and the exact modal analysis. An important case is the analysis of single-mode or few-mode fibers which must be dealt with by using electromagnetic theory. Problems involving coherence or interference phenomena must also be solved with an electromagnetic approach. In addition, a modal analysis is necessary when a knowledge of the field distribution of individual modes is required. This arises, for example when analyzing the excitation of an individual mode or when analyzing the coupling of power between modes at waveguide imperfections (which we shall discuss in Chap. 3).

Another discrepancy between the ray optics approach and the modal analysis occurs when an optical fiber is uniformly bent with a constant radius of curvature. As we shall show in Chap. 3, wave optics correctly predicts that every mode of the curved fiber experiences some radiation loss. Ray optics, on the other hand, erroneously predicts that some ray congruences can undergo total internal reflection at the curve and, consequently, can remain guided without loss.

2-3-3 Step-Index Fiber Structure

We begin our discussion of light propagation in an optical waveguide by considering the step-index fiber. In practical step-index fibers the core of radius a has a refractive index n_1 which is typically equal to 1.48. This is surrounded by a cladding of slightly lower index n_2, where

$$n_2 = n_1(1 - \Delta) \qquad (2-12)$$

The parameter Δ is called the *core-cladding index difference* or simply the *index difference*. Values of n_2 are chosen such that Δ is nominally 0.01. Since the core refractive index is larger than the cladding index, electromagnetic energy at optical frequencies is made to propagate along the fiber waveguide through internal reflection at the core-cladding interface.

2-3-4 Ray Optics Representation

Since the core size of multimode fibers is much larger than the wavelength of the light we are interested in (which is approximately 1 μm), an intuitive picture of the propagation mechanism in an ideal multimode step-index optical waveguide is most easily seen by a simple ray (geometrical) optics representation.[6-9] For simplicity in this analysis we shall consider only a particular ray belonging to a ray congruence which represents a fiber mode. The two types of rays that can propagate in a fiber are meridional rays and skew rays. *Meridional rays* are confined to the meridian planes of the fiber, which are the planes that contain the axis of symmetry of the fiber (the core axis). Since a given meridional ray lies in a single plane, its path is easy to track as it travels along the fiber. Meridional rays can be divided into two general classes: bound rays that are trapped in the core and propagate along the fiber axis according to the laws of geometrical optics, and unbound rays that are refracted out of the fiber core.

Skew rays are not confined to a single plane but, instead, tend to follow a helical type path along the fiber as shown in Fig. 2-10. These rays are more difficult to track as they travel along the fiber since they do not lie in a single plane. Although skew rays constitute a major portion of the total number of guided rays, their analysis is not necessary to obtain a general picture of rays propagating in a

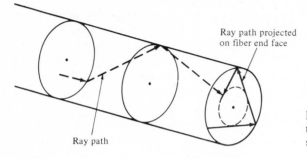

Ray path projected on fiber end face

Ray path

Figure 2-10 Ray optics representation of skew rays traveling in a step-index optical fiber core.

fiber. The examination of meridional rays will suffice for this purpose. However, a detailed inclusion of skew rays will change such expressions as the light acceptance ability of the fiber and power losses of light traveling along a waveguide. Details concerning these parameters are given in the works by Gallawa,[9] Unger,[10] and Kapany.[11]

A greater power loss arises when skew rays are included in the analyses, since many of the skew rays that geometric optics predicts are trapped in the fiber are actually leaky rays.[5,12,13] These leaky rays are only partially confined to the core of the circular optical fiber and attenuate as the light travels along the optical waveguide. This partial reflection of leaky rays cannot be described by pure ray theory alone. Instead, the analysis of radiation loss arising from these types of rays must be described by mode theory. This is explained further in Sec. 2-4.

The meridional ray is shown in Fig. 2-11 for a step-index fiber. The light ray enters the fiber core from a medium of refractive index n at an angle θ_0 with respect to the fiber axis and strikes the core-cladding interface at a normal angle ϕ. If it strikes this interface at such an angle that it is totally internally reflected, the meridional ray follows a zig-zag path along the fiber core passing through the axis of the guide after each reflection.

From Snell's law the minimum angle ϕ_{min} that supports total internal reflection for the meridional ray is

$$\sin(\phi_{min}) = \frac{n_2}{n_1} \tag{2-13}$$

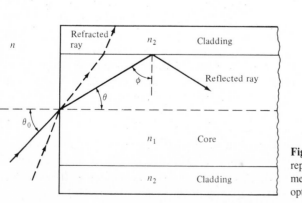

Figure 2-11 Meridional ray optics representation of the propagation mechanism in an ideal step-index optical waveguide.

Rays striking the core-cladding interface at angles less than ϕ_{min} will refract out of the core and be lost in the cladding. The condition of Eq. (2-13) can be related to the maximum entrance angle $\theta_{0,max}$ through the relationship

$$n \sin \theta_{0,max} = n_1 \sin \theta_c = (n_1^2 - n_2^2)^{1/2} \tag{2-14}$$

where θ_c is the critical angle. Thus those rays having entrance angles θ_0 less than $\theta_{0,max}$ will be totally internally reflected at the core-cladding interface.

Equation (2-14) also defines the *numerical aperture* NA of a step-index fiber for meridional rays

$$NA = n \sin \theta_{0,max} = (n_1^2 - n_2^2)^{1/2} \simeq n_1 \sqrt{2\Delta} \tag{2-15}$$

The approximation on the right-hand side is valid for the typical case where Δ as defined by Eq. (2-12) is much less than 1. Since the numerical aperture is related to the maximum acceptance angle, it is commonly used to describe the light acceptance or gathering capability of a fiber and to calculate source-to-fiber optical power coupling efficiencies. This is detailed in Chap. 5. The numerical aperture is a dimensionless quantity which is less than unity, with values normally ranging from 0.14 to 0.50.

2-3-5 Wave Representation

The ray theory appears to allow rays at any angle θ_1 less than the critical angle θ_c to propagate along the fiber. However, when the phase of the plane wave associated with the ray is taken into account, it is seen that only rays at certain discrete angles less than or equal to θ_c are capable of propagating along the fiber.

To see this, consider a light ray in the core incident on the reflective surface at an angle θ as shown in Fig. 2-12. The plane wave associated with this ray is of the form given in Eq. (2-1). As the wave travels it undergoes a phase change δ given by

$$\delta = k_1 s = n_1 k s = \frac{n_1 2\pi s}{\lambda} \tag{2-16}$$

where k_1 = the propagation constant in the medium of refractive index n_1
$k = k_1/n_1$ is the free-space propagation constant
s = the distance traveled along the ray by the wave

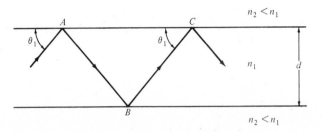

Figure 2-12 Light ray propagating along a fiber waveguide. Phase changes occur both as the wave travels through the fiber medium and at the reflection points.

The phase of the wave changes not only as the wave travels but also upon reflection from a dielectric interface, as shown in Sec. 2-2.

In order for the wave associated with a given ray to propagate along the waveguide shown in Fig. 2-12, the phase of the twice reflected wave must be the same as that of the incident wave. That is, the wave must interfere constructively with itself. If this phase condition is not satisfied, the wave would interfere destructively with itself and just die out. Thus the total phase shift that results when the wave traverses the guide twice (from points A to B to C) and gets reflected twice (at points A and B) must be equal to an integer multiple of 2π rad. Using Eqs. (2-16) and (2-11), we let the phase change that occurs over the distance ABC be $\delta_{AC} = n_1 k \, (2d/\sin \theta_1)$ and the phase changes upon reflection each be (assuming for simplicity that the wave is polarized normal to the reflecting surface)

$$\delta_1 = 2 \arctan \frac{(n^2 \cos^2 \theta_1 - 1)^{1/2}}{n \sin \theta_1} \tag{2-17}$$

where $n = n_1/n_2$. Then the following condition must be satisfied

$$\frac{2n_1 kd}{\sin \theta_1} + 2\delta_1 = 2\pi M \tag{2-18}$$

where M is an integer that determines the allowed ray angles for waveguiding.

2-4 MODE THEORY FOR CIRCULAR WAVEGUIDES

To attain a more detailed understanding of the optical power propagation mechanism in a fiber, it is necessary to solve Maxwell's equations subject to the cylindrical boundary conditions of the fiber. This has been carried out in extensive detail in a number of works.[7-10,14-18] Since a complete treatment is beyond the scope of this book, only a general outline of the analyses will be given here.

Before we progress with our discussion of mode theory in circular optical fibers, let us first qualitatively examine the appearance of modal fields in the planar dielectric slab waveguide shown in Fig. 2-13. This waveguide is composed of a dielectric slab of refractive index n_1 sandwiched between dielectric material of refractive index $n_2 < n_1$, which we shall call the cladding. This represents the simplest form of an optical waveguide and can serve as a model to gain an understanding of wave propagation in optical fibers. In fact, a cross-sectional view of the slab waveguide looks the same as the cross-sectional view of an optical fiber cut along its axis. Figure 2-13 shows the field patterns of several of the lower-order modes (which are solutions of Maxwell's equations for the slab waveguide[7-10]). The *order* of a mode is equal to the number of field maxima across the guide. The order of the mode is also related to the angle that the ray congruence corresponding to this mode makes with the plane of the waveguide (or the axis of a fiber); that is, the steeper the angle, the higher the order of the mode. The plots show that the electric fields of the guided modes are not completely confined to the central

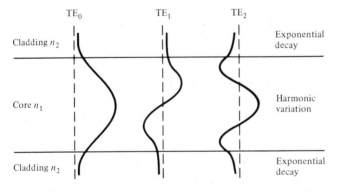

Figure 2-13 Electric field distributions for several of the lower-order guided modes in a symmetrical slab waveguide.

dielectric slab (that is, they do not go to zero at the guide-cladding interface), but, instead, they extend partially into the cladding. The fields vary harmonically in the guiding region of refractive index n_1 and decay exponentially outside of this region. For low-order modes the fields are tightly concentrated near the center of the slab (or the axis of an optical fiber) with little penetration into the cladding region. On the other hand, for higher-order modes the fields are distributed more toward the edges of the guide and penetrate further into the cladding region.

Solving Maxwell's equations shows that, in addition to supporting a finite number of guided modes, the optical fiber waveguide has an infinite continuum of radiation modes that are not trapped in the core and guided by the fiber but are still solutions of the same boundary-value problem. The radiation field basically results from the optical power that is outside the fiber acceptance angle being refracted out of the core. Because of the finite radius of the cladding, some of this radiation gets trapped in the cladding, thereby causing cladding modes to appear. As the core and cladding modes propagate along the fiber, mode coupling occurs between the cladding modes and the higher-order core modes. This coupling occurs because the electric fields of the guided core modes are not completely confined to the core but extend partially into the cladding (see Fig. 2-13) and likewise for the cladding modes. A diffusion of power back and forth between the core and cladding modes thus occurs, which generally results in a loss of power from the core modes. In practice, the cladding modes will be suppressed by a lossy coating which covers the fiber or they will scatter out of the fiber after traveling a certain distance because of roughness on the cladding surface.

In addition to bound and refracted modes, a third category of modes called *leaky modes*[5,7,10,12,13] is present in optical fibers. These leaky modes are only partially confined to the core region, and attenuate by continuously radiating their power out of the core as they propagate along the fiber. This power radiation out of the waveguide results from a quantum mechanical phenomenon known as the *tunnel effect*. Its analysis is fairly lengthy and beyond the scope of this book. However, it is essentially based on the upper and lower bounds that the boundary conditions for the solutions of Maxwell's equations impose on the propagation constant β. As we shall see in Sec. 2-4-3, a mode remains guided as long as β satisfies the condition

$$n_2 k < \beta < n_1 k$$

where n_1 and n_2 are the refractive indices of the core and cladding, respectively, and $k = 2\pi/\lambda$. The boundary between truly guided modes and leaky modes is defined by the cutoff condition $\beta = n_2 k$. As soon as β becomes smaller than $n_2 k$, power leaks out of the core into the cladding region. Leaky modes can carry significant amounts of optical power in short fibers. Most of these modes disappear after a few centimeters, but a few have sufficiently low losses to persist in fiber lengths of a kilometer.

Although the theory of propagation in optical fibers is well understood, a complete description of the guided and radiation modes is rather complex since it involves six-component hybrid electromagnetic fields having very involved mathematical expressions. A simplification[19-23] of these expressions can be carried out in practice since fibers usually are constructed so that the difference in the core and cladding indices of refraction is very small; that is, $n_1 - n_2 \ll 1$. With this assumption only four field components need to be considered and their expressions become significantly simpler. However, the analysis required for these simplifications is fairly involved, and the interested reader is referred to the literature. Here we shall first follow the standard approach of solving Maxwell's equations for a circular step-index waveguide and then describe the resulting solutions for some of the lower-order modes.

2-4-1 Maxwell's Equations

To analyze the optical waveguide we need to consider Maxwell's equations that give the relationships between the electric and magnetic fields. Assuming a linear, isotropic dielectric material having no currents and free charges, these equations take the form[2]

$$\nabla \times \mathbf{E} = -\frac{\partial \mathbf{B}}{\partial t} \tag{2-19a}$$

$$\nabla \times \mathbf{H} = \frac{\partial \mathbf{D}}{\partial t} \tag{2-19b}$$

$$\nabla \cdot \mathbf{D} = 0 \tag{2-19c}$$

$$\nabla \cdot \mathbf{B} = 0 \tag{2-19d}$$

where $\mathbf{D} = \epsilon \mathbf{E}$ and $\mathbf{B} = \mu \mathbf{H}$. The parameter ϵ is the permittivity (or dielectric constant) and μ is the permeability of the medium.

A relationship defining the wave phenomena of the electromagnetic fields can be derived from Maxwell's equations. Taking the curl of Eq. (2-19a) and making use of Eq. (2-19b) yields

$$\nabla \times (\nabla \times \mathbf{E}) = -\mu \frac{\partial}{\partial t}(\nabla \times \mathbf{H}) = -\epsilon\mu \frac{\partial^2 \mathbf{E}}{\partial t^2} \tag{2-20}$$

Using the vector identity (see App. B)

$$\nabla \times (\nabla \times \mathbf{E}) = \nabla(\nabla \cdot \mathbf{E}) - \nabla^2\mathbf{E}$$

and using Eq. (2-19c) (that is, $\nabla \cdot \mathbf{E} = 0$), Eq. (2-20) becomes

$$\nabla^2\mathbf{E} = \epsilon\mu\frac{\partial^2\mathbf{E}}{\partial t^2} \tag{2-21}$$

Similarly, by taking the curl of Eq. (2-19b), it can be shown that

$$\nabla^2\mathbf{H} = \epsilon\mu\frac{\partial^2\mathbf{H}}{\partial t^2} \tag{2-22}$$

Equations (2-21) and (2-22) are the standard *wave equations*.

2-4-2 Waveguide Equations

Consider electromagnetic waves propagating along a cylindrical fiber shown in Fig. 2-14. For this fiber a cylindrical coordinate system $\{r, \phi, z\}$ is defined with the z axis lying along the axis of the waveguide. If the electromagnetic waves are to propagate along the z axis, they will have a functional dependence of the form

$$\mathbf{E} = \mathbf{E}_0(r,\phi)e^{j(\omega t - \beta z)} \tag{2-23}$$

$$\mathbf{H} = \mathbf{H}_0(r,\phi)e^{j(\omega t - \beta z)} \tag{2-24}$$

which are harmonic in time t and coordinate z. The parameter β is the z component of the propagation vector and will be determined by the boundary conditions on the electromagnetic fields at the core-cladding interface described in Sec. 2-4-4.

When Eqs. (2-23) and (2-24) are substituted into Maxwell's curl equations, we have, from Eq. (2-19a),

$$\frac{1}{r}\left(\frac{\partial E_z}{\partial \phi} + jr\beta E_\phi\right) = -j\omega\mu H_r \tag{2-25a}$$

$$j\beta E_r + \frac{\partial E_z}{\partial r} = j\omega\mu H_\phi \tag{2-25b}$$

$$\frac{1}{r}\left[\frac{\partial}{\partial r}(rE_\phi) - \frac{\partial E_r}{\partial \phi}\right] = -j\mu\omega H_z \tag{2-25c}$$

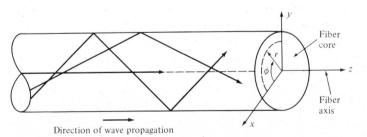

Figure 2-14 Cylindrical coordinate system used for analyzing electromagnetic wave propagation in an optical fiber.

and from Eq. (2-19b)

$$\frac{1}{r}\left(\frac{\partial H_z}{\partial \phi} + jr\beta H_\phi\right) = j\epsilon\omega E_r \qquad (2\text{-}26a)$$

$$j\beta H_r + \frac{\partial H_z}{\partial r} = -j\epsilon\omega E_\phi \qquad (2\text{-}26b)$$

$$\frac{1}{r}\left[\frac{\partial}{\partial r}(rH_\phi) - \frac{\partial H_r}{\partial \phi}\right] = j\epsilon\omega E \qquad (2\text{-}26c)$$

By eliminating variables these equations can be rewritten such that, when E_z and H_z are known, the remaining transverse components E_r, E_ϕ, H_r, and H_ϕ can be determined. For example, E_ϕ or H_r can be eliminated from Eqs. (2-25b) and (2-26b) so that the component H_ϕ or E_r, respectively, can be found in terms of E_z or H_z. Doing so yields

$$E_r = -\frac{j}{q^2}\left(\beta\frac{\partial E_z}{\partial r} + \frac{\mu\omega}{r}\frac{\partial H_z}{\partial \phi}\right) \qquad (2\text{-}27a)$$

$$E_\phi = -\frac{j}{q^2}\left(\frac{\beta}{r}\frac{\partial E_z}{\partial \phi} - \mu\omega\frac{\partial H_z}{\partial r}\right) \qquad (2\text{-}27b)$$

$$H_r = \frac{-j}{q^2}\left(\beta\frac{\partial H_z}{\partial r} - \frac{\omega\epsilon}{r}\frac{\partial E_z}{\partial \phi}\right) \qquad (2\text{-}27c)$$

$$H_\phi = \frac{-j}{q^2}\left(\frac{\beta}{r}\frac{\partial H_z}{\partial \phi} + \omega\epsilon\frac{\partial E_z}{\partial r}\right) \qquad (2\text{-}27d)$$

where $q^2 = \omega^2\epsilon\mu - \beta^2 = k^2 - \beta^2$.

Substitution of Eqs. (2-27c) and (2-27d) into Eq. (2-26c) results in the wave equation in cylindrical coordinates

$$\frac{\partial^2 E_z}{\partial r^2} + \frac{1}{r}\frac{\partial E_z}{\partial r} + \frac{1}{r^2}\frac{\partial^2 E_z}{\partial \phi^2} + q^2 E_z = 0 \qquad (2\text{-}28)$$

and substitution of Eqs. (2-27a) and (2-27b) into Eq. (2-25c) leads to

$$\frac{\partial^2 H_z}{\partial r^2} + \frac{1}{r}\frac{\partial H_z}{\partial r} + \frac{1}{r^2}\frac{\partial^2 H_z}{\partial \phi^2} + q^2 H_z = 0 \qquad (2\text{-}29)$$

It is interesting to note that Eqs. (2-28) and (2-29) each contain either only E_z or H_z. This appears to imply that the longitudinal components of $\mathbf{E}$ and $\mathbf{H}$ are uncoupled and can be chosen arbitrarily provided that they satisfy Eqs. (2-28) and (2-29). However, in general, coupling of E_z and H_z is required by the boundary conditions of the electromagnetic field components described in Sec. 2-4-4. If the boundary conditions do not lead to coupling between the field components, mode solutions can be obtained in which either $E_z = 0$ or $H_z = 0$. When $E_z = 0$ the modes are called *transverse electric* or TE modes, and when $H_z = 0$ *transverse magnetic* or TM modes result. *Hybrid* modes exist if both E_z and H_z are nonzero.

These are designated as HE or EH modes, depending on whether H_z or E_z, respectively, makes a larger contribution to the transverse field. The fact that hybrid modes are present in optical waveguides makes their analysis more complex compared to the simpler case of hollow metallic waveguides where only TE and TM modes are found.

2-4-3 Wave Equations for Step-Index Fibers

We now use the above results to find the guided modes in a step-index fiber. A standard mathematical procedure for solving equations such as Eq. (2-28) is to use the separation-of-variables method, which assumes a solution of the form

$$E_z = AF_1(r)F_2(\phi)F_3(z)F_4(t) \tag{2-30}$$

As was already assumed, the time- and z-dependent factors are given by

$$F_3(z)F_4(t) = e^{j(\omega t - \beta z)} \tag{2-31}$$

since the wave is sinusoidal in time and propagates in the z direction. In addition, because of the circular symmetry of the waveguide, each field component must not change when the coordinate ϕ is increased by 2π. We thus assume a periodic function of the form

$$F_2(\phi) = e^{j\nu\phi} \tag{2-32}$$

The constant ν can be positive or negative, but it must be an integer since the fields must be periodic in ϕ with a period of 2π.

Substituting Eq. (2-32) into Eq. (2-30) the wave equation for E_z [Eq. (2-28)] becomes

$$\frac{\partial^2 F_1}{\partial r^2} + \frac{1}{r}\frac{\partial F_1}{\partial r} + \left(q^2 - \frac{\nu^2}{r^2}\right)F_1 = 0 \tag{2-33}$$

which is a well-known differential equation for Bessel functions.[24-26] An exactly identical equation can be derived for H_z.

For the configuration of the step-index fiber we consider a homogeneous core of refractive index n_1 and radius a, which is surrounded by an infinite cladding of index n_2. The reason for assuming an infinitely thick cladding is that the guided modes in the core have exponentially decaying fields outside the core which must have insignificant values at the outer boundary of the cladding. In practice, optical fibers are designed with claddings that are sufficiently thick so that the guided-mode field does not reach the outer boundary of the cladding. To get an idea of the field patterns, the electric field distributions for several of the lower-order guided modes in a symmetrical slab waveguide were shown in Fig. 2-13. The fields vary harmonically in the guiding region of refractive index n_1 and decay exponentially outside of this region.

Equation (2-33) must now be solved for the regions inside and outside the core. For the inside region the solutions for the guided modes must remain finite as $r \to 0$, whereas on the outside the solutions must decay to zero as $r \to \infty$. Thus for $r < a$

the solutions are Bessel functions of the first kind of order ν. For these functions we use the common designation $J_\nu(ur)$. Here $u^2 = k_1^2 - \beta^2$ with $k_1 = 2\pi n_1/\lambda$. The expressions for E_z and H_z inside the core are thus

$$E_z(r < a) = AJ_\nu(ur)e^{j\nu\phi}e^{j(\omega t - \beta z)} \tag{2-34}$$

$$H_z(r < a) = BJ_\nu(ur)e^{j\nu\phi}e^{j(\omega t - \beta z)} \tag{2-35}$$

where A and B are arbitrary constants.

Outside of the core the solutions to Eq. (2-33) are given by modified Bessel functions of the second kind $K_\nu(wr)$, where $w^2 = \beta^2 - k_2^2$ with $k_2 = 2\pi n_2/\lambda$. The expressions for E_z and H_z outside the core are, therefore,

$$E_z(r > a) = CK_\nu(wr)e^{j\nu\phi}e^{j(\omega t - \beta z)} \tag{2-36}$$

$$H_z(r > a) = DK_\nu(wr)e^{j\nu\phi}e^{j(\omega t - \beta z)} \tag{2-37}$$

with C and D being arbitrary constants.

The definitions of $J_\nu(ur)$ and $K_\nu(wr)$ and various recursion relations are given in App. C. From the definition of the modified Bessel function, it is seen that $K_\nu(wr) \to e^{-wr}$ as $wr \to \infty$. Since $K_\nu(wr)$ must go to zero as $r \to \infty$, it follows that $w > 0$. This, in turn, implies that $\beta \geq k_2$, which represents a cutoff condition. The *cutoff condition* is the point at which a mode is no longer bound to the core region. A second condition on β can be deduced from the behavior of $J_\nu(ur)$. Inside the core the parameter u must be real for F_1 to be real, from which it follows that $k_1 \geq \beta$. The permissible range of β for bound solutions is, therefore,

$$n_2 k = k_2 \leq \beta \leq k_1 = n_1 k \tag{2-38}$$

where $k = 2\pi/\lambda$ is the free-space propagation constant.

2-4-4 Modal Equation

The solutions for β must be determined from the boundary conditions. The boundary conditions require that the tangential components E_ϕ and E_z of $\mathbf{E}$ inside and outside of the dielectric interface at $r = a$ must be the same and, similarly, for the tangential components H_ϕ and H_z. Consider first the tangential components of $\mathbf{E}$. For the z component we have, from Eq. (2-34) at the inner core-cladding boundary ($E_z = E_{z1}$) and from Eq. (2-36) at the outside of the boundary ($E_z = E_{z2}$), that

$$E_{z1} - E_{z2} = AJ_\nu(ua) - CK_\nu(wa) = 0 \tag{2-39}$$

The ϕ component is found from Eq. (2-27b). Inside the core the factor q^2 is given by

$$q^2 = u^2 = k_1^2 - \beta^2 \tag{2-40}$$

where $k_1 = 2\pi n_1/\lambda = \omega\sqrt{\epsilon_1\mu}$, while outside the core

$$w^2 = \beta^2 - k_2^2 \tag{2-41}$$

with $k_2 = 2\pi n_2/\lambda = \omega\sqrt{\epsilon_2\mu}$. Substituting Eqs. (2-34) and (2-35) into Eq.

(2-27b) to find $E_{\phi 1}$, and similarly using Eqs. (2-36) and (2-37) to determine $E_{\phi 2}$, yields at $r = a$

$$E_{\phi 1} - E_{\phi 2} = -\frac{j}{u^2}\left[A\frac{jv\beta}{a}J_\nu(ua) - B\omega\mu u J'_\nu(ua)\right]$$
$$-\frac{j}{w^2}\left[C\frac{jv\beta}{a}K_\nu(wa) - D\omega\mu w K'_\nu(wa)\right] = 0 \quad (2\text{-}42)$$

where the prime indicates differentiation with respect to the argument.

Similarly, for the tangential components of **H** it is readily shown that at $r = a$

$$H_{z1} - H_{z2} = BJ_\nu(ua) - DK_\nu(wa) = 0 \quad (2\text{-}43)$$

and

$$H_{\phi 1} - H_{\phi 2} = -\frac{j}{u^2}\left[B\frac{jv\beta}{a}J_\nu(ua) + A\omega\epsilon_1 u J'_\nu(ua)\right]$$
$$-\frac{j}{w^2}\left[D\frac{jv\beta}{a}K_\nu(wa) + C\omega\epsilon_2 w K'_\nu(wa)\right] = 0 \quad (2\text{-}44)$$

Equations (2-39), (2-42), (2-43), and (2-44) are a set of four equations with four unknown coefficients $A, B, C,$ and D. A solution to these equations exists only if the determinant of these coefficients is zero:

$$\begin{vmatrix} J_\nu(ua) & 0 & -K_\nu(wa) & 0 \\[2mm] \dfrac{\beta v}{au^2}J_\nu(ua) & \dfrac{j\omega\mu}{u}J'_\nu(ua) & \dfrac{\beta v}{aw^2}K_\nu(wa) & \dfrac{j\omega\mu}{w}K'_\nu(wa) \\[2mm] 0 & J_\nu(ua) & 0 & -K_\nu(wa) \\[2mm] -\dfrac{j\omega\epsilon_1}{u}J'_\nu(ua) & \dfrac{\beta v}{au^2}J_\nu(ua) & -\dfrac{j\omega\epsilon_2}{w}K'_\nu(wa) & \dfrac{\beta v}{aw^2}K_\nu(wa) \end{vmatrix} = 0 \quad (2\text{-}45)$$

Evaluation of this determinant yields the following eigenvalue equation for β:

$$(\mathcal{J}_\nu + \mathcal{K}_\nu)(k_1^2\mathcal{J}_\nu + k_2^2\mathcal{K}_\nu) = \left(\frac{\beta v}{a}\right)^2\left(\frac{1}{u^2} + \frac{1}{w^2}\right)^2 \quad (2\text{-}46)$$

where

$$\mathcal{J}_\nu = \frac{J'_\nu(ua)}{uJ_\nu(ua)} \quad \text{and} \quad \mathcal{K}_\nu = \frac{K'_\nu(wa)}{wK_\nu(wa)}$$

Upon solving Eq. (2-46) for β it will be found that only discrete values restricted to the range given by Eq. (2-38) will be allowed. Although Eq. (2-46) is a complicated transcendental equation which is generally solved by numerical techniques, its solution for any particular mode will provide all the characteristics of that mode. We shall now consider this equation for some of the lower-order modes of a step-index waveguide.

2-4-5 Modes in Step-Index Fibers

To help describe the modes we shall first examine the behavior of the J-type Bessel functions. These are plotted in Fig. 2-15 for the first three orders. The J-type Bessel functions are similar to harmonic functions since they exhibit oscillatory behavior for real k as is the case for sinusoidal functions. Because of the oscillatory behavior of J_ν there will be m roots of Eq. (2-46) for a given ν value. These roots will be designated by $\beta_{\nu m}$ and the corresponding modes are either $\text{TE}_{\nu m}$, $\text{TM}_{\nu m}$, $\text{EH}_{\nu m}$, or $\text{HE}_{\nu m}$.

For the dielectric fiber waveguide all modes are hybrid modes except those for which $\nu = 0$. When $\nu = 0$ the right-hand side of Eq. (2-46) vanishes and two different eigenvalue equations result. These are

$$\mathcal{J}_0 + \mathcal{K}_0 = 0 \tag{2-47a}$$

or, using the relations for J'_ν and K'_ν from App. C,

$$\frac{J_1(ua)}{uJ_0(ua)} + \frac{K_1(wa)}{wK_0(wa)} = 0 \tag{2-47b}$$

which corresponds to TM_{0m} modes ($E_z = 0$), and

$$k_1^2 \mathcal{J}_0 + k_2^2 \mathcal{K}_0 = 0 \tag{2-48a}$$

or

$$\frac{k_1^2 J_1(ua)}{uJ_0(ua)} + \frac{k_2^2 K_1(wa)}{wK_0(wa)} = 0 \tag{2-48b}$$

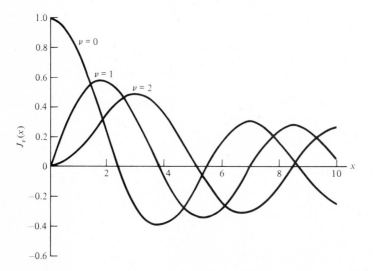

Figure 2-15 Variation of the Bessel function $J_\nu(x)$ for the first three orders ($\nu = 1, 2, 3$) plotted as a function of x.

which corresponds to TE_{0m} modes ($H_z = 0$). The proof of this is left as an exercise (see Prob. 2-16).

When $\nu \neq 0$ the situation is more complex and numerical methods are needed to solve Eq. (2-46) exactly. However, simplified and highly accurate approximations based on the principle that the core and cladding refractive indices are nearly the same have been derived by Snyder[19] and Gloge.[20] The condition that $n_1 - n_2 \ll 1$ was referred to by Gloge as giving rise to *weakly guided* modes. For a treatment of these derivations the interested reader is referred to the literature.[10,19,20]

Let us next examine the cutoff conditions for fiber modes. As was mentioned in relation to Eq. (2-38), a mode is referred to as being cut off when it is no longer bound to the core of the fiber so that its field no longer decays on the outside of the core. The cutoffs for the various modes are found by solving Eq. (2-46) in the limit $w^2 \to 0$. This is, in general, fairly complex so that only the results,[14,16] which are listed in Table 2-1, will be given here.

An important parameter connected with the cutoff condition is the *normalized frequency* V (also called the *V-number* or *V-parameter*) defined by

$$V^2 = (u^2 + w^2)a^2 = \left(\frac{2\pi a}{\lambda}\right)^2 (n_1^2 - n_2^2) \qquad (2\text{-}49)$$

which is a dimensionless number that determines how many modes a fiber can support. The number of modes that can exist in a waveguide as a function of V may be conveniently represented in terms of a *normalized propagation constant b* defined by[20]

$$b = \frac{a^2 w^2}{V^2} = \frac{(\beta/k)^2 - n_2^2}{n_1^2 - n_2^2}$$

A plot of b as a function of V is shown in Fig. 2-16 for a few of the low-order modes. This figure shows that each mode can exist only for values of V that exceed a certain limiting value. The modes are cut off when $\beta/k = n_2$. The HE_{11} mode has no cutoff and ceases to exist only when the core diameter is zero. This is the

Table 2-1 Cutoff conditions for some lower-order modes

ν	Mode	Cutoff condition
0	TE_{0m}, TM_{0m}	$J_0(ua) = 0$
1	HE_{1m}, EH_{1m}	$J_1(ua) = 0$
≥ 2	$EH_{\nu m}$	$J_\nu(ua) = 0$
	$HE_{\nu m}$	$\left(\dfrac{n_1^2}{n_2^2} + 1\right)J_{\nu-1}(ua) = \dfrac{ua}{\nu - 1}J_\nu(ua)$

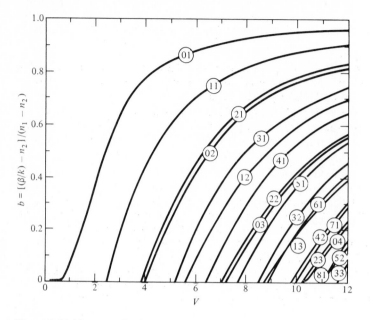

Figure 2-16 The normalized propagation parameter b as a function of the V number. When $\nu \neq 1$, the curve numbers νm designate the $HE_{\nu+1,m}$ and $EH_{\nu-1,m}$ modes. For $\nu = 1$, the curve numbers νm give the HE_{2m}, TE_{0m}, and TM_{0m} modes. *(Reproduced with permission from Gloge.[20])*

principle on which the single-mode fiber is based. By appropriately choosing a, n_1, and n_2 so that

$$V = \frac{2\pi a}{\lambda}(n_1^2 - n_2^2)^{1/2} \geq 2.405 \qquad (2\text{-}50)$$

which is the value at which the lowest-order Bessel function J_0 is zero (see Fig. 2-15), all modes except the HE_{11} mode are cut off.

Single-mode fibers are constructed by letting the dimensions of the core diameter be a few wavelengths (usually 8 to 12) and by having small index differences between the core and the cladding. From Eq. (2-50) with $V = 2.4$, it can be seen that single-mode propagation is possible for fairly large variations in values of the physical core sizes a and the core-cladding index differences Δ. However in practical designs of single-mode fibers,[27] the core-cladding index difference varies between 0.1 and 0.2 percent, and the core diameter should be chosen to be just below the cutoff of the first higher-order mode; that is, for V slightly less than 2.4. For example, a typical single-mode fiber may have a core radius of 3 μm and a numerical aperture of 0.1 at a wavelength of 0.8 μm. From Eqs. (2-15) and (2-49) this yields $V = 2.356$.

The parameter V can also be related to the number of modes M in a multimode fiber when M is large. An approximate relationship for step-index fibers can be derived from ray theory. A ray congruence incident on the end of a fiber will be accepted by the fiber if it lies within an angle θ defined by the numerical aperture as given in Eq. (2-15)

$$NA = \sin \theta = (n_1^2 - n_2^2)^{1/2} \tag{2-51}$$

For practical numerical apertures $\sin \theta$ is small so that $\sin \theta \simeq \theta$. The solid acceptance angle for the fiber is therefore

$$\Omega = \pi \theta^2 = \pi(n_1^2 - n_2^2) \tag{2-52}$$

For electromagnetic radiation of wavelength λ emanating from a laser or a waveguide the number of modes per unit solid angle is given by $2A/\lambda^2$, where A is the area the mode is leaving or entering.[28] The area A in this case is the core cross section πa^2. The factor 2 comes from the fact that the plane wave can have two polarization orientations. The total number of modes M entering the fiber is thus given by

$$M \simeq \frac{2A}{\lambda^2}\Omega = \frac{2\pi^2 a^2}{\lambda^2}(n_1^2 - n_2^2) = \frac{V^2}{2} \tag{2-53}$$

2-4-6 Power Flow in Step-Index Fibers

A final quantity of interest for step-index fibers is the fractional power flow in the core and cladding for a given mode. As is illustrated in Fig. 2-13, the electromagnetic field for a given mode does not go to zero at the core-cladding interface, but changes from an oscillating form in the core to an exponential decay in the cladding. Thus the electromagnetic energy of a guided mode is carried partially in the core and partially in the cladding. The further away a mode is from its cutoff frequency the more concentrated its energy is in the core. As cutoff is approached, the field penetrates further into the cladding region and a greater percentage of the energy travels in the cladding. At cutoff the field no longer decays outside the core and the mode now becomes a fully radiating mode.

The relative amounts of power flowing in the core and the cladding can be obtained by integrating the Poynting vector in the axial direction

$$S_z = \tfrac{1}{2}\mathrm{Re}(\mathbf{E} \times \mathbf{H}^*) \cdot \mathbf{e}_z \tag{2-54}$$

over the fiber cross section. Thus the power in the core and cladding, respectively, is given by

$$P_{\mathrm{core}} = \frac{1}{2}\int_0^a \int_0^{2\pi} r(E_x H_y^* - E_y H_x^*)\,d\phi\,dr \tag{2-55}$$

$$P_{\mathrm{clad}} = \frac{1}{2}\int_a^\infty \int_0^{2\pi} r(E_x H_y^* - E_y H_x^*)\,d\phi\,dr \tag{2-56}$$

where the star denotes the complex conjugate. Gloge[20,29] has shown that, based on the weakly guided mode approximation which has an accuracy on the order of the index difference Δ between the core and cladding, the relative core and cladding powers for a particular mode ν are given by

$$\frac{P_{\text{core}}}{P} = \left(1 - \frac{u^2}{V^2}\right)\left[1 - \frac{J_\nu^2(ua)}{J_{\nu+1}(ua)J_{\nu-1}(ua)}\right] \tag{2-57}$$

and

$$\frac{P_{\text{clad}}}{P} = 1 - \frac{P_{\text{core}}}{P} \tag{2-58}$$

where P is the total power in the mode ν. The relationships between P_{core} and P_{clad} are plotted in Fig. 2-17 in terms of the fractional powers P_{core}/P and P_{clad}/P. In addition, far from cutoff the average total power in the cladding has been derived for fibers in which many modes can propagate. Because of this large number of modes, those few modes that are appreciably close to cutoff can be ignored to a reasonable approximation. The derivation assumes an incoherent source, such as a tungsten filament lamp or a light-emitting diode, which, in general, excites every fiber mode with the same amount of power. The total average cladding power is thus approximated by[20]

$$\left(\frac{P_{\text{clad}}}{P}\right)_{\text{total}} = \tfrac{4}{3} M^{-1/2} \tag{2-59}$$

From Fig. 2-17 and Eq. (2-59) it can be seen that, since M is proportional to V^2, the power flow in the cladding decreases as V increases.

As an example, consider a fiber having a core radius of 25 μm, a core index of 1.48, and $\Delta = 0.01$. At an operating wavelength of 0.84 μm the value of V is

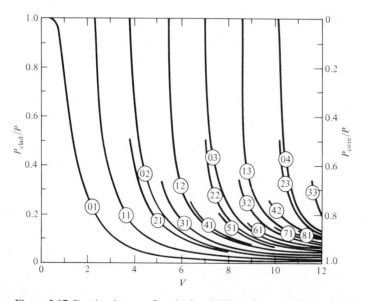

Figure 2-17 Fractional power flow in the cladding of a step-index optical fiber as a function of V. When $\nu \neq 1$, the curve numbers νm designate the HE$_{\nu+1,m}$ and EH$_{\nu-1,m}$ modes. For $\nu = 1$, the curve numbers νm give the HE$_{2m}$, TE$_{0m}$, and TM$_{0m}$ modes. (*Reproduced with permission from Gloge.[20]*)

39 and there are 760 modes in the fiber. From Eq. (2-59) approximately 5 percent of the power propagates in the cladding. If Δ is decreased to, say 0.003, in order to decrease signal dispersion (see Chap. 3), then 242 modes propagate in the fiber and about 9 percent of the power resides in the cladding. For the case of the single-mode fiber, considering the HE_{11} mode in Fig. 2-17, it is seen that for $V = 1$ about 70 percent of the power propagates in the cladding, whereas for $V = 2.405$, which is where the TE_{01} mode begins, approximately 84 percent of the power is now within the core.

2-5 GRADED-INDEX FIBER STRUCTURE

In the graded-index fiber design the core refractive index decreases continuously with radial distance r from the center of the fiber but is generally constant in the cladding. The most commonly used construction for the refractive-index variation in the core is the power law relationship

$$n(r) = \begin{cases} n_1\left[1 - 2\Delta\left(\frac{r}{a}\right)^\alpha\right]^{1/2} & \text{for } 0 \le r \le a \\ n_1(1 - 2\Delta)^{1/2} \simeq n_1(1 - \Delta) = n_2 & \text{for } r \ge a \end{cases} \tag{2-60}$$

Here r is the radial distance from the fiber axis, a is the core radius, n_1 is the refractive index at the core axis, n_2 is the refractive index of the cladding, and the dimensionless parameter α defines the shape of the index profile. The index difference Δ for the graded-index fiber is given by

$$\Delta = \frac{n_1^2 - n_2^2}{2n_1^2} \simeq \frac{n_1 - n_2}{n_1} \tag{2-61}$$

The approximation on the right-hand side of this equation reduces the expression for Δ to that of the step-index fiber given by Eq. (2-12). Thus the same symbol is used in both cases. For $\alpha = \infty$, Eq. (2-60) reduces to the step-index profile $n(r) = n_1$.

2-5-1 Graded-Index Numerical Aperture (NA)

The determination of the NA for graded-index fibers is more complex than for step-index fibers. In graded-index fibers the NA is a function of position across the core end face. This is in contrast to the step-index fiber where the NA is constant across the core. Geometrical optics considerations show that light incident on the fiber core at position r will propagate as a guided mode only if it is within the local numerical aperture $NA(r)$ at that point. The local numerical aperture is defined as[30]

$$NA(r) = \begin{cases} [n^2(r) - n_2^2]^{1/2} \simeq NA(0)\sqrt{1 - (r/a)^\alpha} & \text{for } r \le a \\ 0 & \text{for } r > a \end{cases} \tag{2-62}$$

where the axial numerical aperture is defined as

$$NA(0) = [n^2(0) - n_2^2]^{1/2} = (n_1^2 - n_2^2)^{1/2} \simeq n_1\sqrt{2\Delta} \tag{2-63}$$

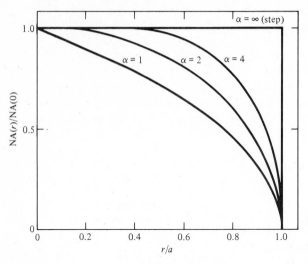

Figure 2-18 A comparison of the numerical apertures for fibers having various α profiles.

From Eq. (2-62) it is clear that the NA of a graded-index fiber decreases from NA(0) to zero as r moves from the fiber axis to the core-cladding boundary. A comparison of the numerical apertures for fibers having various α profiles is given in Fig. 2-18.

2-5-2 Modes in Graded-Index Fibers

A modal analysis of an optical fiber based on solving Maxwell's equations can only be carried out rigorously if the core refractive index is uniform, that is, for step-index fibers. In other cases, such as for graded-index fibers, approximation methods are needed. The most widely used analysis of modes in a graded-index fiber is an approximation based on the WKB method[31,32] (named after Wenzel, Kramers, and Brillouin) which is commonly used in quantum mechanics. The purpose of the WKB method is to obtain an asymptotic representation for the solution of a differential equation containing a parameter that varies slowly over the desired range of the equation. That parameter in this case in the refractive-index profile $n(r)$ which varies only slightly over distances on the order of an optical wavelength.

Analogous to the step-index fiber, Eq. (2-33) for the radial component of the wave equation must be solved:[6,10,29]

$$\frac{d^2F_1}{dr^2} + \frac{1}{r}\frac{dF_1}{dr} + \left[k^2n^2(r) - \beta^2 - \frac{\nu^2}{r^2}\right]F_1 = 0 \qquad (2\text{-}64)$$

where $n(r)$ is given by Eq. (2-60). The general procedure in the WKB method is to let[29]

$$F_1 = Ae^{jkQ(r)} \qquad (2\text{-}65)$$

where the coefficient A is independent of r. Substituting this into Eq. (2-64) gives

$$jkQ'' - (kQ')^2 + \frac{jk}{r}Q' + \left[k^2n^2(r) - \beta^2 - \frac{\nu^2}{r^2}\right] = 0 \qquad (2\text{-}66)$$

where the primes denote differentiation with respect to r. Since $n(r)$ varies slowly over a distance on the order of a wavelength, an expansion of the function $Q(r)$ in powers of λ or, equivalently, in powers of $k^{-1} = \lambda/2\pi$ is expected to converge rapidly. Thus we let

$$Q(r) = Q_0 + \frac{1}{k}Q_1 + \cdots \tag{2-67}$$

where $Q_0, Q_1, \ldots$ are certain functions of r. Substituting Eq. (2-67) into Eq. (2-66) and collecting equal powers of k yield

$$\left\{-k^2(Q_0')^2 + \left[k^2n^2(r) - \beta^2 - \frac{\nu^2}{r^2}\right]\right\} + \left(jkQ_0'' - 2kQ_0'Q_1' + \frac{jk}{r}Q_0'\right)$$

$$+ \text{ terms of order } (k^0, k^{-1}, k^{-2}, \ldots) = 0 \tag{2-68}$$

A sequence of defining relations for the Q_i functions are obtained by setting to zero the terms in equal powers of k. Thus, for the first two terms of Eq. (2-68),

$$-k^2(Q_0')^2 + \left[k^2n^2(r) - \beta^2 - \frac{\nu^2}{r^2}\right] = 0 \tag{2-69}$$

$$jkQ_0'' - 2kQ_0'Q_1' + \frac{jk}{r}Q_0' = 0 \tag{2-70}$$

Integration of Eq. (2-69) yields

$$kQ_0 = \int_{r_1}^{r_2} \left[k^2n^2(r) - \beta^2 - \frac{\nu^2}{r^2}\right]^{1/2} dr \tag{2-71}$$

A mode is bound in the fiber core only if Q_0 is real. For Q_0 to be real, the radical in the integrand must be greater than zero. In general, for a given mode ν, there are two values r_1 and r_2 for which the radical is zero as is indicated by the limits of integration in Eq. (2-71). Note that these values of r are functions of ν. Guided modes exist for r between these two values. For other values of r the function Q_0 is imaginary, which leads to decaying fields.

To help visualize the solutions to Eq. (2-71), consider the cross-sectional projection of a skew ray in a graded-index fiber shown in Fig. 2-19. The path followed by the ray lies completely within the boundaries of two coaxial cylindrical surfaces, known as the *caustic surfaces,* that have inner and outer radii r_1 and r_2, respectively. The radii r_1 and r_2 are those points at which the radical in the integrand of Eq. (2-71) becomes zero. They are called *turning points* since the ray turns from increasing to decreasing values of r or vice versa. To evaluate the turning points, consider the functions

$$k^2n^2(r) - \beta^2$$

and ν^2/r^2 plotted in Fig. 2-19 as solid and dashed curves, respectively. The crossing points of these two curves give the points r_1 and r_2. An oscillating field exists when the solid curve lies above the dashed curve, which indicates bound mode solutions. Evanescent (nonbound decaying mode) fields occur when the solid curve lies below the dashed curve.

To form a bound mode of the graded-index fiber, each wave associated with the ray congruence corresponding to this mode must interfere constructively with

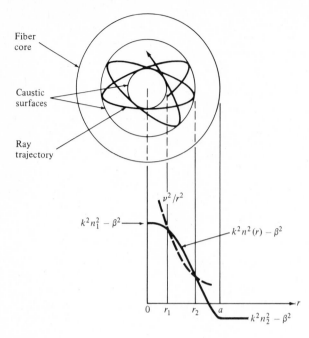

Fiber core

Caustic surfaces

Ray trajectory

Figure 2-19 Cross-sectional projection of a skew ray in a graded-index fiber and the graphical representation of its mode solution from the WKB method. The field is oscillatory between the turning points r_1 and r_2 and is evanescent outside of this region.

itself in such a way as to form a standing-wave pattern in the radial cross-sectional direction. (A full mathematical treatment of this can be found in the literature.[33]) This requirement imposes the condition that the phase function Q_0 between r_1 and r_2 must be a multiple of π (that is, an integer number of half-periods), so that

$$m\pi \simeq \int_{r_1}^{r_2} \left[k^2 n^2(r) - \beta^2 - \frac{\nu^2}{r^2} \right]^{1/2} dr \qquad (2\text{-}72)$$

where $m = 0, 1, 2, \ldots$ is the radial mode number that counts the number of half-periods between the turning points. The total number of bound modes $m(\beta)$ can be found by summing Eq. (2-72) over all ν from 0 to ν_{max}, where ν_{max} is the highest-order bound mode for a given value of β. If ν_{max} is a large number, the sum can be replaced by an integral yielding

$$m(\beta) = \frac{4}{\pi} \int_0^{\nu_{max}} \int_{r_1(\nu)}^{r_2(\nu)} \left[k^2 n^2(r) - \beta^2 - \frac{\nu^2}{r^2} \right]^{1/2} dr \, d\nu \qquad (2\text{-}73)$$

The factor 4 arises from the fact that each combination (m, ν) designates a degenerate group of four modes of different polarization or orientation.[20] If we change the order of integration, the lower limit on r must be $r_1 = 0$ in order to count all the modes, and the upper limit on ν is found from the condition

$$k^2 n^2(r) - \beta^2 - \frac{\nu^2_{max}}{r^2} = 0 \qquad (2\text{-}74)$$

Thus

$$m(\beta) = \frac{4}{\pi} \int_0^{r_2} \int_0^{\nu_{max}} \left[k^2 n^2(r) - \beta^2 - \frac{\nu^2}{r^2} \right]^{1/2} d\nu \, dr \qquad (2\text{-}75)$$

Evaluating the integral over ν with ν_{max} given by Eq. (2-74) yields

$$m(\beta) = \int_0^{r_2} [k^2 n^2(r) - \beta^2] r \, dr \tag{2-76}$$

To evaluate this further, we choose the index profile $n(r)$ given by Eq. (2-60). The upper limit of integration r_2 is determined from the condition that

$$kn(r) = \beta$$

Combining this condition with Eq. (2-60) gives

$$r_2 = a\left[\frac{1}{2\Delta}\left(1 - \frac{\beta^2}{k^2 n_1^2}\right) \right]^{1/\alpha} \tag{2-77}$$

Using Eqs. (2-60) and (2-77), the number of modes is

$$m(\beta) = a^2 k^2 n_1^2 \Delta \frac{\alpha}{\alpha + 2}\left(\frac{k^2 n_1^2 - \beta^2}{2 \Delta k^2 n_1^2} \right)^{(2+\alpha)/\alpha} \tag{2-78}$$

All bound modes in a fiber must have $\beta \geq kn_2$. If this condition does not hold, the mode is no longer perfectly trapped inside the core and loses power by leakage into the cladding. The maximum number of bound modes M is thus found by letting

$$\beta = kn_2 = kn_1(1 - \Delta)$$

where Eq. (2-60) was used for the relationship between n_1 and n_2. Thus

$$M = m(kn_2) = \frac{\alpha}{\alpha + 2} a^2 k^2 n_1^2 \Delta \tag{2-79}$$

gives the total number of bound modes in a graded-index fiber having a refractive-index profile given by Eq. (2-70).

2-6 SUMMARY

In this chapter we have examined the structure of optical fibers and have considered two mechanisms that show how light propagates along these fibers. In its simplest form an optical fiber is a coaxial cylindrical arrangement of two homogeneous dielectric (glass or plastic) materials. This fiber consists of a central core of refractive index n_1 surrounded by a cladding region of refractive index n_2 that is less than n_1. This configuration is referred to as a step-index fiber because the cross-sectional refractive-index profile has a step function at the interface between the core and the cladding.

Graded-index fibers are those in which the refractive-index profile varies as a function of the radial coordinate r in the core but is constant in the cladding. This index profile $n(r)$ is often represented as a power law

$$n(r) = \begin{cases} n_1\left[1 - 2\Delta\left(\frac{r}{a}\right)^\alpha \right]^{1/2} & \text{for } r \leq a \\ n_2 & \text{for } r > a \end{cases}$$

where α defines the shape of the index profile, a is the fiber core radius, and Δ is the relative index difference between the maximum value n_1 on the fiber axis and the value n_2 in the cladding:

$$\Delta = \frac{n_1^2 - n_2^2}{2n_1^2} \simeq \frac{n_1 - n_2}{n_1}$$

A commonly used value of the power law exponent is $\alpha = 2$. This special case is referred to as a parabolic graded-index profile. As we shall show in Chap. 3, a graded-index profile reduces signal distortion and, thus, provides a wider bandwidth than a step-index fiber.

A general picture of light propagation in an optical fiber can be obtained by considering a ray-tracing (or geometrical optics) model in a slab waveguide. The slab consists of a central region of refractive index n_1 which is sandwiched between two material layers having a lower refractive index n_2. Light rays propagate along the slab waveguide by undergoing total internal reflection in accordance with Snell's law at the interfaces of these two materials.

Although the ray model is adequate for an intuitive picture of how light travels along a fiber, a more comprehensive description of light propagation, signal distortion, and power loss in a cylindrical optical fiber waveguide requires a wave theory approach. In this wave approach, electromagnetic fields (at optical frequencies) traveling in the fiber can be expressed as superpositions of elementary field configurations called the modes of the fiber. A mode of monochromatic light of radian frequency ω traveling in the axial (positive z) direction in a fiber can be described by the factor

$$e^{j(\omega t - \beta z)}$$

where β is the propagation constant of the mode. For guided (bound) modes, β can only assume a finite number of possible solutions. These are found by solving Maxwell's equations for a dielectric medium subject to the boundary conditions of the optical fiber. Here is where the analysis becomes rather complex. The boundary conditions at the core-cladding interface lead to a coupling between the longitudinal components of the $\mathbf{E}$ and $\mathbf{H}$ fields. This coupling leads to rather involved hybrid mode solutions. The fact that hybrid modes are present in optical fiber waveguides makes their analysis much more complex compared to the simpler case of hollow metallic waveguides, where only transverse electric and transverse magnetic modes are found.

With this information as a background we are now ready to discuss the signal loss mechanisms in a fiber and how and why a signal is distorted as it propagates along an optical waveguide. These are the subjects of Chap. 3.

PROBLEMS

Section 2-1

2-1 A typical sheet of paper is 0.003 in thick. How many wavelengths of 820-nm light (which is widely used in optical fiber systems) will fit into this distance? How does this compare to a 50-μm-diameter optical fiber?

2-2 A wave is specified by $y = 8 \cos 2\pi(2t - 0.8z)$, where y is expressed in μm and the propagation constant is given in μm^{-1}. Find (a) the amplitude, (b) the wavelength, (c) the angular frequency, and (d) the displacement at time $t = 0$ and $z = 4$ μm.

2-3 What are the energies in electron volts (eV) of light having wavelengths of 820 nm and 1.3 μm? What are the values of the propagation constant k of these two wavelengths?

2-4 Consider the following two waves X_1 and X_2 having the same frequency ω but different amplitudes a_i and phases δ_i:

$$X_1 = a_1 \cos(\omega t - \delta_1)$$

$$X_2 = a_2 \cos(\omega t - \delta_2)$$

According to the principle of superposition, the resultant wave X is simply the sum of X_1 and X_2. Show that X can be written in the form

$$X = A \cos(\omega t - \phi)$$

where

$$A^2 = a_1^2 + a_2^2 + 2a_1 a_2 \cos(\delta_1 - \delta_2)$$

and

$$\tan \phi = \frac{a_1 \sin \delta_1 + a_2 \sin \delta_2}{a_1 \cos \delta_1 + a_2 \cos \delta_2}$$

2-5 Elliptically polarized light can be represented by the two orthogonal waves given by Eqs. (2-2) and (2-3). Show that elimination of the $(\omega t - kz)$ dependence between them yields

$$\left(\frac{E_x}{E_{0x}}\right)^2 + \left(\frac{E_y}{E_{0y}}\right)^2 - 2\frac{E_x}{E_{0x}}\frac{E_y}{E_{0y}} \cos \delta = \sin^2 \delta$$

which is the equation of an ellipse making an angle α with the x axis, where α is given by Eq. (2-5).

Section 2-2

2-6 Light traveling in air strikes a glass plate at an angle $\theta_1 = 33°$, where θ_1 is measured between the incoming ray and the glass surface. Upon striking the glass, part of the beam is reflected and part is refracted. If the refracted and reflected beams make an angle of 90° with each other, what is the refractive index of the glass? What is the critical angle for this glass?

2-7 A point source of light is 12 cm below the surface of a large body of water ($n = 1.33$). What is the radius of the largest circle on the water surface through which the light can emerge?

2-8 A 45°-45°-90° prism is immersed in alcohol ($n = 1.45$). What is the minimum index of refraction the prism must have if a ray incident normally on one of the short faces is to be totally reflected at the long face of the prism?

Section 2-3

2-9 Calculate the numerical aperture of a step-index fiber having $n_1 = 1.48$ and $n_2 = 1.46$. What is the maximum entrance angle $\theta_{0,max}$ for this fiber if the outer medium is air with $n = 1$?

2-10 Derive the approximation on the right-hand side of Eq. (2-15).

2-11 (a) Verify the expressions for the various phase changes that are used to derive Eq. (2-18).

  (b) Show that, for the ray that propagates at the critical angle in Fig. 2-12, the integer M in Eq. (2-18) satisfies the condition

$$M = \frac{2n_1^2 d}{\lambda(n_1^2 - n_2^2)^{1/2}}$$

Section 2-4

2-12 Assume the fields of an electromagnetic wave are of the form

$$\mathbf{E}(\mathbf{x}, t) = \mathbf{e}_1 E_0 \exp[j(\omega t - \mathbf{k} \cdot \mathbf{x})]$$

$$\mathbf{B}(\mathbf{x}, t) = \mathbf{e}_2 B_0 \exp[j(\omega t - \mathbf{k} \cdot \mathbf{x})]$$

where $\mathbf{e}_1$ and $\mathbf{e}_2$ are two constant unit vectors with an unspecified orientation.

(a) Using Maxwell's divergence equations show that $\mathbf{E}$ and $\mathbf{B}$ are both perpendicular to the direction of propagation $\mathbf{k}$. (Such a wave is called a transverse wave.)

(b) From Maxwell's curl equations show that

$$\mathbf{e}_2 = \frac{\mathbf{k} \times \mathbf{e}_1}{k} = \frac{\mathbf{k} \times \mathbf{e}_1}{\omega B_0 / E_0}$$

which shows that $(\mathbf{e}_1, \mathbf{e}_2, \mathbf{k})$ form a set of orthogonal vectors and that $\mathbf{E}$ and $\mathbf{B}$ are in phase.

2-13 Show that Maxwell's equations are satisfied by the solution

$$E_x = A \sin(\omega t + kz) \qquad E_y = 0 \qquad E_z = 0$$

$$H_x = 0 \qquad H_y = -A \sqrt{\frac{\epsilon}{\mu}} \sin(\omega t + kz) \qquad H_z = 0$$

In which plane is the wave polarized and in which direction does it travel?

2-14 Derive Eqs. (2-27a) to (2-27d) from Eqs. (2-25) and (2-26). Show that they lead to Eqs. (2-28) and (2-29).

2-15 Evaluate the determinant given by Eq. (2-45) and show that it yields Eq. (2-46).

2-16 Show that for $\nu = 0$ Eq. (2-47b) corresponds to TM_{0m} modes ($E_z = 0$) and that Eq. (2-48b) corresponds to TE_{0m} modes ($H_z = 0$).

2-17 Determine the normalized frequency at 0.82 μm for a step-index fiber having a 25-μm core radius, $n_1 = 1.48$, and $n_2 = 1.46$. How many modes propagate in this fiber at 0.82 μm? How many modes propagate at a wavelength of 1.3 μm? What percentage of the optical power flows in the cladding in each case?

2-18 Find the core radius necessary for single-mode operation at 820 nm of a step-index fiber with $n_1 = 1.480$ and $n_2 = 1.478$. What is the numerical aperture and maximum acceptance angle of this fiber?

Section 2-5

2-19 Plot the refractive-index profiles from n_1 to n_2 as a function of radial distance $r \leq a$ for graded-index fibers that have α values of 1, 2, 4, 8, and ∞ (step index). Assume the fibers have a 25-μm core radius, $n_1 = 1.48$, and $\Delta = 0.01$.

2-20 Verify the steps leading from Eq. (2-64) to Eq. (2-68).

2-21 Show that Eq. (2-76) results from evaluating the integral over ν in Eq. (2-75).

2-22 Calculate the number of modes at 820 nm and 1.3 μm in a graded-index fiber having a parabolic-index profile ($\alpha = 2$), a 25-μm core radius, $n_1 = 1.48$, and $n_2 = 1.46$. How does this compare to a step-index fiber?

REFERENCES

1. See any general physics book or introductory optics book; for example:

    (a) M. V. Klein, *Optics*, Wiley, New York, 1970.

    (b) E. Hecht and A. Zajac, *Optics*, Addison-Wesley, Reading, Mass., 1974. (Chapter 8 presents a good discussion on polarization.)

    (c) F. A. Jenkins and H. E. White, *Fundamentals of Optics*, McGraw-Hill, New York, 1976.

2. See any introductory electromagnetics book; for example:
   (a) W. H. Hayt, Jr., *Engineering Electromagnetics,* McGraw-Hill, New York, 1974.
   (b) J. D. Kraus and K. R. Carver, *Electromagnetics,* McGraw-Hill, New York, 1973.
   (c) D. Corson and P. Lorrain, *Introduction to Electromagnetic Fields and Waves,* Freeman, San Francisco, 1970.
3. S. E. Miller, E. A. Marcatili, and T. Li, "Research towards optical fiber transmission systems," *Proc. IEEE,* **61,** 1703–1751, Dec. 1973.
4. (a) S. J. Maurer and L. B. Felsen, "Ray-optical techniques for guided waves," *Proc. IEEE,* **55,** 1718–1729, Oct. 1967.
   (b) L. B. Felsen, "Rays and modes in optical fibers," *Electron. Lett.,* **10,** 95–96, Apr. 1974.
5. A. W. Snyder and D. J. Mitchell, "Leaky rays on circular optical fibers," *J. Opt. Soc. Amer.,* **64,** 599–607, May 1974.
6. D. Keck, "Optical fiber waveguides," in *Fundamentals of Optical Fiber Communications,* M. K. Barnoski (Ed.), Academic, New York, 1976; 2d ed., 1982.
7. D. Marcuse, *Theory of Dielectric Optical Waveguides,* Academic, New York, 1974.
8. J. Midwinter, *Optical Fibers for Transmission,* Wiley, New York, 1979.
9. R. L. Gallawa, *A User's Manual for Optical Waveguide Communications,* publication no. OTR76-83, U.S. Dept. of Commerce, Washington, D.C., 1976.
10. H. G. Unger, *Planar Optical Waveguides and Fibres,* Clarendon, Oxford, 1977.
11. N. S. Kapany, *Fiber Optics: Principles and Applications,* Academic, New York, 1967.
12. R. Olshansky, "Leaky modes in graded index optical fibers," *Appl. Opt.,* **15,** 2773–2777, Nov. 1976.
13. J. D. Love, C. Winkler, R. Sammut, and K. Barrell, "Leaky modes and rays on multimode step-index waveguides," *Electron. Lett.,* **14,** 489–490, July 1978.
14. E. Snitzer, "Cylindrical dielectric waveguide modes," *J. Opt. Soc. Amer.,* **51,** 491–498, May 1961.
15. N. S. Kapany and J. J. Burke, *Optical Waveguides,* Academic, New York, 1972.
16. D. Marcuse, *Light Transmission Optics,* Van Nostrand-Reinhold, New York, 1972.
17. R. Olshansky, "Propagation in glass optical waveguides," *Rev. Mod. Phys.,* **51,** 341–367, April 1979.
18. D. Gloge, "The optical fiber as a transmission medium," *Rep. Progr. Phys.,* **42,** 1777–1824, Nov. 1979.
19. A. W. Snyder, "Asymptotic expressions for eigenfunctions and eigenvalues of a dielectric or optical waveguide," *IEEE Trans. Microwave Theory Tech.,* **MTT-17,** 1130–1138, Dec. 1969.
20. D. Gloge, "Weakly guiding fibers," *Appl. Opt.,* **10,** 2252–2258, Oct. 1971.
21. D. Marcuse, "Gaussian approximation of the fundamental modes of graded index fibers," *J. Opt. Soc. Amer.,* **68,** 103–109, Jan. 1978.
22. H. M. DeRuiter, "Integral equation approach to the computation of modes in an optical waveguide," *J. Opt. Soc. Amer.,* **70,** 1519–1524, Dec. 1980.
23. A. W. Snyder, "Understanding monomode optical fibers," *Proc. IEEE,* **69,** 6–13, Jan. 1981.
24. M. Abramowitz and I. A. Stegun, *Handbook of Mathematical Functions,* Dover, New York, 1965.
25. J. Mathews and R. L. Walker, *Mathematical Methods of Physics,* Benjamin, New York, 1964.
26. E. Jahnke, F. Emde, and F. Losch, *Tables of Higher Functions,* McGraw-Hill, New York, 1960.
27. S. E. Miller and A. G. Chynoweth, *Optical Fiber Telecommunications,* Academic, New York, 1979.
28. R. M. Gagliardi and S. Karp, *Optical Communications,* Wiley, New York, 1976, chap. 3.
29. D. Gloge, "Propagation effects in optical fibers," *IEEE Trans. Microwave Theory Tech.,* **MTT-23,** 106–120, Jan. 1975.
30. D. Gloge and E. Marcatili, "Mulitmode theory of graded core fibers," *Bell Sys. Tech. J.,* **52,** 1563–1578, Nov. 1973.
31. E. Merzbacher, *Quantum Mechanics,* Wiley, New York, 1961.
32. G. F. Carrier, M. Krook, and C. E. Pearson, *Functions of a Complex Variable,* McGraw-Hill, New York, 1966.
33. P. M. Morse and H. Feshbach, *Methods of Mathematical Physics,* vol. II, McGraw-Hill, New York, 1953.

THREE

SIGNAL DEGRADATION IN OPTICAL FIBERS

In Chap. 2 we showed the structure of optical fibers and examined the concepts of how light propagates along a cylindrical dielectric optical waveguide. Here we shall continue the discussion of optical fibers by answering two very important questions:

1. What are the loss or signal attenuation mechanisms in a fiber?
2. Why and to what degree do optical signals get distorted as they propagate along a fiber?

Signal attenuation (also known as *fiber loss* or *signal loss*) is one of the most important properties of an optical fiber, because it largely determines the maximum repeaterless separation between a transmitter and a receiver. Since repeaters are expensive to fabricate, install, and maintain, the degree of attenuation in a fiber has a large influence on system cost. Of equal importance is signal distortion. The distortion mechanisms in a fiber cause optical signal pulses to broaden as they travel along a fiber. If these pulses travel sufficiently far, they will eventually overlap with neighboring pulses, thereby creating errors in the receiver output. The signal distortion mechanisms thus limit the information-carrying capacity of a fiber.

Since these two factors are closely tied to how and of what a fiber is constructed, we shall first discuss fiber materials and construction methods. This will be in the form of a very brief overview that defines the terminology and gives the necessary background concepts. A more detailed treatment of fiber materials and fabrication methods is given in Chap. 10. Here we shall mainly concentrate on low-loss glass fibers that are suitable for long-distance information transmission.

3-1 FIBER MATERIALS AND FABRICATION METHODS: AN OVERVIEW

In selecting materials for optical fibers a number of requirements must be satisfied. For example:

1. It must be possible to make long, thin, flexible fibers from the material.
2. The material must be transparent at a particular optical wavelength in order for the fiber to guide light efficiently.
3. Physically compatible materials having slightly different refractive indices for the core and cladding must be available.

Materials satisfying these requirements are glasses and plastics.[1–8]

   The largest category of optically transparent glasses from which optical fibers are made are the oxide glasses. Of these the most common is silica (SiO_2), which has a refractive index of 1.458 at 850 nm. To produce two similar materials having slightly different indices of refraction for the core and cladding, trace amounts of either fluorine or various oxides (referred to as *dopants*) such as B_2O_3, GeO_2, or P_2O_5 are added to the silica. As shown in Fig. 3-1, the addition of GeO_2 or P_2O_5 increases the refractive index, whereas doping the silica with fluorine or B_2O_3 decreases it. When referring to a doped silica glass, notations such as GeO_2-SiO_2 are used, for example, to denote a GeO_2 dopant.

   Two basic techniques[4, 9, 10] are used in the fabrication of all-glass optical waveguides. These are the vapor phase oxidation processes and the direct-melt methods. The direct-melt method follows traditional glass-making procedures in that optical fibers are made directly from the molten state of purified components of silicate glasses. In the vapor phase oxidation processes, highly pure vapors of metal halides (for example, $SiCl_4$ and $GeCl_4$) react with oxygen to form a white powder of SiO_2 particles. The particles are then collected on the surface of a bulk glass by one of three commonly used processes[11–17] (as described in Chap. 10), and are then sintered (transformed to a homogeneous glass mass by heating without melting) by a variety of techniques to form a clear glass rod or tube (depending on the process).

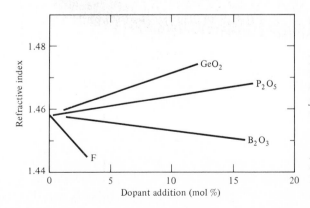

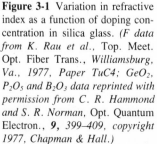

Figure 3-1 Variation in refractive index as a function of doping concentration in silica glass. *(F data from K. Rau et al.*, Top. Meet. Opt. Fiber Trans., *Williamsburg, Va., 1977, Paper TuC4; GeO₂, P₂O₅ and B₂O₃ data reprinted with permission from C. R. Hammond and S. R. Norman,* Opt. Quantum Electron., *9, 399–409, copyright 1977, Chapman & Hall.)*

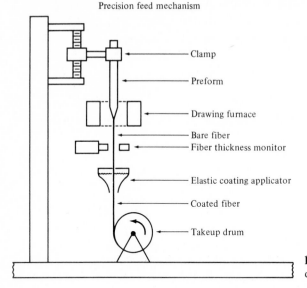

Precision feed mechanism

Clamp

Preform

Drawing furnace

Bare fiber

Fiber thickness monitor

Elastic coating applicator

Coated fiber

Takeup drum

Figure 3-2 Schematic of fiber-drawing apparatus.

This rod or tube is called a *preform*. It is typically around 10 mm in diameter and 60 to 90 cm long. Fibers are made from the preform[18, 19] by using the equipment shown in Fig. 3-2. The preform is precision fed into a circular heater called the *drawing furnace*. Here the preform end is softened to the point where it can be drawn into a very thin filament which becomes the optical fiber.

3-2 ATTENUATION

Attenuation of a light signal as it propagates along a fiber is an important consideration in the design of an optical communication system, since it plays a major role in determining the maximum transmission distance between a transmitter and a receiver. The basic attenuation mechanisms[3, 20–23] in a fiber are absorption, scattering, and radiative losses of the optical energy. Absorption is related to the fiber material, whereas scattering is associated both with the fiber material and with structural imperfections in the optical waveguide. Attenuation owing to radiative effects originates from perturbations (both microscopic and macroscopic) of the fiber goemetry.

In this section we shall first discuss the units in which fiber losses are measured and then present the physical phenomena giving rise to attenuation.

3-2-1 Attenuation Units

Signal attenuation (or *fiber loss*) is defined as the ratio of the optical output power P_{out} from a fiber of length L to the optical input power P_{in}. This power ratio is a function of wavelength, as is shown by the general attenuation curve in Fig. 3-3.

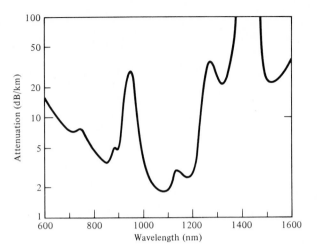

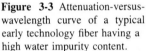

Figure **3-3** Attenuation-versus-wavelength curve of a typical early technology fiber having a high water impurity content.

The symbol α is commonly used to express attenuation in decibels per kilometer (see App. D for a discussion of decibels):

$$\alpha = 10 \log \frac{P_{in}/P_{out}}{L} \qquad (3\text{-}1)$$

An ideal fiber would have no loss so that $P_{out} = P_{in}$. This corresponds to a 0-dB attenuation, which in practice is impossible. An actual low-loss fiber may have a 3-dB/km average loss, for example. This means that the optical signal power would decrease by 50 percent over a 1-km length and would decrease by 75 percent (a 6-dB loss) over a 2-km length since loss contributions expressed in decibels are additive.

3-2-2 Absorption

Absorption is caused by three different mechanisms:

1. Absorption by atomic defects in the glass composition
2. Extrinsic absorption by impurity atoms in the glass material
3. Intrinsic absorption by the basic constituent atoms of the fiber material

Atomic defects are imperfections of the atomic structure of the fiber material such as missing molecules, high-density clusters of atom groups, or oxygen defects in the glass structure. Usually absorption losses arising from these defects are negligible compared to intrinsic and impurity absorption effects. However, they can be significant if the fiber is exposed to intense nuclear radiation levels,[24-27] as might occur inside a nuclear reactor, during a nuclear explosion, or in the earth's Van Allen belts.

The dominant absorption factor in fibers prepared by the direct-melt method is the presence of impurities in the fiber material. Impurity absorption results predominantly from transition metal ions such as iron, chromium, cobalt, and copper, and from OH (water) ions. The transition metal impurities which are present in the starting materials used for direct-melt fibers range between 1 and 10 parts per billion (ppb) causing losses from 1 to 10 dB/km. The impurity levels in vapor phase deposition processes are usually one to two orders of magnitude lower. Impurity absorption losses occur either from electronic transitions between the energy levels associated with the incompletely filled inner subshell of these ions or because of charge transitions from one ion to another. The absorption peaks of the various transition metal impurities tend to be broad, and several peaks may overlap, which further broadens the absorption region.

The presence of OH (water) ion impurities in fiber preforms results mainly from the oxyhydrogen flame used for the hydrolysis reaction of the $SiCl_4$, $GeCl_4$, and $POCl_3$ starting materials. Water impurity concentrations of less than a few parts per billion are required if the attenuation is to be less than 20 dB/km. Early optical fibers had high levels of OH ions which resulted in large absorption peaks occurring at 1400, 950, and 725 nm. These are the first, second, and third overtones, respectively, of the fundamental absorption peak of water near 2.7 μm, as shown in Fig. 3-3. Between these absorption peaks there are regions of low attenuation.

The peaks and valleys in the attenuation curve resulted in the assignment of various "transmission windows" to early optical fibers. Significant progress[28, 29] has been made in reducing the residual OH content of fibers to less than 1 ppb. For example, the loss curve of a fiber prepared by the VAD method[28] with an OH content of less than 0.8 ppb is shown in Fig. 3-4.

Intrinsic absorption is associated with the basic fiber material (for example,[30] pure SiO_2) and is the principal physical factor that defines the transparency window of a material over a specified spectral region. It occurs when the material is in a perfect state with no density variations, impurities, material inhomogeneities, etc. Intrinsic absorption thus sets the fundamental lower limit on absorption for any particular material.

Intrinsic absorption results from electronic absorption bands in the ultraviolet

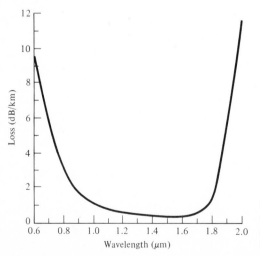

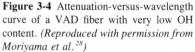

Figure 3-4 Attenuation-versus-wavelength curve of a VAD fiber with very low OH content. (*Reproduced with permission from Moriyama et al.[28]*)

region and from atomic vibration bands in the near-infrared region. The electronic absorption bands are associated with the band gaps of the amorphous glass materials. Absorption occurs when a photon interacts with an electron in the valence band and excites it to a higher energy level, as is described in Sec. 2-1. The ultraviolet edge of the electron absorption bands of both amorphous and crystalline materials follow the empirical relationship[3, 22]

$$\alpha_{uv} = Ce^{E/E_0} \tag{3-2}$$

which is known as Urbach's rule. Here C and E_0 are empirical constants and E is the photon energy. The magnitude and characteristic exponential decay of the ultraviolet absorption are shown in Fig. 3-5.

In the near-infrared region above 1.2 μm, the optical waveguide loss is predominantly determined by the presence of OH ions and the inherent infrared absorption of the constituent material. The inherent infrared absorption is associated with the characteristic vibration frequency of the particular chemical bond between the atoms of which the fiber is composed. An interaction between the vibrating bond and the electromagnetic field of the optical signal results in a transfer of energy from the field to the bond, thereby giving rise to absorption. This absorption is quite strong because of the many bonds present in the fiber.

These mechanisms result in a wedge-shaped spectral-loss characteristic. Within this wedge losses as low as 0.2 dB/km at 1.55 μm in a single-mode fiber have been measured.[32] A comparison[31, 33] of the infrared absorption induced by various doping materials in low-water-content fibers is shown in Fig. 3-6. This

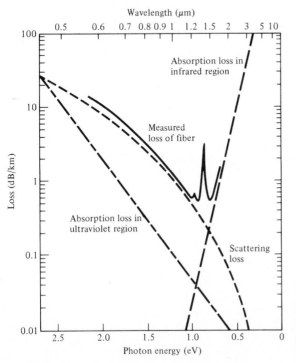

Figure 3-5 Optical fiber attenuation characteristics and their limiting mechanisms for a GeO$_2$-doped low-loss low-OH-content fiber. (*Reproduced with permission from Osanai et al.[31]*)

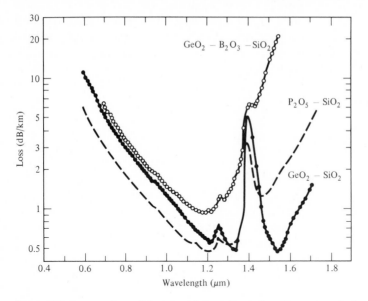

Figure 3-6 A comparison of the infrared absorption induced by various doping materials in low-loss fibers. *(Reproduced with permission from Osanai et al.[31])*

indicates that for operation at longer wavelengths a GeO_2-doped fiber material is the most desirable. Note that the absorption curve shown in Fig. 3-5 is for a GeO_2-doped fiber.

3-2-3 Scattering Losses

Scattering losses in glass arise from microscopic variations in the material density, from compositional fluctuations, and from structural inhomogeneities or defects occurring during fiber manufacture. As we show in Sec. 10-1, glass is composed of a randomly connected network of molecules. Such a structure naturally contains regions in which the molecular density is either higher or lower than the average density in the glass. In addition since glass is made up of several oxides, such as SiO_2, GeO_2, and P_2O_5, compositional fluctuations can occur. These two effects give rise to refractive-index variations which occur within the glass over distances that are small compared to the wavelength. These index variations cause a Rayleigh-type scattering of the light. Rayleigh scattering in glass is the same phenomenon that scatters light from the sun in the atmosphere, thereby giving rise to a blue sky.

The expressions for scattering-induced attenuation are fairly complex owing to the random molecular nature and the various oxide constituents of glass. For single-component glass the scattering loss at a wavelength λ resulting from density fluctuations can be approximated by[20, 34] (in base e units)

$$\alpha_{\text{scat}} = \frac{8\pi^3}{3\lambda^4}(n^2 - 1)^2 k_B T_f \beta_T \tag{3-3}$$

Here n is the refractive index, k_B is Boltzmann's constant, β_T is the isothermal compressibility of the material, and the fictive temperature T_f is the temperature at which the density fluctuations are frozen into the glass as it solidifies (after having been drawn into a fiber). Alternatively the relation[35] (in base e units)

$$\alpha_{\text{scat}} = \frac{8\pi^3}{3\lambda^4} n^8 p^2 k_B T_f \beta_T \tag{3-4}$$

has been derived, where p is the photoelastic coefficient. A comparison of Eqs. (3-3) and (3-4) is given in Prob. 3-2. Note that Eqs. (3-3) and (3-4) are given in units of *nepers* (that is, base e units). To change this to decibels for optical power attenuation calculations, multiply these equations by 10 log e = 4.343.

For multicomponent glasses the scattering is given by[22]

$$\alpha = \frac{8\pi^3}{3\lambda^4} (\delta n^2)^2 \, \delta V \tag{3-5}$$

where the square of the mean-squared refractive-index fluctuation $(\delta n^2)^2$ over a volume of δV is

$$(\delta n^2)^2 = \left(\frac{\partial n}{\partial \rho}\right)^2 (\delta \rho)^2 + \sum_{i=1}^{m} \left(\frac{\partial n^2}{\partial C_i}\right)^2 (\delta C_i)^2 \tag{3-6}$$

Here $\delta \rho$ is the density fluctuation and δC_i is the concentration fluctuation of the ith glass component. The magnitudes of the composition and density fluctuations are generally not known and must be determined from experimental scattering data. Once they are known the scattering loss can be calculated.

Structural inhomogeneities and defects created during fiber fabrication can also cause scattering of light out of the fiber. These defects may be in the form of trapped gas bubbles, unreacted starting materials, and crystallized regions in the glass. In general the preform manufacturing methods that have evolved have minimized these extrinsic effects to the point where scattering resulting from them is negligible compared to the intrinsic Rayleigh scattering.

Since Rayleigh scattering follows a characteristic λ^{-4} dependence, it decreases dramatically with increasing wavelength, as is shown in Fig. 3-5. For wavelengths below about 1 μm it is the dominant loss mechanism in a fiber and gives the attenuation-versus-wavelength plots their characteristic downward trend with increasing wavelength. At wavelengths longer than 1 μm, infrared absorption effects tend to dominate optical signal attenuation.

3-2-4 Radiative Losses

Radiative losses occur whenever an optical fiber undergoes a bend of finite radius of curvature.[36–38] Fibers can be subject to two types of bends: (a) bends having radii that are large compared to the fiber diameter, for example, such as occur when a fiber cable turns a corner, and (b) random microscopic bends of the fiber axis that can arise when the fibers are incorporated into cables.

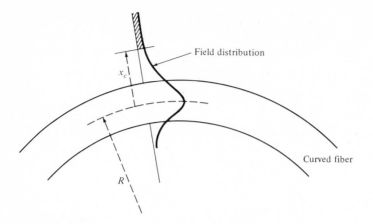

Figure 3-7 Sketch of the fundamental mode field in a curved optical waveguide. *(Reproduced with permission from E. A. J. Marcatili and S. E. Miller, Bell Sys. Tech. J., 48, 2161, Sept. 1969, copyright 1969, The American Telephone and Telegraph Co.)*

Let us first examine large-curvature radiation losses. For slight bends the excess loss is extremely small and is essentially unobservable. As the radius of curvature decreases, the loss increases exponentially until at a certain critical radius the curvature loss becomes observable. If the bend radius is made a bit smaller once this threshold point has been reached, the losses suddenly become extremely large.

Qualitatively these curvature loss effects can be explained by examining the modal electric field distributions shown in Fig. 2-13. Recall that this figure shows that any bound core mode has an evanescent field tail in the cladding which decays exponentially as a function of distance from the core. Since this field tail moves along with the field in the core, part of the energy of a propagating mode travels in the fiber cladding. When a fiber is bent, the field tail on the far side of the center of curvature must move faster to keep up with the field in the core, as is shown in Fig. 3-7 for the lowest-order fiber mode. At a certain critical distance x_c from the center of the fiber the field tail would have to move faster than the speed of light to keep up with the core field. Since this is not possible the optical energy in the field tail beyond x_c radiates away.

The amount of optical radiation from a bent fiber depends on the field strength at x_c and on the radius of curvature R. Since higher-order modes are bound less tightly to the fiber core than lower-order modes, the higher-order modes will radiate out of the fiber first. Thus the total number of modes that can be supported by a curved fiber is less than in a straight fiber. Gloge[38] has derived the following expression for the effective number of modes N_{eff} that are guided by a curved fiber of radius a:

$$N_{\text{eff}} = N_\infty \left\{ 1 - \frac{\alpha + 2}{2\alpha\Delta} \left[\frac{2a}{R} + \left(\frac{3}{2n_2 kR} \right)^{2/3} \right] \right\} \tag{3-7}$$

where α defines the graded-index profile, Δ is the core-cladding index difference,

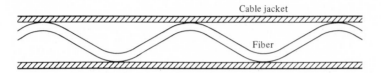

Figure 3-8 Microbends shown as repetitive changes in the radius of curvature of the fiber axis.

n_2 is the cladding refractive index, $k = 2\pi/\lambda$ is the wave propagation constant, and

$$N_\infty = \frac{\alpha}{\alpha + 2}(n_1 ka)^2 \Delta \tag{3-8}$$

is the total number of modes in a straight fiber [see Eq. (2-79)]. As an example, let us find the radius of curvature R at which the number of modes decreases by 50 percent in a graded-index fiber. For this fiber let $\alpha = 2$, $n_2 = 1.5$, $\Delta = 0.01$, $a = 25\ \mu m$, and let the wavelength of the guided light be $1.3\ \mu m$. Solving Eq. (3-7) yields $R = 1.0$ cm.

Another form of radiation loss in optical waveguides results from mode coupling caused by random microbends of the optical fiber.[39-41] *Microbends* are repetitive changes in the radius of curvature of the fiber axis, as is illustrated in Fig. 3-8. They are caused either by nonuniformities in the sheathing of the fiber or by nonuniform lateral pressures created during the cabling of the fiber. The latter effect is often referred to as *cabling* or *packaging losses*. An increase in attenuation results from microbending because the fiber curvature causes repetitive coupling of energy between the guided modes and the leaky or nonguided modes in the fiber.

One method of minimizing microbending losses is by extruding a compressible jacket over the fiber. When external forces are applied to this configuration, the jacket will be deformed but the fiber will tend to stay relatively straight, as shown in Fig. 3-9. For a graded-index fiber having a core radius a, outer radius b (excluding the jacket), and an index difference Δ, the microbending loss α_M of a jacketed fiber is reduced from that of an unjacketed fiber by a factor[42]

$$F(\alpha_M) = \left[1 + \pi\Delta^2 \left(\frac{b}{a}\right)^4 \frac{E_f}{E_j}\right]^{-2} \tag{3-9}$$

Here E_j and E_f are the Young's moduli of the jacket and fiber, respectively. The Young's modulus of common jacket materials ranges from 20 to 500 MPa. The Young's modulus of fused silica glass is about 65 GPa.

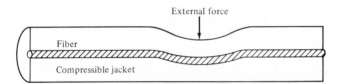

Figure 3-9 A compressible jacket extruded over a fiber minimizes microbending resulting from external forces.

3-2-5 Core and Cladding Losses

Upon measuring the propagation losses in an actual fiber, all the dissipative and scattering losses will be manifested simultaneously. Since the core and cladding have different indices of refraction and, therefore, differ in composition, the core and cladding generally have different attenuation coefficients denoted α_1 and α_2, respectively. If the influence of modal coupling is ignored,[21] the loss for a mode of order (ν, m) for a step-index waveguide is

$$\alpha_{\nu m} = \alpha_1 \frac{P_{core}}{P} + \alpha_2 \frac{P_{clad}}{P} \tag{3-10}$$

where the fractional powers P_{core}/P and P_{clad}/P are shown in Fig. 2-17 for several low-order modes. Using Eq. (2-58), this can be written as

$$\alpha_{\nu m} = \alpha_1 + (\alpha_2 - \alpha_1) \frac{P_{clad}}{P} \tag{3-11}$$

The total loss of the waveguide can be found by summing over all modes weighted by the fractional power in that mode.

For the case of a graded-index fiber the situation is much more complicated. In this case, both the attenuation coefficients and the modal power tend to be functions of the radial coordinate. At a distance r from the core axis the loss is[21]

$$\alpha(r) = \alpha_1 + (\alpha_2 - \alpha_1) \frac{n^2(0) - n^2(r)}{n^2(0) - n_2^2} \tag{3-12}$$

where α_1 and α_2 are the axial and cladding attenuation coefficients, respectively, and the n's are defined by Eq. (2-60). The loss encountered by a given mode is then

$$\alpha_{gi} = \frac{\displaystyle\int_0^\infty \alpha(r)p(r)r \, dr}{\displaystyle\int_0^\infty p(r)r \, dr} \tag{3-13}$$

where $p(r)$ is the power density of that mode at r. The complexity of the multimode waveguide has prevented an experimental correlation with a model. However, it has generally been observed that the loss increases with increasing mode number.[43]

3-3 SIGNAL DISTORTION IN OPTICAL WAVEGUIDES

An optical signal becomes increasingly distorted as it travels along a fiber. This distortion is a consequence of intramodal dispersion and intermodal delay effects. These distortion effects can be explained by examining the behavior of the group velocities of the guided modes, where the *group velocity* is the speed at which energy in a particular mode travels along the fiber.

Intramodal dispersion is pulse spreading that occurs within a single mode. It is a result of the group velocity being a function of the wavelength λ and is, therefore, often referred to as *chromatic dispersion*. Since intramodal dispersion depends on the wavelength, its effect on signal distortion increases with the spectral width of the optical source. This spectral width is the band of wavelengths over which the source emits light. It is normally characterized by the root-mean-square (rms) spectral width σ_λ (see Fig. 4-5). For light-emitting diodes (LEDs) the rms spectral width is approximately 5 percent of a central wavelength. For example, if the peak emission wavelength of an LED source is 850 nm, a typical source spectral width would be 40 nm; that is, the source emits most of its optical power in the 830- to 870-nm wavelength band. Laser diode optical sources have much narrower spectral widths, typical values being 1 to 2 nm.

The two main causes of intramodal dispersion are:

1. *Material dispersion,* which arises from the variation of the refractive index of the core material as a function of wavelength. This causes a wavelength dependence of the group velocity of any given mode.
2. *Waveguide dispersion,* which occurs because the modal propagation constant β is a function of a/λ (the optical fiber dimension relative to the wavelength λ, where a is the core radius).

The other factor giving rise to pulse spreading is *intermodal delay* which is a result of each mode having a different value of the group velocity at a single frequency.

Of these three, waveguide dispersion usually can be ignored in multimode fibers. However, this effect can be significant in single-mode fibers. The full effects of these three distortion mechanisms are seldom observed in practice since they tend to be mitigated by other factors, such as nonideal index profiles, optical power-launching conditions (different amounts of optical power launched into the various modes), nonuniform mode attenuation, mode mixing in the fiber and in splices, and by statistical variations in these effects along the fiber. In this section we shall first discuss the general effects of signal distortion and then examine the various dispersion mechanisms.

3-3-1 Information Capacity Determination

A result of the disperson-induced signal distortion is that a light pulse will broaden as it travels along the fiber. As shown in Fig. 3-10 this pulse broadening will eventually cause a pulse to overlap with neighboring pulses. After a certain amount of overlap has occurred, adjacent pulses can no longer be individually distinguished at the receiver and errors will occur. Thus the dispersive properties determine the limit of the information capacity of the fiber.

A measure of the information capacity of an optical waveguide is usually specified by the *bandwidth-distance product* in MHz · km. For a step-index fiber the various distortion effects tend to limit the bandwidth-distance product to about 20 MHz · km. In graded-index fibers the radial refractive-index profile can be carefully selected so that pulse broadening is minimized at a specific operating wavelength. This had led to bandwidth-distance products as high as 2.5 GHz · km. Single-mode

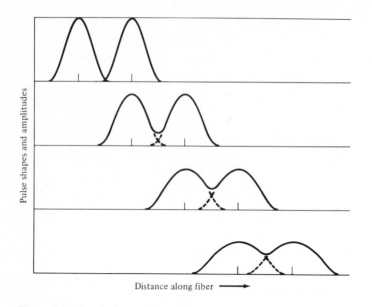

Figure 3-10 Broadening and attenuation of two adjacent pulses as they travel along a fiber.

fibers can have capacities well in excess of this. A comparison of the information capacity of an 800-MHz · km optical fiber with the capacities of typical coaxial cables used for UHF and VHF transmission is shown in Fig. 3-11. The curves are shown in terms of signal attenuation versus data rate. The fiber has a 6-dB/km low-frequency attenuation. The flatness of the attenuation curve for this fiber extends up to the microwave spectrum.

   The information-carrying capacity can be determined by examining the deformation of short light pulses propagating along the fiber.[44] The following discussion

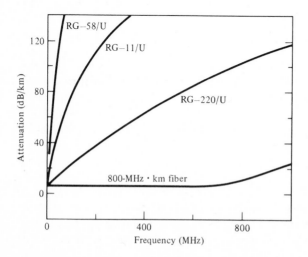

Figure 3-11 A comparison of the attenuation as a function of frequency or data rate of various coaxial cables and an 800 MHz · km optical waveguide.

on signal distortion is thus carried out primarily from the standpoint of pulse broadening, which is representative of digital transmission.

3-3-2 Group Delay

Let us examine a signal that modulates an optical source. We shall assume that the modulated optical signal excites all modes equally at the input end of the fiber. Each mode thus carries an equal amount of energy through the fiber. Furthermore, each mode contains all of the spectral components in the wavelength band over which the source emits. The signal may be considered as modulating each of these spectral components in the same way. As the signal propagates along the fiber, each spectral component can be assumed to travel independently and to undergo a time delay or *group delay* per unit length in the direction of propagation given by[45, 46]

$$\frac{t_g}{L} = \frac{1}{V_g} = \frac{1}{c}\frac{d\beta}{dk} = -\frac{\lambda^2}{2\pi c}\frac{d\beta}{d\lambda} \tag{3-14}$$

Here L is the distance traveled by the pulse, β is the propagation constant along the fiber axis, $k = 2\pi/\lambda$, and the *group velocity*

$$V_g = c\left(\frac{d\beta}{dk}\right)^{-1}$$

is the velocity at which the energy in a pulse travels along a fiber.

Since the group delay depends on the wavelength, each spectral component of any particular mode takes a different amount of time to travel a certain distance. As a result of this difference in time delays, the optical signal pulse spreads out with time as it is transmitted over the fiber. The quantity we are thus interested in is the amount of pulse spreading that arises from the group delay variation.

If the spectral width of the optical source is not too wide, the delay difference per unit wavelength along the propagation path is approximately $dt_g/d\lambda$. For spectral components which are $\Delta\lambda$ apart and which lie $\Delta\lambda/2$ above and below a central wavelength λ_0, the total delay difference τ over a distance L is

$$\tau = \frac{dt_g}{d\lambda}\Delta\lambda \tag{3-15}$$

If the spectral width $\Delta\lambda$ of an optical source is characterized by its rms value σ_λ (see Fig. 4-5), then the pulse spreading can be approximated by

$$\tau_g = \frac{dt_g}{d\lambda}\sigma_\lambda = -\frac{L\sigma_\lambda}{2\pi c}\left(2\lambda\frac{d\beta}{d\lambda} + \lambda^2\frac{d^2\beta}{d\lambda^2}\right) \tag{3-16}$$

The factor

$$D = \frac{1}{L}\frac{dt_g}{d\lambda} \tag{3-17}$$

is designated as the dispersion. It defines the pulse spread as a function of wavelength and is measured in nanoseconds per kilometer per nanometer. It is a result of material and waveguide dispersion. In many theoretical treatments of intramodal dispersion it is assumed for simplicity that material dispersion and waveguide dispersion can be calculated separately and then added to give the total dispersion of the mode. In reality these two mechanisms are intricately related since the dispersive properties of the refractive index (which gives rise to material dispersion) also affects the waveguide dispersion. However, an examination[47] of the interdependence of material and waveguide dispersion has shown that, unless a very precise value is desired, a good estimate of the total intramodal dispersion can be obtained by calculating the effect of signal distortion arising from one type of dispersion in the absence of the other, and then adding the results. Material dispersion and waveguide dispersion are, therefore, considered separately in the next two sections.

3-3-3 Material Dispersion

Material dispersion occurs because the index of refraction varies as a nonlinear function of the optical wavelength. This is exemplified in Fig. 3-12 for silica.[48] As a consequence, since the group velocity V_g of a mode is a function of the index of refraction, the various spectral components of a given mode will travel at different speeds, depending on the wavelength.[49] Material dispersion is, therefore, an intramodal dispersion effect, and is of particular importance for single-mode waveguides and for LED systems (since an LED has a broader output spectrum than a laser diode).

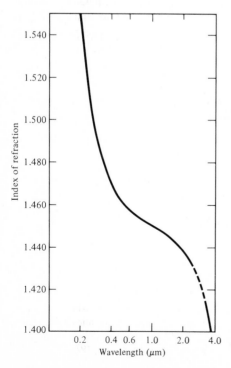

Figure 3-12 Variation in the index of refraction as a function of the optical wavelength for silica. *(Reproduced with permission from I. H. Malitson, J. Opt. Soc. Amer., **55**, 1205–1209, Oct. 1965.)*

To calculate material-induced dispersion, we consider a plane wave propagating in an infinitely extended dielectric medium that has a refractive index $n(\lambda)$ equal to that of the fiber core. The propagation constant β is thus given by

$$\beta = \frac{2\pi n(\lambda)}{\lambda} \tag{3-18}$$

Substituting this expression for β into Eq. (3-14) with $k = 2\pi/\lambda$ yields the group delay t_{mat} resulting from material dispersion

$$t_{mat} = \frac{L}{c}\left(n - \lambda\frac{dn}{d\lambda}\right) \tag{3-19}$$

Using Eq. (3-16), the pulse spread τ_{mat} for a source of spectral width σ_λ is found by differentiating this group delay with respect to wavelength and multiplying by σ_λ to yield

$$\tau_{mat} = \frac{dt_{mat}}{d\lambda}\sigma_\lambda = -\frac{L}{c}\lambda\frac{d^2n}{d\lambda^2}\sigma_\lambda \tag{3-20}$$

A plot of Eq. (3-20) for unit length L and unit optical source spectral width σ_λ is given in Fig. 3-13 for the silica material shown in Fig. 3-12. From Eq. (3-20) and Fig 3-13 it can be seen that material dispersion can be reduced either by choosing sources with narrower spectral output widths (reducing σ_λ) or by operating at longer wavelengths. As an example, consider a typical GaAlAs LED having a spectral width of 40 nm at an 800-nm peak output so that $\sigma_\lambda/\lambda = 5$ percent. As can be seen from Fig. 3-13 and Eq. (3-20) this produces a pulse spread of 4.4 ns/km. Note that material dispersion goes to zero at 1.27 μm for pure silica.

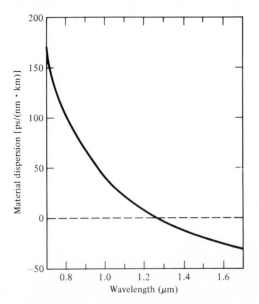

Figure 3-13 Material dispersion as a function of optical wavelength for silica. *(Reproduced with permission from J. W. Fleming, Elec-tron. Lett., **14**, 326, May 1978.)*

3-3-4 Waveguide Dispersion

The effect of waveguide dispersion on pulse spreading can be approximated by assuming that the refractive index of the material is independent of wavelength. Let us first consider the group delay, that is, the time required for a mode to travel along a fiber of length L. To make the results independent of fiber configuration,[49] we shall express the group delay in terms of the normalized propagation constant b defined by

$$b = 1 - \left(\frac{ua}{V}\right)^2 = \frac{\beta^2/k^2 - n_2^2}{n_1^2 - n_2^2} \tag{3-21}$$

For small values of the index difference $\Delta = (n_1 - n_2)/n_1$, Eq. (3-21) can be approximated by

$$b = \frac{\beta/k - n_2}{n_1 - n_2} \tag{3-22}$$

Solving Eq. (3-22) for β, we have

$$\beta = n_2 k(b\Delta + 1) \tag{3-23}$$

With this expression for β and using the assumption that n_2 is not a function of wavelength, we find that the group delay t_{wg} arising from waveguide dispersion is

$$t_{wg} = \frac{L}{c}\frac{d\beta}{dk} = \frac{L}{c}\left[n_2 + n_2\Delta\frac{d(kb)}{dk}\right] \tag{3-24}$$

The modal propagation constant β is obtained from the eigenvalue equation expressed by Eq. (2-46), and is generally given in terms of the normalized frequency V defined by Eq. (2-49). We shall therefore use the approximation

$$V = ka(n_1^2 - n_2^2)^{1/2} \simeq kan_2\sqrt{2\Delta}$$

which is valid for small values of Δ, to write the group delay in Eq. (3-24) in terms of V instead of k, yielding

$$t_{wg} = \frac{L}{c}\left[n_2 + n_2\Delta\frac{d(Vb)}{dV}\right] \tag{3-25}$$

The first term in Eq. (3-25) is a constant and the second term represents the group delay arising from waveguide dispersion. The factor $d(Vb)/dV$ can be expressed as[49]

$$\frac{d(Vb)}{dV} = b\left[1 - \frac{2J_\nu^2(ua)}{J_{\nu+1}(ua)J_{\nu-1}(ua)}\right]$$

where u is defined by Eq. (2-40) and a is the fiber radius. This factor is plotted in Fig. 3-14 as a function of V. The plots show that, for a fixed value of V, the group delay is different for every guided mode. When a light pulse is launched into a fiber, it is distributed among many guided modes. These various modes arrive at the fiber end at different times depending on their group delay, so that a pulse spreading results. For multimode fibers the waveguide dispersion is generally very small compared to material dispersion and can therefore be neglected.

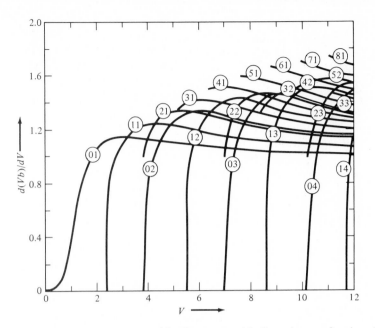

Figure 3-14 The group delay arising from waveguide dispersion as a function of the V number for a step-index fiber. When $\nu \neq 1$, the curve numbers νm designate the $HE_{\nu+1,m}$ and $EH_{\nu-1,m}$ modes. For $\nu = 1$, the curve numbers νm give the HE_{2m}, TE_{0m}, and TM_{0m} modes. *(Reproduced with permission from Gloge.[44])*

For single-mode fibers, however, waveguide dispersion is of importance and can be of the same order of magnitude as material dispersion. To see this, let us compare the two dispersion factors. The pulse spread τ_{wg} occurring over a distribution of wavelengths σ_λ is obtained from the derivative of the group delay with respect to wavelength[49]

$$\tau_{wg} = \sigma_\lambda \frac{dt_{wg}}{d\lambda} = -\frac{V}{\lambda} \sigma_\lambda \frac{dt_{wg}}{dV} = -\frac{n_2 L}{c\lambda} \Delta\sigma_\lambda \, V \frac{d^2(Vb)}{dV^2} \qquad (3\text{-}26)$$

The factor $Vd^2(Vb)/dV^2$ is plotted as a function of V in Fig. 3-15 for the fundamental mode shown in Fig. 3-14. This factor reaches a maximum at $V = 1.2$, but runs between 0.2 and 0.1 for a practical single-mode operating range of $V = 2.0$ to 2.4. Thus for values of $\Delta = 0.01$ and $n_2 = 1.5$,

$$\frac{\tau_{wg}}{L} \simeq -\frac{0.003\sigma_\lambda}{c\lambda} \qquad (3\text{-}27)$$

Comparing this with the material-dispersion-induced pulse spreading from Eq. (3-20) for $\lambda = 900$ nm, where

$$\frac{\tau_{mat}}{L} \simeq -\frac{0.02\sigma_\lambda}{c\lambda} \qquad (3\text{-}28)$$

it is clear that material dispersion dominates at lower wavelengths. However, at

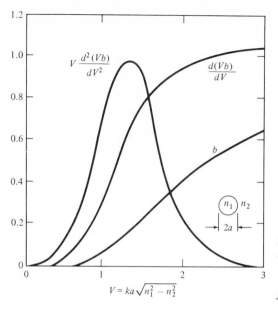

Figure 3-15 The waveguide parameter *b* and its derivatives $d(Vb)/dV$ and $Vd^2(Vb)/dV^2$ plotted as a function of the *V* number. *(Reproduced with permission from Gloge.[49])*

longer wavelengths such as at 1.3 μm, which is the spectral region of extremely low material dispersion in silica, waveguide dispersion can become the dominating pulse-distorting mechanism. Examples of the magnitudes of material and waveguide dispersions are given in Fig. 3-16 for a fused-silica-core single-mode fiber having $V = 2.4$. In this figure the approximation that material and waveguide dispersions are additive was used.

Figure 3-16 shows that in single-mode fibers the total dispersion can be reduced to zero at a particular wavelength through the mutual cancellation of material and waveguide dispersions.[50–56] The particular wavelength at which the total dispersion is reduced to a minimum can be selected anywhere in the 1.3- to 1.7-μm spectral range. For GeO_2-doped fibers this is achieved by varying the amount of GeO_2 dopant to obtain different material behavior, and by controlling the waveguide effects through variations in core diameter and core-cladding index difference. For example, fibers designed for zero dispersion in the minimum-attenuation region of 1.55 μm could have core diameters of approximately 4.0 to 4.8 μm, core dopant concentrations of about 13 mol % GeO_2, and index differences ranging from 0.55 to 1.8 percent.

3-3-5 Intermodal Dispersion

The final factor giving rise to signal distortion is intermodal dispersion, which is a result of different values of the group delay for each individual mode at a single frequency. To see this pictorially, consider the meridional ray picture given for the step-index fiber in Fig. 2-11. The steeper the angle of propagation of the ray congruence, the higher is the mode number and, consequently, the slower the axial group velocity. This variation in the group velocities of the different modes results in a group delay spread or intermodal dispersion. This distortion mechanism is

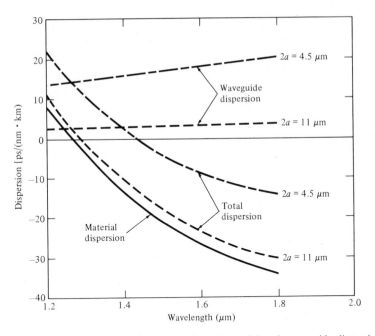

Figure 3-16 Examples of the magnitudes of material and waveguide dispersions as a function of optical wavelength for a single-mode fused-silica-core fiber. *(Reproduced with permission from White and Nelson.[50])*

eliminated by single-mode operation, but is important in multimode fibers. The pulse broadening arising from intermodal dispersion is the difference in travel time between the longest ray congruence paths (the highest-order mode) and the shortest ray congruence paths (the fundamental mode). This is simply obtained from ray tracing and is given by

$$\tau_{\text{mod}} = T_{\text{max}} - T_{\text{min}} = \frac{n_1 \, \Delta L}{c} \tag{3-29}$$

For values of $n_1 = 1.5$ and $\Delta = 0.01$ the modal spread is $\tau_{\text{mod}} = 0.015L/c$. Comparing this with Eqs. (3-27) and (3-28) with a relative spectral width of $\sigma_\lambda/\lambda = 4$ percent for a light-emitting diode, we have $\tau_{wg} = 1.2 \times 10^{-4}L/c$ and $\tau_{\text{mat}} = 8 \times 10^{-4}L/c$, which shows that τ_{mod} dominates the pulse spreading by about an order of magnitude in step-index fibers. The relative dominance of τ_{mod} is even greater if a laser diode light source, which has a narrower spectral output width than an LED, is used.

3-4 PULSE BROADENING IN GRADED-INDEX WAVEGUIDES

The analysis of pulse broadening in graded-index waveguides is more involved owing to the radial variation in core refractive index. The feature of this grading of the refractive-index profile is that it offers multimode propagation in a relatively large core together with the possibility of very low intermodal delay distortion. This combination allows the transmission of high data rates over long distances while still maintaining a reasonable degree of light launching and coupling ease. The reason for this low intermodal distortion can be seen by examining the light ray congruence propagation paths shown in Fig. 2-9. Since the index of refraction is lower at the outer edges of the core, light rays will travel faster in this region than in the center of the core where the refractive index is higher. This can be seen from the fundamental relationship $v = c/n$, where v is the speed of light in a medium of refractive index n. Thus the ray congruence characterizing the higher-order mode will tend to travel further than the fundamental ray congruence, but at a faster rate. The higher-order mode will thereby tend to keep up with the lower-order mode, which, in turn, reduces the spread in the modal delay.

The root-mean-square (rms) pulse broadening σ in a graded-index fiber can be obtained from the sum[57]

$$\sigma = (\sigma_{\text{intermodal}}^2 + \sigma_{\text{intramodal}}^2)^{1/2} \qquad (3\text{-}30)$$

where $\sigma_{\text{intermodal}}$ is the rms pulse width resulting from intermodal delay distortion and $\sigma_{\text{intramodal}}$ is the rms pulse width resulting from pulse broadening within each mode. To find the intermodal delay distortion, we use the relationship connecting intermodal delay to pulse broadening derived by Personick[57]

$$\sigma_{\text{intermodal}} = (<\tau_g^2> - <\tau_g>^2)^{1/2} \qquad (3\text{-}31)$$

where the group delay τ_g of a mode is given by Eq. (3-14) and the quantity $<A>$ is defined as the average of the variable $A_{\nu m}$ over the mode distribution, that is, it is given by

$$<A> = \sum_{\nu,m} \frac{P_{\nu m} A_{\nu m}}{M} \qquad (3\text{-}32)$$

where $P_{\nu m}$ is the power contained in the mode of order (ν, m).

The group delay

$$\tau_g = \frac{L}{c} \frac{\partial \beta}{\partial k} \qquad (3\text{-}33)$$

is the time it takes energy in a mode having a propagation constant β to travel a distance L. To evaluate τ_g we solve Eq. (2-78) for β, which yields

$$\beta = \left[k^2 n_1^2 - 2\left(\frac{\alpha + 2}{\alpha} \frac{m}{a^2}\right)^{\alpha/(\alpha+2)} (n_1^2 k^2 \Delta)^{2/(\alpha+2)} \right]^{1/2} \qquad (3\text{-}34)$$

or, equivalently,

$$\beta = kn_1 \left[1 - 2\Delta \left(\frac{m}{M}\right)^{\alpha/(\alpha+2)} \right]^{1/2}$$

where m is the number of guided modes having propagation constants between $n_1 k$ and β, and M is the total number of possible guided modes given by Eq. (2-79).

Substituting Eq. (3-34) into Eq. (3-33), keeping in mind that n_1 and Δ also depend on k, we obtain

$$\tau = \frac{L}{c}\frac{kn_1}{\beta}\left[N_1 - \frac{4\Delta}{\alpha + 2}\left(\frac{m}{M}\right)^{\alpha/(\alpha+2)}\left(N_1 + \frac{n_1 k}{2\Delta}\frac{\partial\Delta}{\partial k}\right)\right]$$

$$= \frac{LN_1}{c}\frac{kn_1}{\beta}\left[1 - \frac{\Delta}{\alpha + 2}\left(\frac{m}{M}\right)^{\alpha/(\alpha+2)}(4 + \epsilon)\right] \tag{3-35}$$

where we have used Eq. (2-79) for M and have defined the quantities

$$N_1 = n_1 + k\frac{\partial n_1}{\partial k} \tag{3-36a}$$

$$\epsilon = \frac{2n_1 k}{N_1\Delta}\frac{\partial\Delta}{\partial k} \tag{3-36b}$$

As we noted in Eq. (2-38), guided modes only exist for values of β lying between kn_2 and kn_1. Since n_1 differs very little from n_2, that is

$$n_2 = n_1(1 - \Delta)$$

where $\Delta \ll 1$ is the core-cladding index difference, it follows that $\beta \simeq n_1 k$. Thus we can use the relationship

$$y = \Delta\left(\frac{m}{M}\right)^{\alpha/(\alpha+2)} \ll 1 \tag{3-37}$$

in order to expand Eq. (3-35) in a power series in y. Using the approximation

$$\frac{kn_1}{\beta} = (1 - 2y)^{-1/2} \simeq 1 + y + \frac{3y^2}{2} \tag{3-38}$$

we have that

$$\tau = \frac{N_1 L}{c}\left[1 + \frac{\alpha - 2 - \epsilon}{\alpha + 2}\Delta\left(\frac{m}{M}\right)^{\alpha/(\alpha+2)}\right.$$

$$\left. + \frac{3\alpha - 2 - 2\epsilon}{2(\alpha + 2)}\Delta^2\left(\frac{m}{M}\right)^{2\alpha/(\alpha+2)} + 0(\Delta^3)\right] \tag{3-39}$$

Equation (3-39) shows that to first order in Δ, the group delay difference between modes is zero if

$$\alpha = 2 + \epsilon \tag{3-40}$$

Since ϵ is generally small, this indicates that minimum intermodal distortion will result from core refractive-index profiles which are nearly parabolic, that is, $\alpha \simeq 2$.

If we assume that all modes are equally excited, that is, $P_{vm} = P$ for all modes, and if the number of fiber modes is assumed to be large, then the summation in Eq.

(3-32) can be replaced by an integral. Using these assumptions, Eq. (3-39) can be substituted into Eq. (3-31) to yield[34, 58, 59]

$$\sigma_{\text{intermodal}} = \frac{LN_1\Delta}{2c} \frac{\alpha}{\alpha + 1} \left(\frac{\alpha + 2}{3\alpha + 2}\right)^{1/2}$$

$$\times \left[c_1^2 + \frac{4c_1c_2(\alpha + 1)\Delta}{2\alpha + 1} + \frac{16\Delta^2 c_2^2(\alpha + 1)^2}{(5\alpha + 2)(3\alpha + 2)} \right]^{1/2} \quad (3\text{-}41)$$

where we have used the abbreviations

$$c_1 = \frac{\alpha - 2 - \epsilon}{\alpha + 2}$$

$$c_2 = \frac{3\alpha - 2 - 2\epsilon}{2(\alpha + 2)} \quad (3\text{-}42)$$

To find the intramodal pulse broadening, we use the definition[34, 58]

$$\sigma_{\text{intramodal}}^2 = L^2 \left(\frac{\sigma_\lambda}{\lambda}\right)^2 \left\langle \left(\lambda \frac{d\tau_g}{d\lambda}\right)^2 \right\rangle \quad (3\text{-}43)$$

where σ_λ is the rms spectral width of the optical source. Equation (3-39) can be used to evaluate the term in parentheses. If we neglect all terms of second and higher order in Δ, we obtain

$$\lambda \frac{d\tau_g}{d\lambda} = -\frac{L}{c} \lambda^2 \frac{d^2 n_1}{d\lambda^2} + \frac{N_1 L \Delta}{c} \frac{\alpha - 2 - \epsilon}{\alpha + 2} \frac{2\alpha}{\alpha + 2} \left(\frac{m}{M}\right)^{\alpha/(\alpha+2)} \quad (3\text{-}44)$$

Here we have kept only the largest terms, that is, terms involving factors such as $d\Delta/d\lambda$ and $\Delta \, dn_1/d\lambda$ are negligibly small. Both terms in Eq. (3-44) contribute to $\lambda \, d\tau_g/d\lambda$ for large values of α, since $\lambda^2 \, d^2 n_1/d\lambda^2$ and Δ are the same order of magnitude. However, the second term in Eq. (3-44) is small compared to the first term when α is close to 2.

To evaluate $\sigma_{\text{intramodal}}$ we again assume that all modes are equally excited and that the summation in Eq. (3-32) can be replaced by an integral. Thus substituting Eq. (3-44) into Eq. (3-43) we have[34, 58]

$$\sigma_{\text{intramodal}} = \frac{L}{c} \frac{\sigma_\lambda}{\lambda} \left[\left(-\lambda^2 \frac{d^2 n_1}{d\lambda^2}\right)^2 - N_1 c_1 \Delta \left(2\lambda^2 \frac{d^2 n_1}{d\lambda^2} \frac{\alpha}{\alpha + 1} - N_1 \Delta \frac{2\alpha}{3\alpha + 2}\right) \right]^{1/2}$$

$$(3\text{-}45)$$

Olshansky and Keck[58] have evaluated σ as a function of α at $\lambda = 900$ nm for a titania-doped silica fiber having a numerical aperture of 0.16. This is shown in Fig. 3-17. Here the uncorrected curve assumes $\epsilon = 0$ and includes only intermodal dispersion (no material dispersion). The inclusion of the effect of ϵ shifts the curve to higher values of α. The effect of the spectral width of the optical source on the rms pulse width is clearly demonstrated in Fig. 3-17. The light sources shown are an LED, an injection laser diode, and a distributed-feedback laser having rms

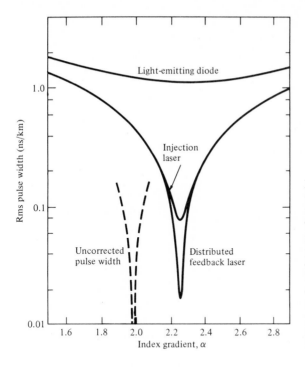

Figure 3-17 Calculated rms pulse spreading in a graded-index fiber versus the index parameter alpha at 900 nm. The uncorrected pulse curve is for $\epsilon = 0$ and assumes mode dispersion only. The other curves include material dispersion for an LED, an injection laser diode, and a distributed-feedback laser having spectral widths of 15, 1, and 0.2 nm, respectively. *(Reproduced with permission from Olshansky and Keck.[58])*

spectral widths of 15, 1, and 0.2 nm, respectively. The data transmission capacities of these sources are approximately 0.13, 2, and 10 $(\text{Gb} \cdot \text{km})/\text{s}$, respectively.

The value of α which minimizes pulse distortion depends strongly on wavelength. To see this, let us examine the structure of a graded-index fiber. A simple model of this structure is to consider the core to be composed of concentric cylindrical layers of glass, each of which has a different material composition. For each layer the refractive index has a different variation with wavelength λ since the glass composition is different in each layer. Consequently, a fiber with a given index profile α will exhibit different pulse spreading according to the source wavelength used. This is generally called *profile dispersion*. An example of this is given in Fig. 3-18 for a GeO_2-SiO_2 fiber.[60] This shows that the optimum value of α decreases with increasing wavelength. Suppose one wishes to transmit at 900 nm. A fiber having an optimum profile α_{opt} at 900 nm should exhibit a sharp bandwidth peak at that wavelength. Fibers with undercompensated profiles, characterized by $\alpha > \alpha_{\text{opt}}$ (900 nm), tend to have a peak bandwidth at a shorter wavelength. On the other hand, overcompensated fibers which have an index profile $\alpha < \alpha_{\text{opt}}$ (900 nm) become optimal at a longer wavelength.

If the effect of material dispersion is ignored (that is, for $dn_1/d\lambda = 0$), an expression for the optimum index profile can be found from the minimum of Eq. (3-41) as a function of α. This occurs at[58]

$$\alpha_{\text{opt}} = 2 + \epsilon - \Delta \frac{(4 + \epsilon)(3 + \epsilon)}{5 + 2\epsilon} \tag{3-46}$$

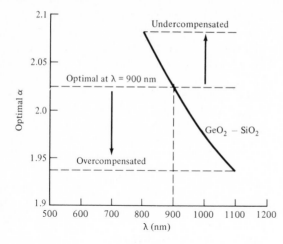

Figure 3-18 Profile dispersion effect on the optimum value of α as a function of wavelength for a GeO_2-SiO_2 graded-index fiber. *(Reproduced with permission from Cohen, Kaminow, Astle, and Stulz,[60] copyright 1978, IEEE.)*

If we take $\epsilon = 0$ and $dn_1/d\lambda = 0$, then Eq. (3-41) reduces to

$$\sigma_{opt} = \frac{n_1 \, \Delta^2 L}{20\sqrt{3}c} \tag{3-47}$$

This can be compared with the dispersion in a step-index fiber by setting $\alpha = \infty$ and $\epsilon = 0$ in Eq. (3-41), yielding

$$\sigma_{step} = \frac{n_1 \, \Delta L}{c} \frac{1}{2\sqrt{3}} \left(1 + 3\Delta + \frac{12\Delta^2}{5}\right)^{1/2} \simeq \frac{n_1 \, \Delta L}{2\sqrt{3}c} \tag{3-48}$$

Thus, under the assumptions made in Eqs. (3-47) and (3-48),

$$\frac{\sigma_{step}}{\sigma_{opt}} = \frac{10}{\Delta} \tag{3-49}$$

Hence, since typical values of Δ are 0.01, Eq. (3-49) indicates that the capacity of

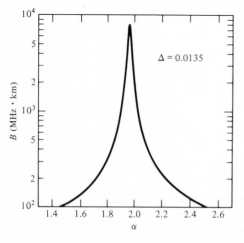

Figure 3-19 Variations in bandwidth resulting from slight deviations in the refractive-index profile for a graded-index fiber with $\Delta = 0.0135$. *(Reproduced with permission from Marcuse and Presby.[61])*

a graded-index fiber is about three orders of magnitude larger than that of a step-index fiber. For $\Delta = 1$ percent the rms pulse spreading in a step-index fiber is about 14 ns/km, whereas that for a graded-index fiber is calculated to be 0.014 ns/km. In practice these values are greater because of manufacturing difficulties. For example, it has been shown[61] that, although theory predicts a bandwidth of about 8 GHz·km, in practice very slight deviations of the refractive-index profile from its optimum shape, owing to unavoidable manufacturing tolerances, can decrease the fiber bandwidth dramatically. This is illustrated[61] in Fig. 3-19 for a fiber with $\Delta = 0.0135$. A change in α of a few percent can decrease the bandwidth by an order of magnitude.

3-5 MODE COUPLING

In real systems pulse distortion will increase less rapidly after a certain initial length of fiber because of mode coupling and differential mode loss.[62-64] In this initial length of fiber, coupling of energy from one mode to another arises because of structural imperfections, fiber diameter and refractive-index variations, and cabling-induced microbends. The mode coupling tends to average out the propagation delays associated with the modes, thereby reducing intermodal dispersion. Associated with this coupling is an additional loss, which is designated by h and which has units of dB/km. The result of this phenomenon is that, after a certain coupling length L_c, the pulse distortion will change from an L dependence to a $(L_c L)^{1/2}$ dependence.

The improvement in pulse spreading caused by mode coupling over the distance $Z < L_c$ is related to the excess loss hZ incurred over this distance by the equation

$$hZ\left(\frac{\sigma_c}{\sigma_0}\right)^2 = C \tag{3-50}$$

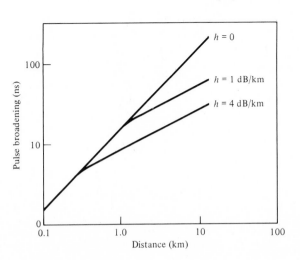

Figure 3-20 Mode-coupling effects on pulse distortion in long fibers for various coupling losses.

Here C is a constant, σ_0 is the pulse width increase in the absence of mode coupling, σ_c is the pulse broadening in the presence of strong mode coupling, and hZ is the excess attenuation resulting from mode coupling. The constant C in Eq. (3-50) is independent of all dimensional quantities and refractive indices. It depends only on the fiber profile shape, the mode-coupling strength, and the modal attenuations.

The effect of mode coupling on pulse distortion can be significant for long fibers, as is shown in Fig. 3-20 for various coupling losses in a graded-index fiber. The parameters of this fiber are $\Delta = 1$ percent, $\alpha = 4$, and $C = 1.1$. The coupling loss h must be determined experimentally, since a calculation would require a detailed knowledge of the mode coupling introduced by the various waveguide perturbations. Measurements[62, 65, 66] of bandwidth as a function of distance have produced values of L_c ranging from about 100 to 550 m.

3-6 SUMMARY

In this chapter we have discussed various engineering aspects of making and using optical fiber waveguides. Glasses and plastics are the principal materials of which fibers are made. Of these, silica-based glass is the most widely used material for the following reasons:

1. It is possible to make long, thin, flexible fibers from this material (and also from plastic).
2. Pure silica glass is highly transparent in the 800- to 1600-nm wavelength range. This is necessary in order for the fiber to guide light efficiently. Plastic fibers are less widely used because of their higher attenuation compared with highly pure glass fibers.
3. By adding trace amounts of certain elements (known as dopants) to the silica, physically compatible materials having slightly different refractive indices for the core and cladding are made available.

All-glass optical fibers are generally made either by a vapor phase oxidation process or by a direct-melt method. The direct-melt method follows traditional glass-making procedures in that optical fibers are made directly from the molten state of purified glasses. Although this method has the advantage of being a continuous process, great care must be taken during the melting process to avoid contaminants (which can give rise to high optical attenuation). In the vapor phase oxidation process, highly pure vapors of metal halides react with oxygen to form a white powder of SiO_2 particles. These particles are collected on the surface of a bulk glass by one of three processes and are sintered to form a clear glass rod or tube (depending on the process). This rod or tube is called a *preform*. Once the preform is made, one end is softened to the point where it can be drawn into a very thin filament which becomes the optical fiber.

Attenuation of a light signal as it propagates along a fiber is an important consideration in the design of an optical communication system, since it plays a major role in determining the maximum transmission distance between a transmitter and a receiver. The basic attenuation mechanisms are absorption, scattering, and radiative losses of optical energy. The major causes of absorption are extrinsic absorption by impurity atoms and intrinsic absorption by the basic constituent atoms of the fiber material. Intrinsic absorption, which sets the fundamental lower limit on attenuation for any particular material, results from electronic absorption bands in the ultraviolet region and from atomic vibration bands in the near-infrared region. Scattering losses arise from microscopic variations in the material density, from compositional fluctuations, and from structural inhomogeneities or defects occurring during fiber manufacture. Scattering follows a Rayleigh λ^{-4} dependence, which gives attenuation-versus-wavelength plots their characteristic downward trend with increasing wavelength. Radiative losses occur whenever an optical fiber undergoes a bend of finite radius of curvature. This can arise from macroscopic bends, for example, such as when a fiber cable turns a corner or from microscopic bends (microbends) of the fiber axis occurring during fiber cabling. Of these, microbends are the most troublesome so that special care must be taken during the cabling process to minimize them.

In addition to being attenuated, an optical signal becomes increasingly distorted as it travels along a fiber. This distortion is a consequence of intramodal and intermodal dispersion effects. Intramodal dispersion is pulse spreading that occurs within a single mode. Its two main causes are:

1. Material dispersion, which arises form the variation with wavelength of the refractive index of the core material
2. Waveguide dispersion, which occurs because the modal propagation constant β is a function of the ratio between the core radius and the wavelength

In multimode fibers, pulse distortions also occur since each mode travels at a different group velocity (which is known as intermodal delay distortion). Mode delay is the dominant pulse-distorting mechanism in step-index fibers. On the other hand, intermodal delay distortion can be made very small by careful tailoring of the core refractive-index profile. Material dispersion thus tends to be the dominant pulse-distorting effect in graded-index fibers.

PROBLEMS

Section 3-2

3-1 A certain optical fiber has an attenuation of 3.5 dB/km at 850 nm. If 0.5 mW of optical power is initially launched into the fiber, what is the power level in mW after 4 km?

3-2 An optical signal has lost 85 percent of its power after traversing 500 m of fiber. What is the loss in dB/km of this fiber?

3-3 The optical power loss resulting from Rayleigh scattering in a fiber can be calculated from either Eq. (3-3) or Eq. (3-4). Compare these two equations for silica ($n = 1.460$ at 630 nm), given that the fictive temperature T_f is 1400 K, the isothermal compressibility β_T is 6.8×10^{-12} cm^2/dyn, and the photoelastic coefficient is 0.286. How does this agree with measured values ranging from 3.9 to 4.8 dB/km at 633 nm?

3-4 Consider graded-index fibers having index profiles $\alpha = 2.0$, cladding refractive indices $n_2 = 1.50$, and index differences $\Delta = 0.01$. Using Eq. (3-7), plot the ratio N_{eff}/N_∞ for bend radii less than 10 cm at $\lambda = 1$ μm for fibers having core radii of 4, 25, and 100 μm.

3-5 Three common fiber jacket materials are Elvax® 265 ($E_j = 21$ MPa) and Hytrel® 4056 ($E_j = 58$ MPa), both made by DuPont, and Versalon® 1164 ($E_j = 104$ MPa) made by General Mills. If the Young's modulus of a glass fiber is 64 GPa, plot the reduction in microbending loss as a function of the index difference Δ when fibers are coated with these materials. Make these plots for Δ values ranging from 0.1 to 1.0 percent and for a fiber cladding-to-core ratio of $b/a = 2$.

Section 3-3

3-6 Derive Eq. (3-19) from Eq. (3-18).

3-7 (a) Using Eqs. (2-40), (2-41), and (2-49) show that the normalized propagation constant defined by Eq. (3-21) can be written in the form

$$b = \frac{\beta^2/k^2 - n_2^2}{n_1^2 - n_2^2}$$

(b) For small core-cladding refractive-index differences show that the expression for b derived in (a) reduces to

$$b = \frac{\beta/k - n_2}{n_1 - n_2}$$

from which it follows that

$$\beta = n_2 k (b\Delta + 1)$$

3-8 Using the approximation $V \simeq k a n_2 \sqrt{2\Delta}$ show that Eq. (3-25) results from Eq. (3-24).

3-9 For wavelengths less than 1.0 μm the refractive index n satisfies a Sellmeier relation of the form[67]

$$n^2 = 1 + \frac{E_0 E_d}{E_0^2 - E^2}$$

where $E = hc/\lambda$ is the photon energy and E_0 and E_d are, respectively, material oscillator energy and dispersion energy parameters. In SiO_2 glass $E_0 = 13.4$ eV and $E_d = 14.7$ eV. Show that, for wavelengths between 0.20 and 1.0 μm, the values of n found from the Sellmeier relation are in good agreement with those shown in Fig. 3-12.

3-10 Derive Eq. (3-29) by using a ray-tracing method.

Section 3-4

3-11 Verify that Eq. (3-41) reduces to Eq. (3-48) for the case $\alpha = \infty$ and $\epsilon = 0$.

3-12 Show that, when the effect of material dispersion is ignored and for $\epsilon = 0$, Eq. (3-41) reduces to Eq. (3-47).

3-13 Make a plot on log-log paper of the rms pulse broadening in a parabolic graded-index fiber ($\alpha = 2$) as a function of the optical source spectral width σ_λ in the range 0.10 to 100 nm for peak operating wavelengths of 850 nm and 1300 nm. Let $\Delta = 0.01$, $N_1 = 1.46$, and $\epsilon = 0$ at both wavelengths. Assume the factor $\lambda^2 d^2 n/d\lambda^2$ is 0.025 at 850 nm and 0.004 at 1300 nm.

3-14 Repeat Prob. 3-13 for a graded-index single-mode fiber with $\Delta = 0.001$.

3-15 Derive Eq. (3-35) by substituting Eq. (3-34) into Eq. (3-33).

3-16 Using the approximation given by Eq. (3-38), show that Eq. (3-35) can be rewritten as Eq. (3-39).

3-17 Derive Eq. (3-44) from Eq. (3-39).

REFERENCES

1. G. W. Morey, *The Properties of Glass,* Reinhold, New York, 1954.
2. C. J. Phillips, *Glass: Its Industrial Applications,* Reinhold, New York, 1960.
3. B. C. Bagley, C. R. Kurkjian, J. W. Mitchell, G. E. Peterson, and A. R. Tynes, "Materials, properties, and choices," in *Optical Fiber Telecommunications,* S. E. Miller and A. G. Chynoweth (Eds.), Academic, New York, 1979.
4. P. C. Schultz, "Progress in optical waveguide process and materials," *Appl. Opt.,* **18,** 3684–3693, Nov. 1979.
5. D. A. Pinnow, A. L. Gentile, A. G. Standler, A. T. Timper, and L. M. Holbrook, "Polycrystalline fiber optical waveguides for infrared transmission," *Appl. Phys. Lett.,* **33,** 28–29, July 1978.
6. L. G. Van Uitert and S. H. Wemple, "ZnCl₂ glass: A potential ultralow-loss optical fiber material," *Appl. Phys. Lett.,* **33,** 57–59, July 1978.
7. S. Mitachi, S. Shibata, and T. Manabe, "Teflon FEP-clad fluoride glass fiber," *Electron. Lett.,* **17,** 128–129, Feb. 1981.
8. P. Kaiser, A. C. Hart, Jr., and L. L. Blyler, Jr., "Low-loss FEP-clad silica fibers," *Appl. Opt.,* **14,** 156–162, Jan. 1975.
9. P. C. Schultz, "Vapor phase materials and processes for glass optical waveguides," in *Fiber Optics,* B. Bendow and S. S. Mitra (Eds.), Plenum, New York, 1979.
10. W. G. French, R. E. Jaeger, J. B. MacChesney, S. R. Nagel, K. Nassau, and A. D. Pearson, "Fiber preform preparation," in *Optical Fiber Telecommunications,* S. E. Miller and A. G. Chynoweth (Eds.), Academic, New York, 1979.
11. F. P. Kapron, D. B. Keck, and R. D. Maurer, "Radiation losses in glass optical waveguides," *Appl. Phys. Lett.,* **17,** 423–425, Nov. 1970.
12. P. C. Schultz, "Fabrication of optical waveguides by the outside vapor deposition process," *Proc. IEEE,* **68,** 1187–1190, Oct. 1980.
13. T. Izawa and N. Inagaki, "Materials and processes for fiber preform fabrication—vapor-phase axial deposition," *Proc. IEEE,* **68,** 1184–1187, Oct. 1980.
14. J. B. MacChesney, P. B. O'Conner, and H. M. Presby, "A new technique for preparation of low-loss and graded-index optical fibers," *Proc. IEEE,* **62,** 1280–1281, Sept. 1974.
15. J. B. MacChesney, "Materials and processes for preform fabrication—modified chemical vapor deposition and plasma chemical vapor deposition," *Proc. IEEE,* **68,** 1181–1184, Oct. 1980.
16. K. J. Beales, C. R. Day, A. G. Dunn, and S. Partington, "Multicomponent glass fibers for optical communications," *Proc. IEEE,* **68,** 1191–1194, Oct. 1980.
17. J. E. Midwinter, *Optical Fibers for Transmission,* Wiley, New York, 1979.
18. D. L. Meyers and F. P. Partus, "Preform fabrication and fiber drawing by Western Electric Product Engineering Control Center," *Bell Sys. Tech. J.,* **57,** 1735–1744, July–Aug. 1978.
19. R. E. Jaeger, A. D. Pearson, J. C. Williams, and H. M. Presby, "Fiber drawing and control," in *Optical Fiber Telecommunications,* S. E. Miller and A. G. Chynoweth (Eds.), Academic, New York, 1979.
20. R. Maurer, "Glass fibers for optical communications," *Proc. IEEE,* **61,** 452–462, Apr. 1973.
21. D. Gloge, "Propagation effects in optical fibers," *IEEE Trans. Microwave Theory Tech.,* **MTT-23,** 106–120, Jan. 1975.
22. R. Olshansky, "Propagation in glass optical waveguides," *Rev. Mod. Phys.,* **51,** 341–367, Apr. 1979.
23. D. Gloge, "The optical fibre as a transmission medium," *Rpts. Prog. Phys.,* **42,** 1777–1824, Nov. 1979.
24. J. E. Golob, P. B. Lyons, and L. D. Looney, "Transient radiation effects in low-loss optical waveguides," *IEEE Trans. Nuc. Sci.,* **NS-24,** 2164–2168, Dec. 1977.
25. P. L. Mattern, L. M. Watkins, C. D. Skoog, J. R. Brandon, and E. H. Barsis, "The effects of radiation on the absorption and luminescence of fiber optic waveguides and materials," *IEEE Trans. Nuc. Sci.,* **NS-21,** 81–95, Dec. 1974.
26. J. C. Blackburn, "Only with fibers: Space radiation testing," *Opt. Spectra,* **14,** 46–50, May 1980.

27. E. J. Friebele, G. H. Sigel, and M. E. Gingerich, "Radiation response of fiber optic waveguides in the 0.4 to 1.8 micron region," *IEEE Trans. Nuc. Sci.*, **NS-25**, 1261–1266, Dec. 1978.

28. (a) T. Moriyama, O. Fukuda, K. Sanada, K. Inada, T. Edahiro, and K. Chida, "Ultimately low OH content VAD optical fibers," *Electron. Lett.*, **16**, 699–700, Aug. 1980.
    (b) F. Hanawa, S. Sudo, M. Kawachi, and M. Nakahara, "Fabrication of completely OH-free VAD fibre," *ibid.*, 700–701.

29. J. Stone and C. A. Burrus, "Reduction of the 1.38 μm water peak in optical fibers by deuterium-hydrogen exchange," *Bell Sys. Tech. J.*, **59**, 1541–1548, Oct. 1980.

30. T. Izawa, N. Shibata, and A. Takeda, "Optical attenuation in pure and doped fused silica in the ir wavelength region," *Appl. Phys. Lett.*, **31**, 33–35, July 1977.

31. H. Osanai, T. Shioda, T. Moriyama, S. Araki, M. Horiguchi, T. Izawa, and H. Takata, "Effects of dopants on transmission loss of low OH content optical fibers," *Electron. Lett.*, **12**, 549–550, Oct. 1976.

32. T. Miya, Y. Terunuma, T. Hosaka, and T. Miyashita, "An ultimate low loss single mode fiber at 1.55 μm," *Electron. Lett.*, **15**, 106–108, Feb. 1979.

33. M. Horiguchi and H. Osanai, "Spectral losses of low OH content fibers," *Electron. Lett.*, **12**, 310–312, June 1976.

34. D. B. Keck, "Optical fiber waveguides," in *Fundamentals of Optical Fiber Communications*, M. K. Barnoski (Ed.), Academic, New York, 1976; 2d ea., 1982.

35. D. A. Pinnow, T. C. Rich, F. W. Ostermeyer, and M. DiDomenico, Jr., "Fundamental optical attenuation limits in the liquid and glassy state with application to fiber optical waveguide material," *Appl. Phys. Lett.*, **22**, 527–529, May 1973.

36. D. Marcuse, "Curvature loss formula for optical fibers," *J. Opt. Soc. Amer.*, **66**, 216–220, Mar. 1976.

37. I. A. White, "Radiation from bends in optical waveguides: the volume-current method," *Microwaves, Opt., Acous. (IEE)*, **3**, 186–188, Sept. 1979.

38. D. Gloge, "Bending loss in multimode fibers with graded and ungraded core index," *Appl. Opt.*, **11**, 2506–2512, Nov. 1972.

39. W. B. Gardener, "Microbending loss in optical fibers," *Bell Sys. Tech. J.*, **54**, 457–465, Feb. 1975.

40. J. Sakai and T. Kimura, "Practical microbending loss formula for single mode optical fibers," *IEEE J. Quantum Electron.*, **QE-15**, 497–500, June 1979.

41. K. Furuya and Y. Suematsu, "Random-bend loss in single-mode and parabolic-index multimode optical fiber cables," *Appl. Opt.*, **19**, 1493–1500, May 1980.

42. D. Gloge, "Optical fiber packaging and its influence on fiber straightness and loss," *Bell Sys. Tech. J.*, **54**, 245–262, Feb. 1975.

43. D. Keck, "Spatial and temporal transfer measurements on low loss optical waveguides," *Appl. Opt.*, **13**, 1882–1888, Aug. 1974.

44. D. Gloge, E. A. J. Marcatili, D. Marcuse, and S. D. Personick, "Dispersion properties of fibers," in *Optical Fiber Telecommunications*, S. E. Miller and A. G. Chynoweth (Eds.), Academic, New York, 1979.

45. S. E. Miller, E. A. J. Marcatili, and T. Li, "Research toward optical fiber transmission systems," *Proc. IEEE*, **61**, 1703–1751, Dec. 1973.

46. R. B. Dyott and J. R. Stern, "Group delay in glass fiber waveguide," *Electron. Lett.*, **7**, 82–84, Feb. 1971.

47. D. Marcuse, "Interdependence of waveguide and material dispersion," *Appl. Opt.*, **18**, 2930–2932, Sept. 1979.

48. R. P. Kapron and D. B. Keck, "Pulse transmission through a dielectric optical waveguide," *Appl. Opt.*, **10**, 1519–1523, July 1971.

49. D. Gloge, "Weakly guiding fibers," *Appl. Opt.*, **10**, 2252–2258, Oct. 1971; "Dispersion in weakly guiding fibers," *ibid.*, 2442–2445, Nov. 1971.

50. K. I. White and B. P. Nelson, "Zero total dispersion in step index monomode fibers at 1.30 and 1.55 μm," *Electron. Lett.*, **15**, 396–397, June 1979.

51. W. A. Gambling, H. Matsumara, and C. M. Ragdale, "Zero total dispersion in graded index single mode fibers," *Electron. Lett.,* **15,** 474–476, July 1971; "Mode dispersion, material dispersion, and profile dispersion in graded index single mode fibers," *Microwaves, Opt., and Acoustics,* **3,** 239–246, Nov. 1979.
52. C. T. Chang, "Minimum dispersion at 1.55 μm for single mode step index fibers," *Electron. Lett.,* **15,** 765–767, Nov. 1979.
53. L. Jeunhomme, "Dispersion minimization in single mode fibers between 1.3 and 1.7 μm," *Electron. Lett.,* **15,** 478–479, July 1979.
54. L. G. Cohen, C. Lin, and W. G. French, "Tailoring zero chromatic dispersion into the 1.5–1.6 μm low loss spectral region of single mode fibers," *Electron. Lett.,* **15,** 334–335, June 1979.
55. J. Yamada, S. Machida, T. Mukai, and T. Kimura, "800 Mbit/s optical transmission experiments with dispersion free fibers at 1.5 μm," *Electron. Lett.,* **16,** 115–117, Feb. 1980.
56. P. S. M. Pires, D. A. Rogers, E. J. Bochove, and R. F. Souza, "Prediction of laser wavelength for minimum total dispersion in single-mode step-index fibers," *IEEE Trans. Microwave Theory Tech.,* **MTT-30,** 131–139, Feb. 1982.
57. S. D. Personick, "Receiver design for digital fiber optic communication systems," *Bell Sys. Tech. J.,* **52,** 843–874, July–Aug. 1973.
58. R. Olshansky and D. Keck, "Pulse broadening in graded index optical fibers," *Appl. Opt.,* **15,** 483–491, Feb. 1976.
59. H. G. Unger, *Planar Optical Waveguides and Fibers,* Clarendon Press, Oxford, 1977.
60. L. Cohen, I. Kaminow, H. Astle, and L. Stulz, "Profile dispersion effects on transmission bandwidths in graded index optical fibers," *IEEE J. Quantum Electron.,* **QE-14,** 37–41, Jan. 1978.
61. D. Marcuse and H. M. Presby, "Effects of profile deformation on fiber bandwidth," *Appl. Opt.,* **18,** 3758–3763, Nov. 1979; *ibid.,* **19,** 188, Jan. 1980.
62. R. Olshansky, "Mode coupling effects in graded index optical fibers," *Appl. Opt.,* **14,** 935–945, Apr. 1975.
63. D. Marcuse, "Losses and impulse response of a parabolic index fiber with random bends," *Bell Sys. Tech. J.,* **52,** 1423–1437, Oct. 1973.
64. D. Gloge, "Impulse response of clad optical multimode fibers," *Bell Sys. Tech. J.,* **52,** 801–816, July 1973.
65. S. Zemon and D. Fellows, "Characterization of the approach to steady-state and the steady-state properties of multimode optical fibers using LED excitation," *Opt. Commun.,* **13,** 198–202, Feb. 1975.
66. E. L. Chinnock, L. G. Cohen, W. S. Holden, R. D. Stanley, and D. B. Keck, "The length dependence of pulse spreading in the CGW-Bell-10 optical fiber," *Proc. IEEE,* **61,** 1499–1500, Oct. 1973.
67. M. DiDomenico, Jr., "Material dispersion in optical fiber waveguides," *Appl. Opt.,* **11,** 652–654, Mar. 1972.

FOUR

OPTICAL SOURCES

The principal light sources used for fiber optic communications applications are heterojunction-structured semiconductor *laser diodes* (also referred to as *injection laser diodes* or ILDs) and *light-emitting diodes* (LEDs). A *heterojunction* consists of two adjoining semiconductor materials with different band-gap energies. These devices are suitable for fiber transmission systems because they have adequate output power for a wide range of applications, their optical power output can be directly modulated by varying the input current to the device, they have a high efficiency, and their dimensional characteristics are compatible with those of the optical fiber. Comprehensive treatments of the major aspects of LEDs and laser diodes are presented in the books by Kressel and Butler,[1] Casey and Panish,[2] and Thompson.[3] Shorter reviews covering the operating principles of these devices are also available[4-8] to which the reader is referred for details. The intent of this chapter is to give an overview of the pertinent characteristics of fiber-compatible luminescent sources. The first two sections in this chapter present the output and operating characteristics of LEDs and laser diodes, respectively. This is followed by sections discussing the temperature responses of optical sources, their linearity characteristics, and their reliability under various operating conditions. Since the material in this chapter assumes a rudimentary knowledge of semiconductor physics, various relevant definitions are given in App. E to refresh the reader's memory.

We shall show in this chapter that the light-emitting region of both LEDs and laser diodes consists of a *pn* junction constructed of direct-band-gap III-V semiconductor materials. When this junction is forward-biased, electrons and holes are injected into the *p* and *n* regions, respectively. These injected minority carriers can recombine either radiatively, in which case a photon of energy $h\nu$ is emitted, or nonradiatively, whereupon the recombination energy is dissipated in the form of heat. This *pn* junction is thus known as the *active* or *recombination region*.

A major difference between LEDs and laser diodes is that the optical output from an LED is incoherent, whereas that from a laser diode is coherent. In a coherent source the optical energy is produced in an optical resonant cavity, as is described in Sec. 4-2-1. The optical energy released from this cavity has spatial and temporal coherence, which means it is highly monochromatic and that the output beam is very directional. In an incoherent LED source no optical cavity exists for wavelength selectivity. The output radiation has a broad spectral width since the emitted photon energies range over the energy distribution of the recombining electrons and holes which usually lie between 1 and $2k_B T$ (k_B is Boltzmann's constant and T is the absolute temperature at the *pn* junction). In addition, the incoherent optical energy is emitted into a hemisphere according to a cosine power distribution and, thus, has a large beam divergence.

In choosing an optical source compatible with the optical waveguide, various characteristics of the fiber such as its geometry, its attenuation as a function of wavelength, its group delay distortion (bandwidth), and its modal characteristics must be taken into account. The interplay of these factors with the optical source power, spectral width, radiation pattern, and modulation capability needs to be considered. The spatially directed coherent optical output from a laser diode can be coupled into either single-mode or multimode fibers. However, the incoherent optical power from an LED can only be coupled into a multimode fiber in sufficient quantities to be useful.

4-1 LIGHT-EMITTING DIODES (LEDs)

For optical communication systems requiring bit rates less than approximately 50 Mb/s together with multimode fiber-coupled optical power in the tens of micro-watts, semiconductor light-emitting diodes (LEDs) are usually the best light source choice. LEDs require less complex drive circuitry than laser diodes since no thermal or optical stabilization circuits are needed (see Sec. 4-2-4), and they can be fabricated less expensively with higher yields.

4-1-1 LED Structures

To be useful in fiber transmission applications an LED must have a high radiance output, a fast emission response time, and a high quantum efficiency. Its *radiance* (or *brightness*) is a measure in watts of the optical power radiated into a unit solid angle per unit area of the emitting surface. High radiances are necessary to couple sufficiently high optical power levels into a fiber, as we show in detail in Chap. 5. The emission response time is the time delay between the application of a current pulse and the onset of optical emission. As we discuss in Secs. 4-1-4 and 4-2-3, this time delay is the factor limiting the bandwidth with which the source can be modulated directly by varying the injected current. The quantum efficiency is related to the fraction of injected electron-hole pairs that recombine radiatively. This is defined and described in detail in Sec. 4-1-3.

Function	Material	Thickness	
Metal contact layer			
Used for better metal contact	p-type GaAs	$\sim 1\ \mu m$	
Light guiding and carrier confinement	p-type $Ga_{1-x}As_xP$	$\sim 1\ \mu m$	Holes
Recombination region	n-type $Ga_{1-y}As_yP$	$\sim 0.3\ \mu m$	$h\nu$
Light guiding and carrier confinement	n-type $Ga_{1-x}As_xP$	$\sim 1\ \mu m$	Electrons
Substrate	n-type GaAs		
Metal contact layer			

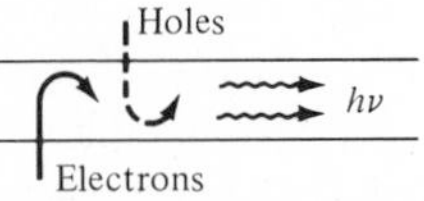

Figure 4-1 Cross-sectional drawing (not to scale) of a typical GaAlAs double-heterostructure light emitter. In this structure $x > y$ to provide for both carrier confinement and optical guiding.

To achieve a high radiance and a high quantum efficiency, the LED structure must provide a means of confining the charge carriers and the stimulated optical emission to the active region of the *pn* junction where radiative recombination takes place. Carrier confinement is used to achieve a high level of radiative recombination in the active region of the device, which yields a high quantum efficiency (see Sec. 4-1-3). Optical confinement is of importance for preventing absorption of the emitted radiation by the material surrounding the *pn* junction.

To achieve carrier and optical confinement, LED configurations such as homojunctions and single and double heterojunctions have been widely investigated. The most effective of these predominantly in use at this time is the configuration shown in Fig. 4-1. This is referred to as a *double-heterostructure* (or *heterojunction*) device because of the two different alloy layers on each side of the active region. This configuration evolved from earlier studies on laser diodes. By means of this sandwich structure of differently composed alloy layers, both the carriers and the optical field are confined in the central active layer. The band-gap differences of adjacent layers confine the charge carriers, while the differences in the indices of refraction of adjoining layers confine the optical field to the central active layer. This dual confinement leads to both high efficiency and high radiance. Other parameters influencing the device performance include optical absorption in the active region (self-absorption), carrier recombination at the heterostructure interfaces, doping concentration of the active layer, injection carrier density, and active-layer thickness. We shall see the effects of these parameters in the following sections.

The two basic LED configurations being used for fiber optics are *surface emitters*[9] (also called *Burrus* or *front emitters*) and *edge emitters*.[10] In the surface emitter the plane of the active light-emitting region is oriented perpendicularly to the axis of the fiber, as shown in Fig. 4-2. In this configuration a well is etched through the substrate of the device, into which a fiber is then cemented in order to accept the emitted light. The circular active area in practical surface emitters is nominally 50 μm in diameter and up to 2.5 μm thick. The emission pattern is essentially isotropic with a 120° half-power beam width.

The edge emitter depicted in Fig. 4-3 consists of an active junction region,

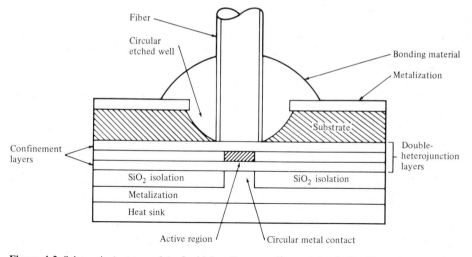

Figure 4-2 Schematic (not to scale) of a high-radiance surface-emitting LED. The active region is limited to a circular section having an area compatible with the fiber core end face.

which is the source of the incoherent light, and two guiding layers. The guiding layers both have a refractive index which is lower than that of the active region but higher than the index of the surrounding material. This structure forms a waveguide channel that directs the optical radiation toward the fiber core. To match the typical fiber core diameters (50 to 100 μm), the contact stripes for the edge emitter are 50 to 70 μm wide. Lengths of the active regions usually range from 100 to 150 μm. The emission pattern of the edge emitter is more directional than that of the surface emitter, as is illustrated in Fig. 4-3. In the plane parallel to the junction where there

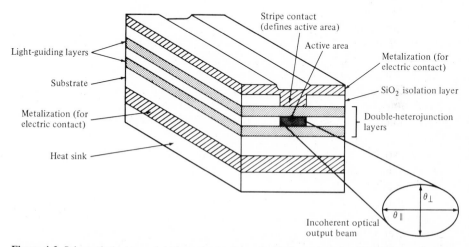

Figure 4-3 Schematic (not to scale) of an edge-emitting double-heterojunction LED. The output beam is lambertian in the plane of the *pn* junction ($\theta_{\parallel} = 120°$) and highly directional perpendicular to the *pn* junction ($\theta_{\perp} \simeq 30°$).

is no waveguide effect, the emitted beam is lambertian (varying as cos θ) with a half-power width of $\theta_\parallel = 120°$. In the plane perpendicular to the junction the half-power beam width $\theta_\perp$ has been made as small as 25 to 35° by a proper choice of the waveguide thickness.[10]

4-1-2 Light Source Materials

The semiconductor material that is used for the active layer of an optical source must have a direct band gap. As we show in App. E, in a direct band gap semiconductor electrons and holes can recombine directly across the band gap without needing a third particle to conserve momentum. Only in direct-band-gap material is the radiative recombination sufficiently high to produce an adequate level of optical emission. Although none of the normal single-element semiconductors are direct-gap materials, many binary compounds are. The most important of these are the so-called III-V materials. These are made from compounds of a group III element (such as Al, Ga, or In) and a group V element (such as P, As, or Sb). Various ternary and quaternary combinations of binary compounds of these elements are also direct-gap materials and are suitable candidates for optical sources.

For operation in the 800- to 900-nm spectrum the principal material used is the ternary alloy $Ga_{1-x}Al_xAs$. The ratio x of aluminum arsenide to gallium arsenide determines the band gap of the alloy and, correspondingly, the wavelength of the peak emitted radiation. This is illustrated in Fig. 4-4. The value of x for the active-area material is usually chosen to give an emission wavelength of 800 to 850

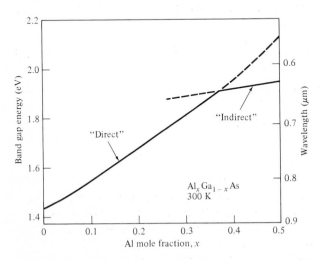

Figure 4-4 Band-gap energy and output wavelength as a function of aluminum mole fraction x for $Al_xGa_{1-x}As$ at room temperature. *(Reproduced with permission from Miller, Marcatili, and Lee, Proc. IEEE, **61**, 1703–1751, Dec. 1973.)*

nm. An example of the emission spectrum of a $Ga_{1-x}Al_xAs$ LED with $x = 0.08$ is shown in Fig. 4-5. The peak output power occurs at 810 nm and the half-power spectral width σ_λ is 36 nm.

At longer wavelengths the quaternary alloy $In_{1-x}Ga_xAs_yP_{1-y}$ is one of the primary material candidates. By varying the mole fractions x and y in the active area, LEDs with peak output powers at any wavelength between 1.0 and 1.7 μm can be constructed. For simplicity the notations GaAlAs and InGaAsP are generally used unless there is an explicit need to know the values of x and y. Other notations such as AlGaAs, (Al,Ga)As, (GaAl)As, GaInPAs, and $In_xGa_{1-x}As_yP_{1-y}$ are also found in the literature. From the last notation it is obvious that depending on the preference of the particular author, the values of x and $1 - x$ for the same material could be interchanged in different articles in the literature.

The alloys GaAlAs and InGaAsP are chosen to make semiconductor light sources, because it is possible to match the lattice parameters of the heterostructure interfaces by using a proper combination of binary, ternary, and quaternary materials. A very close match between the crystal lattice parameters of the two adjoining heterojunctions is required to reduce interfacial defects and to minimize strains in the device as the temperature varies. These factors directly affect the radiative efficiency and lifetime of a light source. Using the fundamental quantum-mechanical relationship between energy E and frequency ν

$$E = h\nu = \frac{hc}{\lambda}$$

the peak emission wavelength λ in micrometers can be expressed as a function of the band gap energy E_g in electron volts by the equation

$$\lambda \ (\mu m) = \frac{1.240}{E_g \ (eV)} \tag{4-1}$$

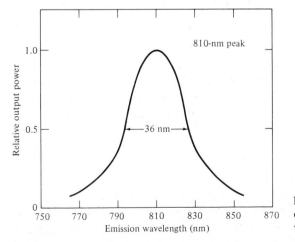

Figure 4-5 Spectral emission pattern of a representative $Ga_{1-x}Al_xAs$ LED with $x = 0.08$.

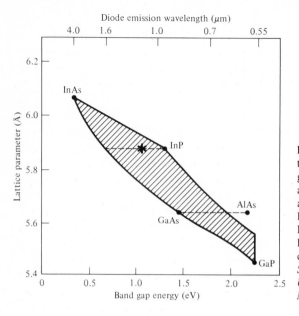

Figure 4-6 Relationships between the crystal lattice spacing, energy gap, and diode emission wavelength at room temperature. The shaded area is for the quaternary alloy InGaAsP. The star ⋆ is for $In_{0.8}Ga_{0.2}As_{0.35}P_{0.65}$ ($E_g \simeq 1.1$ eV) lattice-matched in InP. *(From Optical Fibre Communications by Tech. Staff of CSELT, copyright 1980. Used with the permission of McGraw-Hill Book Co.)*

The relationships between the band gap energy E_g and the crystal lattice spacing (or lattice constant) a_0 for various III-V compounds is plotted in Fig. 4-6.

A heterojunction with matching lattice parameters is created by choosing two material compositions having the same lattice constant but different band gap energies (the band-gap differences are used to confine the charge carriers). In the ternary alloy GaAlAs the band gap energy E_g and the crystal lattice spacing a_0 are determined by the dashed line in Fig. 4-6, connecting the materials GaAs ($E_g = 1.43$ eV and $a_0 = 5.64$ Å) and AlAs ($E_g = 2.16$ eV and $a_0 = 5.66$ Å). The energy gap in electron volts for values of x between zero and 0.37 (the direct-band-gap region) can be found from the empirical equation[1]

$$E_g = 1.424 + 1.266x + 0.266x^2 \qquad (4-2)$$

Given the value of E_g in electron volts, the peak emission wavelength in micrometers is found from Eq. (4-1). For example, $Ga_{0.93}Al_{0.07}As$ has $E_g = 1.51$ eV and emits at $\lambda = 0.82$ μm.

The band gap energy and lattice constant range for the quaternary alloy InGaAsP is much larger, as shown by the shaded area in Fig. 4-6. These materials are generally grown on an InP substrate so that lattice-matched configurations are obtained by selecting a compositional point along the top dashed line in Fig. 4-6 which passes through the InP point. Along this line the compositional parameters x and y follow the relationship[1] $y \simeq 2.20x$ with $0 \le x \le 0.47$. As an illustration, the alloy $In_{0.74}Ga_{0.26}As_{0.56}P_{0.44}$ emits at 1.3 μm ($E_g = 0.96$ eV).

4-1-3 Internal Quantum Efficiency

An excess of electrons and holes in p- and n-type material, respectively, (referred to as *minority carriers*) is created in a semiconductor light source by carrier injection at the device contacts. The excess density of electrons Δn and holes Δp is equal, since the injected carriers are formed and recombine in pairs in accordance with the requirement for charge neutrality in the crystal. When carrier injection stops, the carrier density returns to the equilibrium value. In general, the excess carrier density decays exponentially with time according to the relation

$$\Delta n = \Delta n_0 e^{-t/\tau} \tag{4-3}$$

where Δn_0 is the initial injected excess electron density and the time constant τ is the carrier lifetime. This lifetime is one of the most important operating parameters of an electro-optic device. Its value can range from milliseconds to fractions of a nanosecond depending on material composition and device defects.

As noted earlier, the excess carriers can recombine either radiatively or non-radiatively. In radiative recombination a photon of energy $h\nu$, which is approximately equal to the band gap energy, is emitted. When an electron-hole pair recombines nonradiatively, the energy is released in the form of heat (lattice vibrations). The *internal quantum efficiency* in the active region is the fraction of electron-hole pairs that recombine radiatively. If the radiative recombination rate per unit volume is R_r and the nonradiative recombination rate is R_{nr}, the internal quantum efficiency η_0 is the ratio of the radiative recombination rate to the total recombination rate[8]

$$\eta_0 = \frac{R_r}{R_r + R_{nr}} \tag{4-4}$$

For exponential decay of excess carriers, the radiative recombination lifetime is $\tau_r = \Delta n/R_r$ and the nonradiative recombination lifetime is $\tau_{nr} = \Delta n/R_{nr}$. Thus the internal quantum efficiency is

$$\eta_0 = \frac{1}{1 + (\tau_r/\tau_{nr})} = \frac{\tau}{\tau_r} \tag{4-5}$$

where the *bulk recombination lifetime* τ is

$$\frac{1}{\tau} = \frac{1}{\tau_r} + \frac{1}{\tau_{nr}} \tag{4-6}$$

In a heterojunction structure, nonradiative recombination at the boundaries of the different semiconductor layers resulting from crystal lattice mismatches tends to decrease this lifetime, which, in turn, decreases the internal efficiency. Before we examine this, let us first define some terms. In a semiconductor the flow of electrons or holes gives rise to an electric current i. This is given by

$$i_e = qD_e\frac{\partial(\Delta n)}{\partial x} \qquad \text{and} \qquad i_h = qD_h\frac{\partial(\Delta p)}{\partial x}$$

for electrons and holes, respectively. This current flow is a result of nonuniform carrier distributions in the material and flows even in the absence of an applied electric field. The constants D_e and D_h are the electron and hole diffusion coefficients (or constants), respectively, which are expressed in units of centimeters squared per second. As the charge carriers diffuse through the material, some will disappear by recombination. On the average, they move a distance L_e or L_h for electrons and holes, respectively. This distance is known as the *diffusion length,* and is determined from the diffusion coefficient and the lifetime of the material through the relationship

$$L_D = (D\tau)^{1/2}$$

Here we denote a general diffusion length by L_D and its corresponding diffusion coefficient by D.

A heterojunction is commonly characterized by the quantity SL_D/D, which is the ratio of the interfacial recombination velocity S (expressed in centimeters per second) and the bulk diffusion velocity D/L_D. For a completely reflecting boundary $S = 0$, an ohmic contact is characterized by $S = \infty$, and $S = D/L_D$ describes an interface which cannot be distinguished from a continuation of the bulk material. The criterion for high efficiency in an LED is that S should be less than 10^4 cm/s at the heterojunction interfaces. Experimental data show that $S \simeq 5 \times 10^3$ cm/s in practical heterostructure devices.[12]

The reduction in the bulk lifetime owing to nonradiative heterointerface recombination can be found from the solution to the one-dimensional steady-state continuity equation[11, 12]

$$D\frac{d^2[\Delta n(x)]}{dx^2} - \frac{\Delta n(x)}{\tau} = 0 \tag{4-7}$$

where $\Delta n(x)$ is the density of the excess electrons per cubic centimeter at the position x in the active layer of thickness d as measured from the *pn* junction (see Fig. 4-7). Assuming the same surface recombination velocity S at both heterointerfaces the boundary conditions are [12]

$$\left.\frac{d(\Delta n)}{dx}\right|_{x=0} = -\frac{J}{qD} + \frac{S}{D}\Delta n(0) \tag{4-8}$$

and

$$\left.\frac{d(\Delta n)}{dx}\right|_{x=d} = -\frac{S}{D}\Delta n(d) \tag{4-9}$$

The term $J/(qD)$ gives the number of carriers injected across the *pn* junction at $(x = 0)$, and $S\,\Delta n/D$ represents the number of carriers recombining at the inter-

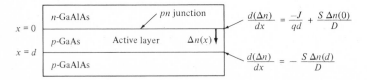

Figure 4-7 Boundary conditions at the *pn* junction for a double-heterostructure LED.

face. The parameter J is the current density. Assuming a solution of the form

$$\Delta n(x) = Ae^{x/L_D} + Be^{-x/L_D}$$

where A and B are constants and $L_D = \sqrt{D\tau}$, Eqs. (4-8) and (4-9) yield

$$\Delta n(x) = \frac{JL_D}{qD}\left\{\frac{\cosh[(d-x)/L_D] + (L_DS/D)\sinh[(d-x)/L_D]}{[(L_DS/D)^2 + 1]\sinh(d/L_D) + (2L_DS/D)\cosh(d/L_D)}\right\} \quad (4\text{-}10)$$

In the active region the average electron density is given by[12]

$$\overline{\Delta n} = \frac{1}{d}\int_0^d \Delta n(x)\,dx = \frac{J}{q}\frac{\tau_{\text{eff}}}{d} \quad (4\text{-}11)$$

Here

$$\tau_{\text{eff}} = \tau\left\{\frac{\sinh(d/L_D) + (L_DS/D)(\cosh d/L_D - 1)}{[(L_DS/D)^2 + 1]\sinh(d/L_D) + (2L_DS/D)\cosh(d/L_D)}\right\} \quad (4\text{-}12)$$

is the average effective carrier lifetime when surface recombination is important. When interface recombination is the dominant nonradiative recombination process, the surface recombination velocity S is much smaller than the bulk diffusion velocity D/L_D ($= L_D/\tau$). Using this condition, that is, $L_DS/D \ll 1$, and the fact that the active-layer thickness d is equal to or smaller than the diffusion length L_D, Eq. (4-12) reduces to[12]

$$\frac{1}{\tau_{\text{eff}}} = \frac{1}{\tau} + \frac{2S}{d} \quad (4\text{-}13)$$

Equation (4-13) gives the lifetime reduction caused by interfacial recombination. This lifetime reduction, in turn, decreases the internal quantum efficiency.

A further reduction of the lifetime and the internal efficiency results from self-absorption in the recombination region. This internal absorption of the luminescense is a result of strong energy-dependent absorption near the band gap of a direct-gap semiconductor. The self-absorption becomes important when the interfacial recombination rate is small and the doping in the active region is low. A high efficiency can be achieved in double-heterostructure LEDs because the self-absorption is reduced as a result of the thin active regions that are characteristic of these devices. Self-absorption is more severe in p-type material than in n-type material. Thus its effects can also be minimized by gathering light out of a device from the n side.

If α_λ is the absorption coefficient at a wavelength λ of the active-layer material, then the peak optical power in the pn junction at this wavelength is given by[12]

$$P = \frac{hc}{\lambda\tau_r}\int_0^d \Delta n(x)e^{-\alpha_\lambda x}\,dx \quad (4\text{-}14)$$

Substituting Eq. (4-10) into Eq. (4-14) yields

$$P = \frac{hc}{\lambda q}\eta_i^{dh}J \quad (4\text{-}15)$$

where

$$\eta_i^{dh} = \frac{\eta_0}{2} \left\{ \left[\left(\frac{L_D S}{D} \right)^2 + 1 \right] \sinh \frac{d}{L_D} + \frac{2 L_D S}{D} \cosh \frac{d}{L_D} \right\}^{-1}$$
$$\cdot \left(\frac{1 + L_D S/D}{1 + \alpha_\lambda L_D} \left\{ 1 - \exp \left[\frac{-d(1 + \alpha_\lambda L_D)}{L_D} \right] \right\} e^{d/L_D} \right.$$
$$\left. - \frac{1 - L_D S/D}{1 - \alpha_\lambda L_D} \left\{ 1 - \exp \left[\frac{d(1 - \alpha_\lambda L_D)}{L_D} \right] \right\} e^{-d/L_D} \right) \qquad (4\text{-}16)$$

is the reduced internal quantum efficiency when interface recombinations and self-absorption are significant. The factor η_0 is given by Eq. (4-5), which holds for $S = 0$ and $\alpha_\lambda = 0$. The superscript dh emphasizes that this expression holds for a double-heterostructure LED with the light being gathered from the n-side passive layer. Similar expressions for homojunction and single-heterostructure devices can be found in the literature.[12, 13]

Similar to the derivation of Eq. (4-12), the total recombination lifetime $\tau_{\text{eff}}(\alpha_\lambda)$ when self-absorption is taken into account can be found by considering the average electron density in the active region

$$\overline{\Delta n(\alpha_\lambda)} = \frac{1}{d} \int_0^d \Delta n(x) e^{-\alpha_\lambda x} \, dx = \frac{J}{q} \frac{\tau_{\text{eff}}(\alpha_\lambda)}{d} \qquad (4\text{-}17)$$

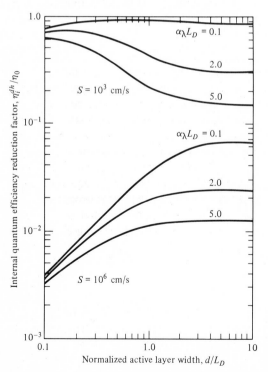

Figure 4-8 Examples of the relative internal quantum efficiency reduction as a function of the normalized active-layer width for two values of the surface recombination velocity. (*Reproduced with permission from Lee and Dentai.*[12])

A comparison of Eqs. (4-14), (4-15), and (4-17) shows that

$$\tau_{\text{eff}}(\alpha_\lambda) = \eta_i^{dh} \tau_r \qquad (4\text{-}18)$$

where η_i^{dh} is given by Eq. (4-16).

Examples of the relative internal quantum efficiency reduction η_i^{dh}/η_0 determined from Eq. (4-16) are shown in Fig. 4-8 as a function of the normalized active-layer width d/L_D for two values of the surface recombination velocity and with $D = 80$ cm²/s in GaAs. The values of α_λ and L_D depend on the active-area doping concentration and were chosen as follows:[12] $\alpha_\lambda = 10^3$ cm^{-1} and $L_D = 10^{-4}$ cm for a 10^{19} cm^{-3} p-type doping; $\alpha_\lambda = 4 \times 10^3$ cm^{-1} and $L_D = 5 \times 10^{-4}$ cm for a 10^{18} cm^{-3} doping; and $\alpha_\lambda = 7 \times 10^3$ cm^{-1} and $L_D = 7 \times 10^{-4}$ cm for a 10^{17} cm^{-3} doping. This results in the $\alpha_\lambda L_D$ products of 0.1, 2, and 5 shown in Fig. 4-8. These curves show that surface recombination severely degrades both the efficiency and the total recombination lifetime. Interfacial recombination dominates for $S \geq 10^5$ cm/s, while for $S \leq 10^3$ cm/s self-absorption becomes significant. An optimum active-region thickness for surface emitters occurs between 2.0 and 2.5 μm for $S = 10^4$ cm/s, whereas thinner values of d are desirable for edge emitters.[5]

4-1-4 Modulation Capability

The frequency response of an LED is limited by its diffusion capacitance because of the storage of injected carriers in the active region of the diode. If the drive current is modulated at a frequency ω, the intensity of the optical output will vary as[14]

$$I(\omega) = I_0[1 + (\omega\tau_{\text{eff}})^2]^{-1/2} \qquad (4\text{-}19)$$

where I_0 is the intensity emitted at zero modulation frequency and τ_{eff} is the effective carrier lifetime given by Eq. (4-12). Parasitic diode space charge capacitance can cause a delay of the carrier injection into the active junction, and consequently could delay the optical output.[15,16] This delay is negligible if a small, constant forward bias is applied to the diode. Under this condition Eq. (4-19) is valid and the modulation response is limited only by the carrier recombination time.

The modulation bandwidth of an LED is defined in electrical terms as the 3-dB bandwidth of the detected electric power resulting from the modulated portion of the optical signal. Since the detected electric signal power $p(\omega)$ is proportional to $I^2(\omega)$, the modulation bandwidth is defined as the frequency band over which $p(\omega) = p(0)/2$. This is equivalent to setting $I^2(\omega) = I^2(0)/2$. Using Eq. (4-19), the 3-dB modulation bandwidth $\Delta\omega$ is given by

$$\Delta\omega = \frac{1}{\tau_{\text{eff}}} \qquad (4\text{-}20)$$

Sometimes the modulation bandwidth of an LED is given in terms of the 3-dB bandwidth of the modulated optical power, that is, $I(\omega) = \frac{1}{2}I(0)$. This naturally gives an apparent but erroneous increase in modulation bandwidth by a factor of $3^{1/2}$.

For relatively lightly doped active regions (2×10^{17} cm^{-3}) and active-region thicknesses d less than or equal to a carrier diffusion length L_D, the nonradiative recombination process is dominated by interfacial recombination. In this case the bulk lifetime τ in the active region can be approximated by the radiative lifetime τ_r. Under this condition Eq. (4-13) becomes

$$\frac{1}{\tau_{\text{eff}}} = \frac{1}{\tau_r} + \frac{2S}{d} \tag{4-21}$$

An expression for the radiative lifetime can be found from the following considerations. The radiative carrier lifetime is related to the sum of the initial carrier concentration $n_0 + p_0$ (where n_0 and p_0 are the electron and hole concentrations, respectively, at thermal equilibrium) and the injected electron-hole pair density Δn through the expression

$$\tau_r = [B_r(n_0 + p_0 + \Delta n)]^{-1} \tag{4-22}$$

The radiative recombination coefficient B_r is a material characteristic which depends on the doping concentration. Its value can range from 0.46×10^{-10} to 7.2×10^{-10} cm^3/s. In the steady-state condition the average injected electron concentration Δn is given by

$$\Delta n = \frac{J\tau_r}{qd} \tag{4-23}$$

where J is the injected current density.
Substituting this into Eq. (4-22) for Δn and solving for τ_r yield

$$\tau_r = \frac{[(n_0 + p_0)^2 + 4J/B_r qd]^{1/2} - (n_0 + p_0)}{2J/qd} \tag{4-24}$$

When the carrier injection is low compared to the background concentration, that is, $\Delta n \ll n_0 + p_0$, Eq. (4-24) reduces to

$$\tau_r \simeq [B_r(n_0 + p_0)]^{-1} \tag{4-25}$$

In this case the lifetime is a constant independent of J.
In the other extreme case, at very high carrier injection levels $\Delta n \gg n_0 + p_0$ and the lifetime becomes

$$\tau_r = \left(\frac{qd}{B_r}\right)^{1/2} J^{-1/2} \tag{4-26}$$

At low doping levels (approximately 2×10^{17} cm^{-3}) high carrier injection increases both the efficiency and the modulation speed. For surface emitters the upper current density limit of 7.5 kA/cm^2 at which the LED can safely operate is determined by thermal heating.[12] For a 2-μm-thick active layer the modulation bandwidth is approximately 25 MHz. This increases to about 60 MHz for an 0.3-μm active layer. A higher modulation bandwidth at the same drive level can occur for edge emitters, which operate under the condition where the lifetime is predominantly controlled by the injected carrier density.[5]

4-1-5 Transient Response

A closed-form approximation of the transient response of high-radiance double-heterojunction LEDs has been derived for practical engineering analyses of optical fiber communication systems.[17, 18] The basic assumption of this approximation is that the junction space charge capacitance C_s varies much more slowly with current than the diffusion capacitance C_d and can, therefore, be considered constant. Typical values of C_s range from 350 to 1000 pF for low to moderately high currents. Under this assumption, the rise time to the half-current point (which is also the half-power point) of the LED is

$$t_{1/2} = \frac{C_s}{\beta I_p} \ln \frac{I_p}{I_s} + \tau \ln 2 \qquad (4\text{-}27)$$

and the 10 to 90 percent rise time is

$$t_{10\text{-}90} = \left(\frac{2C_s}{\beta I_p} + \tau \right) \ln 9 \qquad (4\text{-}28)$$

In these expressions $\beta = q/(2k_B T)$, I_p is the amplitude of the current step function used to drive the LED, I_s is the diode saturation current, and τ is the minority carrier lifetime. These expressions show that the rise time decreases with increasing current. In the high-current limit the rise times depend only on the carrier lifetime so that $t_{1/2} = \tau \ln 2$ and $t_{10\text{-}90} = \tau \ln 9$. The rise time can be reduced considerably by external means such as *current peaking*.[15, 17] If the current rises to a point in excess of the desired level for a very short time and then decays back to the desired level, as is shown in Fig. 4-9, a reduction in rise time to $t_{10\text{-}90} = 0.55\tau$ has been achieved.[17] Similarly the 90 to 10 percent fall time can be made considerably shorter than $\tau \ln 9$ by applying a negative bias for a short time after the pulse to sweep out the injected carriers. In addition, as noted in Sec. 4-1-2, the application of a small, constant forward bias on the LED minimizes the delay time of the onset of optical output.[15, 16]

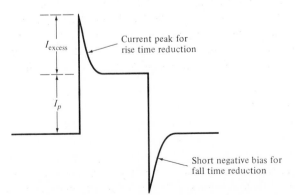

I_{excess}

I_p

Current peak for rise time reduction

Short negative bias for fall time reduction

Figure 4-9 Current waveform showing a short positive peak for rise time reduction and a short negative bias for fall time reduction.

4-1-6 Power-Bandwidth Product

An important parameter to consider for LEDs is the *power-bandwidth product*. This can be found by multiplying both sides of Eq. (4-15) by $\Delta\omega = 1/\tau_{\text{eff}}$:

$$\Delta\omega P = \frac{1}{\tau_{\text{eff}}} \frac{hc}{q\lambda} \eta_i^{dh} J \qquad (4\text{-}29)$$

Using Eq. (4-18), this becomes

$$\Delta\omega P = \frac{hc}{q\lambda} \frac{1}{\tau_r} J \qquad (4\text{-}30)$$

which is constant for a given current injection level. For example, suppose the doping in the active layer is increased. This decreases the effective carrier lifetime, which results in an increase in LED bandwidth. However, this bandwidth increase is accompanied by a decrease in power by the same proportionality factor since η_i^{dh} is equal to τ_{eff}/τ_r. Thus, for a fixed injection level, the net power-bandwidth product remains unchanged. This means that faster LEDs generally emit less power than slow ones.

4-2 LASER DIODES

Lasers come in many forms with dimensions ranging from the size of a grain of salt to one that will occupy an entire room. The lasing medium can be a gas, a liquid, a crystal (solid state), or a semiconductor. For optical fiber systems the laser sources used almost exclusively are semiconductor laser diodes. They are similar to other lasers, such as the conventional solid-state and gas lasers in that the emitted radiation has spatial and temporal coherence; that is, the output radiation is highly monochromatic and the light beam is very directional.

Despite their differences the basic principle of operation is the same for each type of laser. Laser action is the result of three key processes. These are photon absorption, spontaneous emission, and stimulated emission. These three processes are represented by the simple two-energy-level diagrams in Fig. 4-10, where E_1 is the ground-state energy and E_2 is the excited-state energy. According to Planck's law, a transition between these two states involves the absorption or emission of a photon of energy $h\nu_{12} = E_2 - E_1$. Normally the system is in the ground state. When a photon of energy $h\nu_{12}$ impinges on the system, an electron in state E_1 can absorb the photon energy and be excited to state E_2, as shown in Fig. 4-10a. Since this is an unstable state, the electron will shortly return to the ground state, thereby emitting a photon of energy $h\nu_{12}$. This occurs without any external stimulation and is called *spontaneous emission*. These emissions are isotropic and of random phase and, thus, appear as a narrowband gaussian output.

The electron can also be induced to make a downward transition from the excited level to the ground-state level by an external stimulation. As shown in Fig. 4-10c, if a photon of energy $h\nu_{12}$ impinges on the system while the electron is still

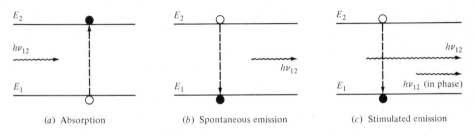

E_2 E_2 E_2

$h\nu_{12}$ $h\nu_{12}$ $h\nu_{12}$

E_1 E_1 E_1 $h\nu_{12}$ (in phase)

(a) Absorption   (b) Spontaneous emission   (c) Stimulated emission

Figure 4-10 The three key transition processes involved in laser action. The open circle represents the initial state of the electron and the heavy dot represents the final state. Incident photons are shown on the left of each diagram and emitted photons are shown on the right.

in its excited state, the electron is immediately stimulated to drop to the ground state and give off a photon of energy $h\nu_{12}$. This emitted photon is in phase with the incident photon, and the resultant emission is known as *stimulated emission*.

In thermal equilibrium the density of excited electrons is very small. Most photons incident on the system will therefore be absorbed, so that stimulated emission is essentially negligible. Stimulated emission will exceed absorption only if the population of the excited states is greater than that of the ground state. This condition is known as *population inversion*. Since this is not an equilibrium condition, population inversion is achieved by various "pumping" techniques. In a semiconductor laser, population inversion is accomplished by injecting electrons into the material at the device contacts to fill the lower energy states of the conduction band.

4-2-1 Laser Diode Structures and Threshold Conditions

For optical fiber communication systems requiring bandwidths greater than approximately 50 MHz, the semiconductor injection laser diode is preferred over the LED. Laser diodes typically have response times less than 1 ns, have optical bandwidths of 2 nm or less, and, in general, are capable of coupling several milliwatts of useful luminescent power into optical fibers with small cores and small numerical apertures. Virtually all laser diodes in use and under investigation at present are multilayered heterojunction devices. As mentioned in Sec. 4-1, the double-heterojunction LED configuration evolved from the successful demonstration of both carrier and optical confinement in heterojunction injection laser diodes. The more rapid evolvement and utilization of LEDs as compared to laser diodes lies in the inherently simpler construction, the smaller temperature dependence of the emitted optical power, and the absence of catastrophic degradation in LEDs (see Sec. 4-4). The construction of laser diodes is more complicated, mainly because of the additional requirement of current confinement in a small lasing cavity.

Stimulated emission in semiconductor lasers arises from optical transitions between distributions of energy states in the valence and conduction bands. This differs from gas and solid-state lasers in which radiative transitions occur between discrete isolated atomic or molecular levels. The radiation in the laser diode is

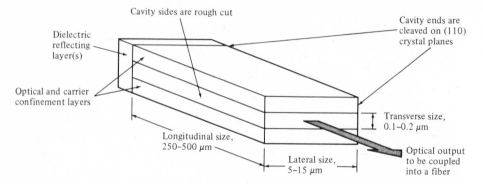

Figure 4-11 Fabry-Perot resonator cavity for a laser diode. The cleaved crystal ends function as partially reflecting mirrors. The unused end can be coated with a dielectric reflector to reduce optical loss in the cavity.

generated within a Fabry-Perot resonator cavity,[1-3, 7] shown in Fig. 4-11, as in most other types of lasers. However, this cavity is much smaller, the sizes being approximately 250 to 500 μm long, 5 to 15 μm wide, and 0.1 to 0.2 μm thick. These dimensions are commonly referred to as the *longitudinal, lateral,* and *transverse dimensions* of the cavity, respectively.

In the laser diode Fabry-Perot resonator a pair of flat, partially reflecting mirrors are directed toward each other to enclose the cavity. The mirror facets are constructed by making two parallel cleaves along natural cleavage planes of the semiconductor crystal. In another laser diode type, commonly referred to as the *distributed-feedback* (DFB) *laser,* [1, 2, 7] the cleaved facets are not required for optical feedback. A typical DFB laser configuration is given in Fig. 4-12. The fabrication of this device is similar to the Fabry-Perot types, except that the lasing action is obtained from Bragg reflectors (gratings) or periodic variations of the refractive index (called *distributed-feedback corrugations*) which are incorporated into the multilayer structure along the length of the diode.

In general, the full optical output is only needed from the front facet of the laser, that is, the one to be aligned with an optical fiber. In this case a dielectric

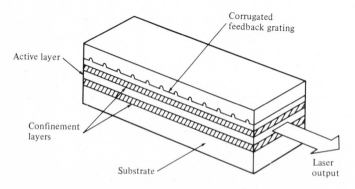

Figure 4-12 Structure of a distributed-feedback (DFB) laser diode.

reflector can be deposited on the rear laser facet to reduce the optical loss in the cavity, to reduce the threshold current density (the point at which lasing starts), and to increase the external quantum efficiency. Reflectivities greater than 98 percent have been achieved with a six-layer reflector.[19]

The optical radiation within the resonance cavity of a laser diode sets up a pattern of electric and magnetic field lines called the *modes of the cavity* (see Sec. 2-3 and 2-4 for details on modes). These can conveniently be separated into two independent sets of transverse electric (TE) and transverse magnetic (TM) modes. Each set of modes can be described in terms of the longitudinal, lateral, and transverse half-sinusoidal variations of the electromagnetic fields along the major axes of the cavity. The *longitudinal modes* are related to the length L of the cavity and determine the principal structure of the frequency spectrum of the emitted optical radiation. Since L is much larger than the lasing wavelength of approximately 1 μm, many longitudinal modes can exist. *Lateral modes* lie in the plane of the *pn* junction. These modes depend on the side wall preparation and the width of the cavity, and determine the shape of the lateral profile of the laser beam. *Transverse modes* are associated with the electromagnetic field and beam profile in the direction perpendicular to the plane of the *pn* junction. These modes are of great importance since they largely determine such laser characteristics as the radiation pattern (the angular distribution of the optical output power) and the threshold current density.

Lasing is the condition at which light amplification becomes possible in the laser diode. The requirement for lasing is that a population inversion be achieved. This condition can be understood by considering the fundamental relationship between the optical field intensity I, the absorption coefficient α_λ, and the gain coefficient g in the Fabry-Perot cavity. The stimulated emission rate into a given mode is proportional to the intensity of the radiation in that mode. The radiation intensity at a photon energy $h\nu$ varies exponentially with the distance z that it traverses along the lasing cavity according to the relationship

$$I(z) = I(0) \exp \{[g(h\nu) - \overline{\alpha}(h\nu)]z\} \tag{4-31}$$

where $\overline{\alpha}$ is the effective absorption coefficient of the material in the optical path.

Optical amplification of selected modes is provided by the feedback mechanism of the optical cavity. In the repeated passes between the two partially reflecting parallel mirrors, a portion of the radiation associated with those modes having the highest optical gain coefficient is retained and further amplified during each trip through the cavity.

Lasing occurs when the gain of one or several guided modes is sufficient to exceed the optical loss during one roundtrip through the cavity, that is, for $z = 2L$. During this roundtrip only the fractions R_1 and R_2 of the optical radiation are reflected from the two laser ends 1 and 2, respectively, where R_1 and R_2 are the mirror reflectivities. Thus Eq. (4-31) becomes

$$I(2L) = I(0)R_1R_2 \exp \{2L[g(h\nu) - \overline{\alpha}(h\nu)]\} \tag{4-32}$$

At the lasing threshold $I(2L) = I(0)$, so that the lasing threshold optical gain coefficient g_{th} is

$$g_{th} = (2L)^{-1} \ln \left[(R_1 R_2)^{-1} \right] + \overline{\alpha} \qquad (4\text{-}33)$$

For GaAs $R_1 = R_2 = 0.32$ for uncoated facets (that is, 32 percent of the radiation is reflected at a facet) and $\overline{\alpha} \simeq 10 \text{ cm}^{-1}$. This yields $g_{th} = 33 \text{ cm}^{-1}$ for a laser diode length of $L = 500 \ \mu\text{m}$.

The mode that satisfies Eq. (4-33) reaches threshold first. Theoretically, at the onset of this condition, all additional energy introduced into the laser should augment the growth of this particular mode. In practice, various phenomena lead to the excitation of more than one mode.[1] Studies on the conditions of longitudinal single-mode operation have shown that important factors are thin active regions and a high degree of temperature stability.[20]

The relationship between optical output power and diode drive current is presented in Fig. 4-13. At low diode currents only spontaneous radiation is emitted. Both the spectral range and the lateral beam width of this emission are broad like that of an LED. A dramatic and sharply defined increase in the power output occurs at the lasing threshold. As this transition point is approached, the spectral range and the beam width both narrow with increasing drive current. The final spectral width of approximately 1 nm and the fully narrowed lateral beam width of nominally 5 to 10° are reached just past the threshold point. The *threshold current* I_{th} is conventionally defined by extrapolation of the lasing region of the power-versus-current curve, as shown in Fig. 4-13. At high power outputs the slope of the curve decreases because of junction heating.

The *external differential quantum efficiency* η_{ext} is defined as the number of photons emitted per radiative electron-hole pair recombination above threshold. Under the assumption that above threshold the gain coefficient remains fixed at g_{th},

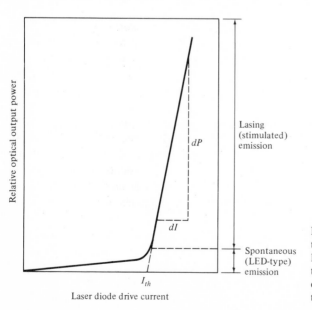

Figure 4-13 Relationship between optical output power and laser diode drive current. Below the lasing threshold the optical output is a spontaneous LED-type emission.

η_{ext} is given by[1]

$$\eta_{\text{ext}} = \frac{\eta_i(g_{th} - \overline{\alpha})}{g_{th}} \qquad (4\text{-}34)$$

Here η_i is the internal quantum efficiency. This is not a well-defined quantity in laser diodes, but most measurements show that $\eta_i \simeq 0.6$ to 0.7 at room temperature. Experimentally η_{ext} is calculated from the straight-line portion of the curve for the emitted optical power P versus drive current I, which gives

$$\eta_{\text{ext}} = \frac{q}{E_g}\frac{dP}{dI} = 0.8065\lambda \ (\mu\text{m}) \frac{dP \ (\text{mW})}{dI \ (\text{mA})} \qquad (4\text{-}35)$$

where E_g is the band gap energy in electron volts, dP is the incremental change in the emitted optical power in milliwatts for an incremental change dI in the drive current (in milliamperes), and λ is the emission wavelength in micrometers. For standard semiconductor lasers, external differential quantum efficiencies of 15 to 20 percent per facet are typical. High-quality devices have differential quantum efficiencies of 30 to 40 percent.

4-2-2 Modal Properties and Radiation Patterns

A basic requirement for efficient operation of laser diodes is that, in addition to transverse optical and carrier confinement between heterojunction layers, the current flow must be restricted laterally to a narrow stripe along the length of the laser. Numerous novel methods of achieving this with varying degrees of success have been proposed,[1-3,7] but all strive for the same goals of limiting the number of lateral modes so that lasing is confined to a single filament, stabilizing the lateral gain, and ensuring a relatively low threshold current.

In a double-heterojunction laser the highest-order transverse mode that can be excited depends on the waveguide thickness and on the refractive-index differentials at the waveguide boundaries.[1] If the refractive-index differentials are kept at approximately 0.08, then only the fundamental transverse mode will propagate if the active area is thinner than 1 μm.

In designing the width and thickness of the optical cavity, a tradeoff must be made between current density and output beam width. As either the width or the thickness of the active region is increased, a narrowing occurs of the lateral or transverse beam widths, respectively, but at the expense of an increase in the threshold current density. Detailed curves showing these tradeoffs are presented in chap. 7 of Ref. 1. In practical continuous wave (CW) laser diodes the transverse and lateral half-power beam widths shown in Fig. 4-4 are about $\theta_\perp \simeq 30$ to $50°$ and $\theta_\parallel \simeq 5$ to $10°$, respectively.

4-2-3 Modulation of Laser Diodes

The two principal methods used to vary the optical output from laser diodes are pulse modulation used in digital systems and amplitude modulation used for analog data transmission. The basic limitation on the modulation rate of laser diodes depends on the carrier and photon lifetime parameters associated with the operation of the laser. These are the spontaneous (radiative) and stimulated carrier lifetimes and the photon lifetime. The *spontaneous lifetime* τ_{sp} is discussed in Sec. 4-1 and is a function of the semiconductor band structure and the carrier concentration. At room temperature the radiative lifetime τ_r is about 1 ns in GaAs-based materials for dopant concentrations on the order of 10^{19} cm^{-3}. The *stimulated carrier lifetime* τ_{st} depends on the optical density in the lasing cavity and is on the order of 10 ps. The *photon lifetime* τ_{ph} is the average time that the photon resides in the lasing cavity before being lost either by absorption or by emission through the facets. In a Fabry-Perot cavity the photon lifetime is[1]

$$\tau_{ph}^{-1} = \frac{c}{n}\left(\bar{\alpha} + \frac{1}{2L}\ln\frac{1}{R_1 R_2}\right) = \frac{c}{n}g_{th} \qquad (4\text{-}36)$$

For a typical value of $g_{th} = 50$ cm^{-1} and a refractive index in the lasing material of $n = 3.5$, the photon lifetime is approximately $\tau_{ph} = 2$ ps. This value sets the upper limit to the modulation capability of the laser diode.

A laser diode can readily be pulse modulated since the photon lifetime is much smaller than the carrier lifetime. If the laser is completely turned off after each pulse, the spontaneous carrier lifetime will limit the modulation rate. This is because at the onset of a current pulse of amplitude I_p, a period of time t_d given by [21] (see Prob. 4-15)

$$t_d = \tau \ln \frac{I_p}{I_p + (I_B - I_{th})} \qquad (4\text{-}37)$$

is needed to achieve the population inversion necessary to produce a gain that is sufficient to overcome the optical losses in the lasing cavity. In Eq. (4-37) the parameter I_B is the bias current and τ is the average lifetime of the carriers in the recombination region when the total current $I = I_p + I_B$ is close to I_{th}. From Eq. (4-37) it is clear that the delay time can be eliminated by dc-biasing the diode at the lasing threshold current. Pulse modulation is then carried out by modulating the laser only in the operating region above threshold. In this region the carrier lifetime is now shortened to the stimulated emission lifetime so that high modulation rates are possible.

In addition to turnon delay the ability of laser diodes to be modulated at high pulse rates can be limited by other factors,[1, 21-24] such as relaxation oscillations, interaction between pulses (pattern effect), self-sustaining oscillations, spectral broadening during nanosecond pulses, and Q-switching effects. Various fabrication techniques are being investigated to minimize these effects.[25-27]

Analog modulation of laser diodes is carried out by making the drive current above threshold proportional to the baseband information signal. A requirement for this modulation scheme is that a linear relation exists between the light output and the current input. However, signal degradation resulting from nonlinearities that are a consequence of the transient response characteristics of laser diodes makes the implementation of analog intensity modulation susceptible to both intermodulation and cross-modulation effects. This can be alleviated by using pulse code modulation or through special distortion compensation techniques, as is discussed in more detail in Sec. 4-3.

4-2-4 Temperature Effects

An important factor to consider in the application of laser diodes is the temperature dependence of the threshold current $I_{th}(T)$. This parameter increases with temperature in all types of semiconductor lasers because of various complex temperature-dependent factors.[28] The complexity of these factors prevents the formulation of a single equation holding for all devices and temperature ranges. However, the temperature variation of I_{th} can be approximated by the empirical expression[29, 30]

$$I_{th}(T) = I_z e^{T/T_0} \tag{4-38}$$

where T_0 is a measure of the relative temperature insensitivity and I_z is a constant. For a conventional stripe geometry laser diode T_0 is typically 120 to 165°C in the vicinity of room temperature. An example of a laser diode with $T_0 = 135°C$ and $I_z = 52$ mA is shown in Fig. 4-14. The variation in I_{th} with temperature is 0.8 percent/°C, as is shown in Fig. 4-15. Smaller dependences of I_{th} on temperature have been demonstrated for GaAlAs quantum-well heterostructure lasers.[29] For these lasers T_0 can be as high as 437°C. The temperature dependence of I_{th} for this device is also shown in Fig. 4-15. The threshold variation for this particular laser type is 0.23 percent/°C.

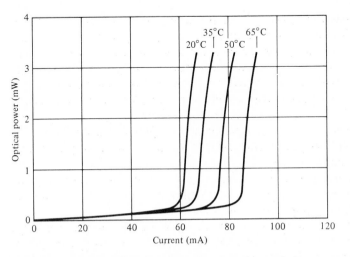

Figure 4-14 Temperature-dependent behavior of the laser optical output power as a function of the bias current for a particular laser diode.

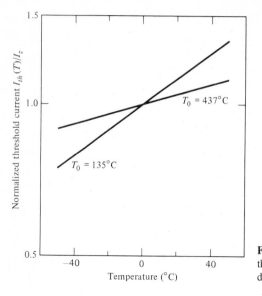

Figure 4-15 Variation with temperature of the threshold current I_{th} for two types of laser diodes.

For the laser diode shown in Fig. 4-14 the threshold current increases by a factor of about 1.4 between 20 and 60°C. In addition, the lasing threshold can change as the laser ages. Consequently, if a constant optical output power level is to be maintained as the temperature of the laser changes or as the laser ages, it is necessary to adjust the dc bias current level. Possible methods for achieving this automatically are optical feedback schemes,[31,34] temperature-matching transistors,[35] and threshold-sensing circuits.[36]

Optical feedback can be carried out by using a photodetector to either sense the variation in optical power emitted from the rear facet of the laser or by tapping off and monitoring a small portion of the fiber-coupled power emitted from the front facet. The photodetector compares the optical power output with a reference level and adjusts the dc bias current level automatically to maintain a constant peak light output relative to the reference. The photodetector used must have a stable long-term responsivity which remains constant over a wide temperature range. For operation in the 800- to 900-nm region, a silicon *pin* photodiode generally exhibits these characteristics (see Chap. 6).

An example of a feedback-stabilizing circuit[31] that can be used for a digital transmitter is shown in Fig. 4-16. In this scheme the light emerging from the rear facet of the laser is monitored by a *pin* photodiode. With this circuit the electric input signal pattern is compared to the optical output level of the laser diode. This effectively prevents the feedback circuit from erroneously raising the bias current level during long sequences of digital zeros or during a period in which there is no input signal on the channel. In this circuit the dc reference through resistor R_1 sets the bias current at the proper operating point during long sequences of zeros. When this bias current is added to the laser drive current, the desired peak output power from the laser is obtained. The resistor R_2 balances the signal reference current against the *pin* photocurrent for a 50 percent duty ratio at 25°C. As the lasing

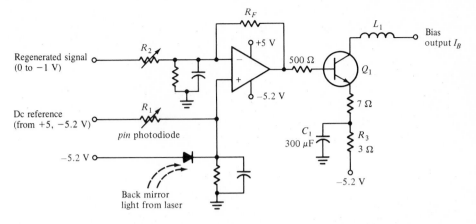

Figure 4-16 Example of a bias circuit that provides feedback stabilization of laser output power. *(Reproduced with permission from Shumate, Chen, and Dorman,[31] copyright 1978, The American Telephone and Telegraph Company.)*

threshold changes because of aging or temperature variations, the bias current I_B is automatically adjusted to maintain a balance between the data reference and the *pin* photocurrent. A more sophisticated version of this circuit[32] simultaneously and independently controls both the bias current and the modulation current. The interested reader is referred to the literature for further details.

4-3 LIGHT SOURCE LINEARITY

High-radiance LEDs and laser diodes are well-suited optical sources for wideband analog applications provided a method is implemented to compensate for any nonlinearities of these devices. In an analog system the time-varying electric analog signal $s(t)$ is used to modulate directly an optical source about a bias current point I_B, as shown in Fig. 4-17. With no signal input the optical power output is P_t. When the signal $s(t)$ is applied, the optical output power $P(t)$ is

$$P(t) = P_t[1 + ms(t)] \qquad (4\text{-}39)$$

Here m is the *modulation index* (or *modulation depth*) defined by

$$m = \frac{\Delta I}{I'_B} \qquad (4\text{-}40)$$

where $I'_B = I_B$ for LEDs and $I'_B = I_B - I_{th}$ for laser diodes. The parameter ΔI is the variation in current about the bias point. To prevent distortions in the output signal, the modulation must be confined to the linear region of the curve for optical output versus drive current. Furthermore if ΔI is greater than I'_B (that is, m is greater than 100 percent), the lower portion of the signal gets cut off and severe distortion will result.

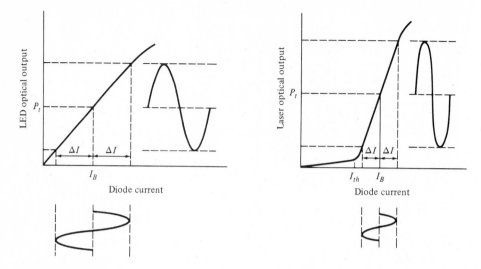

Figure 4-17 Bias point and amplitude modulation range for analog applications of LEDs and laser diodes.

In analog applications any device nonlinearities will create frequency components in the output signal that were not present in the input signal.[37] Two important nonlinear effects are harmonic and intermodulation distortions. If the signal input to a nonlinear device is a simple cosine wave $x(t) = A \cos \omega t$, the output will be

$$y(t) = A_0 + A_1 \cos \omega t + A_2 \cos 2\omega t + A_3 \cos 3\omega t + \cdots \qquad (4\text{-}41)$$

That is, the output signal will consist of a component at the input frequency ω plus spurious components at zero frequency, at the second harmonic frequency 2ω, at the third harmonic frequency 3ω, and so on. This effect is known as *harmonic distortion*. The amount of nth-order distortion in decibels is given by

$$n\text{th-order harmonic distortion} = 20 \log \frac{A_n}{A_1} \qquad (4\text{-}42)$$

To determine *intermodulation distortion*, the modulating signal of a nonlinear device is taken to be the sum of two cosine waves $x(t) = A_1 \cos \omega_1 t + A_2 \cos \omega_2 t$. The output signal will then be of the form

$$y(t) = \sum_{m,n} B_{mn} \cos (m\omega_1 + n\omega_2) \qquad (4\text{-}43)$$

where m and $n = 0, \pm1, \pm2, \pm3, \ldots$.

This signal includes all the harmonics of ω_1 and ω_2 plus cross-product terms such as $\omega_2 - \omega_1$, $\omega_2 + \omega_1$, $\omega_2 - 2\omega_1$, $\omega_2 + 2\omega_1$, etc. The sum and difference frequencies give rise to the intermodulation distortion. The sum of the absolute

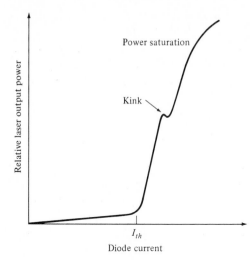

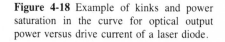

Figure 4-18 Example of kinks and power saturation in the curve for optical output power versus drive current of a laser diode.

values of the coefficients m and n determines the order of the intermodulation distortion. For example, the second-order intermodulation products are at $\omega_1 \pm \omega_2$ with amplitude B_{11}; the third-order intermodulation products are at $\omega_1 \pm 2\omega_2$ and $2\omega_1 \pm \omega_2$ with amplitudes B_{12} and B_{21}; and so on. (Harmonic distortions are also present wherever either $m \neq 0$ and $n = 0$ or when $m = 0$ and $n \neq 0$. The corresponding amplitudes are B_{m0} and B_{0n}, respectively.) In general, the odd-order intermodulation products having $m = n \pm 1$ (such as $2\omega_1 - \omega_2$, $2\omega_2 - \omega_1$, $3\omega_1 - 2\omega_2$, etc.,) are the most troublesome since they may fall within the bandwidth of the channel. Of these only the third-order terms are usually important since the amplitudes of higher-order terms tend to be significantly smaller. If the operating frequency band is less than an octave, all other intermodulation products will fall outside the passband and can be eliminated with appropriate filters in the receiver.

Nonlinear distortions in LEDs are due to effects depending on the carrier injection level, radiative recombination, and other subsidiary mechanisms, as is described in detail by Asatani and Kimura.[38] In certain laser diodes there often are nonlinearities in the curve for optical power output versus diode current as is illustrated in Fig. 4-18. These nonlinearities are a result of inhomogeneities in the active region of the device and also arise from power switching between the dominant lateral modes in the laser. They are generally referred to as "kinks." Power saturation (as indicated by a downward curving of the output-versus-current curve) can occur at high output levels because of active-layer heating.

Total harmonic distortions[39–42] in GaAlAs LEDs and laser diodes tend to be in the range of 30 to 40 dB below the output at the fundamental modulation frequency for modulation depths around 0.5. The second- and third-order harmonic distortions as a function of bias current for several modulation frequencies are shown in Fig. 4-19 for a GaAlAs double-heterojunction LED.[39] The harmonic distortions decrease with increasing bias current but become large at higher modulation fre-

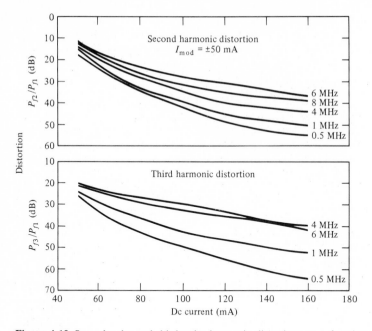

Figure 4-19 Second-order and third-order harmonic distortions as a function of bias current in a GaAlAs LED for several modulation frequencies. The distortion is given in terms of the power P_{fn} at the nth harmonic relative to the power P_{f1} at the modulation frequency $f1$. *(Reproduced with permission from Dawson.*[39]*)*

quencies. The intermodulation distortion curves (not shown) follow the same characteristics as those in Fig. 4-19, but are 5 to 8 dB worse.

A number of compensation techniques for linearization of optical sources in analog communication systems have been investigated. These methods include circuit techniques such as complementary distortion,[38, 42] negative feedback,[43] selective harmonic compensation,[44] and quasi-feedforward compensation,[45] and the use of pulse position modulation (PPM) schemes.[46] One of the most successful circuit design techniques is the quasi-feedforward method with which a 30- to 40-dB reduction in total harmonic distortion has been achieved.

4-4 MODAL AND REFLECTION NOISE

Two significant problems affecting the linearity of laser diodes arise when using them for high-speed digital and analog applications. These phenomena are called *modal* or *speckle noise*[47–49] and reflection noise.[50, 51] These two factors are important because they can introduce receiver output noise which can be particularly serious for analog systems.

Modal or speckle noise is associated with intensity fluctuations in the longitudinal modes of a laser diode. The output from a laser diode generally can come from more than one longitudinal mode. The optical output may arise from all of the modes simultaneously or it may switch from one mode to another randomly in time. Intensity fluctuations can occur among the various modes in a multimode laser even when the total optical output does not vary. Since the output pattern of a laser diode is highly directional, the light from these fluctuating modes can be coupled into a multimode fiber with a high coupling efficiency. Each of the longitudinal modes that is simultaneously coupled into the fiber has a different attenuation and time delay, as was shown in Sec. 3-3. If the time delay difference between any two modes is shorter than the coherence time of the optical source (the time between mode fluctuations), then a speckled light pattern will arise across the end face of the output fiber (hence the name speckle noise). The speckle pattern depends on what type of fiber is used. Since the bandwidth of graded-index multimode fibers is generally larger than that of step-index multimode fibers, the speckle pattern will persist after much longer transmission distances in graded-index multimode fibers than in step-index multimode fibers. As a result of the fiber output fluctuations, the power level and signal arrival times at the optical receiver can vary as the laser mode output pattern changes. The result is a degradation in receiver performance.

Reflection noise[50,51] is associated with laser diode output linearity distortion caused by some of the light output being reflected back into the laser cavity from fiber joints. This reflected power couples with the lasing modes, thereby causing their phases to vary. This produces a periodically modulated noise spectrum that is peaked on the low-frequency side of the intrinsic noise profile. The fundamental frequency of this noise is determined by the roundtrip delay of the light from the laser to the reflecting point and back again. Depending on the roundtrip time, these reflections can create noise peaks in the frequency region where optical fiber data transmission systems operate, even though the lasers themselves are very noise-free at these frequencies. Reflection noise problems can be greatly reduced by using optical isolators between the laser diode and the optical fiber transmission line or by using index-matching fluid in the gaps at fiber-to-fiber joints to eliminate reflections at the fiber-to-air interfaces.

4-5 RELIABILITY CONSIDERATIONS

The reliability of double-heterojunction LEDs and laser diodes is of importance since these devices are of greatest interest in optical communications systems. The lifetimes of these sources are affected by both operating conditions and fabrication techniques. Thus it is important to understand the relationships between light source operation characteristics, degradation mechanisms, and system reliability requirements. A comprehensive review of the reliability of GaAlAs laser diodes has been presented by Ettenberg and Kressel[52] to which the reader is referred for further details and an extensive list of references. The extension of these results to LEDs is straightforward.[53-56]

Lifetime tests of optical sources are done either at room temperature or at elevated temperatures to accelerate the degradation process. A commonly used elevated temperature is 70°C. At present there is no standard method for determining the lifetime of an optical source. The two most popular techniques either maintain the light output constant by increasing the bias current automatically or keep the current constant and monitor the optical output level. In the first case, the end of life of the device is assumed to be reached when the source can no longer put out a specified power at the maximum current value for CW (continuous-wave) operation. In the second case, the lifetime is determined by the time taken for the optical output power to decrease by 3 dB.

Degradation of light sources can be divided into three basic categories: internal damage and ohmic contact degradation, which hold for both lasers and LEDs, and damage to the facets of laser diodes.

The limiting factor on LED and laser diode lifetime is internal degradation. This effect arises from the migration of crystal defects into the active region of the light source. These defects decrease the internal quantum efficiency and increase the optical absorption. Fabrication steps which can be taken to minimize internal degradation include the use of substrates with low surface dislocation densities (less than 2×10^3 dislocations/cm^2), keeping work-damaged edges out of the diode current path, and minimizing stresses in the active region (to less than 10^8 dyn/cm^2).

For high-quality sources having lifetimes which follow a slow internal-degradation mode, the optical power P decreases with time according to the exponential relationship

$$P = P_0 e^{-t/\tau_m} \qquad (4\text{-}44)$$

Here P_0 is the initial optical power at time $t = 0$ and τ_m is a time constant for the degradation process which is approximately twice the -3-dB mean time to failure. Since the operating lifetime depends on both the current density J and the junction temperature T, internal degradation can be accelerated by increasing either one of these parameters.

The operating lifetime τ_s has been found experimentally to depend on the current density J through the relation

$$\tau_s \propto J^{-n} \qquad (4\text{-}45)$$

where $1.5 \leq n \leq 2.0$. For example, by doubling the current density, the lifetime decreases by a factor of 3 to 4. Since the degradation rate of optical sources increases with temperature, an Arrhenius relationship of the form

$$\tau_s = K e^{E_A/k_B T} \qquad (4\text{-}46)$$

has been sought. Here E_A is an activation energy characterizing the lifetime τ_s, k_B is Boltzmann's constant, T is the absolute temperature at which τ_s was evaluated, and K is a constant. The problem in establishing such an expression is that several competing factors are likely to contribute to the degradation, thereby making it difficult to estimate the activation energy E_A. Activation energies for laser degradation reported in the literature have ranged from 0.3 to 1.0 eV. For practical calculations, a value of 0.7 eV is generally used. However, this value is subject to change as more long-term statistical data at various temperatures are obtained.

Equations (4-45) and (4-46) indicate that, to increase the light source lifetime,

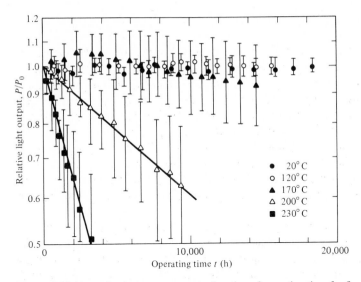

Figure 4-20 Normalized output power as a function of operating time for five ambient temperatures. P_0 is the initial optical output power. *(Reproduced with permission from Yamakoshi et al.[53])*

it is advantageous to operate these devices at as low a current and temperature as is practicable. Examples[55] of the luminescent output of InGaAsP LEDs as a function of time for different temperatures are shown in Fig. 4-20. At temperatures below 120°C the output power remains almost constant over the entire measured 15,000-h (1.7-yr) operating time. At higher temperatures the power output drops as a function of time. For example, at 230°C the optical power has dropped to one-half its initial value (a 3-dB decrease) after approximately 3000 h (4.1 months) of operation. The activation energy of these lasers is about 1.0 eV.

A second fabrication-related degradation mechanism is ohmic contact deterioration. In LEDs and laser diodes the thermal resistance of the contact between the light source chip and the device heat sink occasionally increases with time. This effect is a function of the solder used to bond the chip to the heat sink, the current density through the contact, and the contact temperature. An increase in the thermal resistance results in a rise in the junction temperature for a fixed operating current. This, in turn, leads to a decrease in the optical output power. However, careful designs and implementations of high-quality bonding procedures have minimized effects resulting from contact degradation.

Facet damage is a degradation problem that exists for laser diodes. This degradation reduces the laser mirror reflectivity and increases the nonradiative carrier recombination at the laser facets. The two types of facet damage that can occur are generally referred to as *catastrophic facet degradation* and *facet erosion*. Catastrophic facet degradation is a mechanical damage of the facets which may arise after short operating times of laser diodes at high optical power densities. This damage tends to reduce greatly the facet reflectivity, thereby increasing the threshold current and decreasing the external quantum efficiency. The fundamental cause of catastrophic facet degradation has not yet been determined, but it has been observed to be a function of the optical power density and the pulse length.

Facet erosion is a gradual degradation occurring over a longer period of time compared to catastrophic facet damage. The decrease in mirror reflectivity and the increase in nonradiative recombination at the facets owing to facet erosion lower the internal quantum efficiency of the laser and increase the threshold current. In GaAlAs lasers facet erosion arises from oxidation of the mirror surface. It is speculated that the oxidation process is stimulated by the optical radiation emitted from the laser. Facet erosion is minimized by depositing a half-wavelength-thick Al_2O_3 film on the facet. This type of coating acts as a moisture barrier and does not affect the mirror reflectivity or the lasing threshold current.

A comparison[52] of two definitions of failure for laser diodes operating at 70°C is shown in Fig. 4-21. The lower curve shows the time required for the laser output to drop to one-half its initial value when a constant current passes through the device. This is the "3-dB life."

The "end-of-life" failure is given by the top trace in Fig. 4-21. This condition is defined as the time at which the device can no longer emit a fixed power level (1.25 mW in this case) at the 70°C heat sink temperature. The mean operating times (time for 50 percent of the lasers to fail) are 3800 h and 1900 h for the end-of-life and 3-dB-life conditions respectively. The right-hand ordinate of Fig. 4-21 gives an estimate of the operating time at 22°C, assuming an activation energy of 0.7 eV.

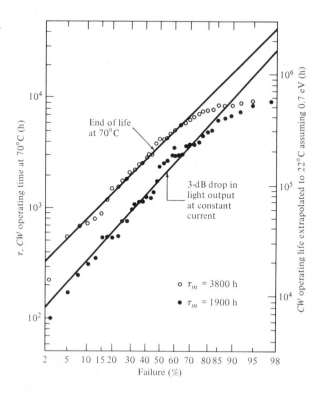

Figure 4-21 Time-to-failure plot on log-normal coordinates for 40 low-threshold ($\simeq$ 50 mA) oxide-defined stripe lasers at a 70°C heat sink temperature. τ_m is the time it took for 50 percent of the lasers to fail for the two types of failure mechanisms. *(Reproduced with permission from Kressel and Ettenberg.[47])*

4-6 SUMMARY

In this chapter we have presented the basic operating characteristics of heterojunction-structured semiconductor light-emitting diodes (LEDs) and laser diodes. We first discussed the structures of these light sources, which is a sandwich-type construction of different semiconductor materials. These layers serve to confine the electrical and optical carriers to yield optical sources with high output and high efficiencies. The principal materials of which these layers are composed include the ternary alloy GaAlAs for operation in the 800- to 900-nm wavelength region and the quaternary alloy InGaAsP for use between 1100 and 1600 nm.

We then discussed in detail the internal quantum efficiency and the various factors that can adversely reduce this efficiency. The two principal factors are:

1. Nonradiative recombination of charge carriers at the boundaries of the hetero-structure layers resulting from crystal lattice mismatches. This effect can be minimized by choosing materials with closely matched lattice spacings for adjacent heterojunction layers.
2. Optical absorption in the active region of the device. This effect, which is known as self-absorption, is minimized by using thin active regions.

In conjunction with the quantum efficiency we addressed the modulation capability of optical sources and their response when subjected to transient current pulses. We noted in particular that the time delay between the application of a current pulse and the onset of optical power output can be reduced by applying a small dc bias to the source. This bias reduces the parasitic diode space charge capacitance that could cause a delay of the carrier injection into the active region, which, in turn, could delay the optical output.

We next turned our attention to semiconductor laser diode optical sources. When deciding whether to choose an LED or a laser diode source, a tradeoff must be made between the advantages and drawbacks of each type of device. The advantages that a laser diode has over an LED are:

1. A faster response time, so that much greater modulation rates (data transmission rates) are possible with a laser diode.
2. A narrower spectral width of the output which implies less dispersion-induced signal distortion.
3. A much higher optical power level can be coupled into a fiber with a laser diode, thus allowing greater transmission distances.

Some of the drawbacks of laser diodes are:

1. Their construction is more complicated mainly because of the requirement of current confinement in a small lasing cavity.

2. The optical output power level is strongly dependent on temperature. This increases the complexity of the transmitter circuitry. If a laser diode is to be used over a wide temperature range, then either a cooling mechanism must be used to maintain the laser at a constant temperature or a threshold-sensing circuit can be implemented to adjust the bias current with changes in temperature.

3. They are susceptible to catastrophic facet degradation which greatly reduces the device lifetime. This is a mechanical damage of the facets that may arise after short operating times at high optical power densities.

High-radiance LEDs and laser diodes are well-suited for wideband analog applications, provided a method is used to compensate for any nonlinearities in the output of these devices. These nonlinearities will create frequency components in the output signal that were not present in the input signal. Two important nonlinear effects are harmonic and intermodulation distortions. For intermodulation distortion, only the third-order terms are usually important. All other intermodulation products can be filtered out in the receiver if the operating bands are less than an octave. Total harmonic distortions in GaAlAs LEDs and laser diodes (which operate in the 800- to 900-nm range) tend to be about 30 to 40 dB below the output of the fundamental modulation frequency for modulation depths of around 0.5. Special circuit design techniques can be implemented for linearization of optical sources in analog communciation systems. These designs can reduce total harmonic distortions by 30 to 40 dB.

An important issue in any application is device reliability. Degradation of light sources can be divided into three basic categories: internal damage and contact degradation, which hold for both laser diodes and LEDs, and damage to the facets of laser diodes. Lifetime tests of optical sources are often done at elevated temperatures (e.g., 70°C) to accelerate the degradation process. Since optical sources are adversely affected by high currents and high temperatures, it is recommended that in order to increase the light source lifetime, they be operated at as low a current and temperature as is practical in a system.

PROBLEMS

Section 4-1

4-1 An engineer has two $Ga_{1-x}Al_xAs$ light-emitting diodes, one with $x = 0.02$ and the other with $x = 0.09$. Find the band gap energy and the peak emission wavelengths of these two devices.

4-2 The lattice spacing of $In_{1-x}Ga_xAs_yP_{1-y}$ has been shown to obey Vegard's law.[57] This states that for quaternary alloys of the form $A_{1-x}B_xC_yD_{1-y}$, where A and B are group-III elements (e.g., Al, In, and Ga) and C and D are group-V elements (e.g., As, P, and Sb), the lattice spacing $a(x, y)$ of the quaternary alloy can be approximated by

$$a(x, y) = xya(BC) + x(1 - y)a(BD) + (1 - x)ya(AC) + (1 - x)(1 - y)a(AD)$$

where the $a(IJ)$ are the lattice spacings of the binary compounds IJ.

(a) Show that for $In_{1-x}Ga_xAs_yP_{1-y}$ with

$$a(GaAs) = 5.6536 \text{ Å}$$

$$a(GaP) = 5.4512 \text{ Å}$$

$$a(InAs) = 6.0590 \text{ Å}$$

$$a(InP) = 5.8696 \text{ Å}$$

the quaternary lattice spacing becomes

$$a(x, y) = 0.1894y - 0.4184x + 0.0130xy + 5.8696 \text{ Å}$$

(b) For quaternary alloys that are lattice-matched to InP the relation between x and y can be determined by letting $a(x, y) = a(InP)$. Show that since $0 \le x \le 0.47$ the resulting expression can be approximated by $y \simeq 2.20x$.

(c) A simple empirical relation that gives the band gap energy in terms of x and y is[57]

$$E_g(x, y) = 1.35 + 0.668x - 1.17y + 0.758x^2 + 0.18y^2 - 0.069xy - 0.322x^2y + 0.03xy^2 \quad \text{eV}$$

Find the band gap energy and the peak emission wavelength of $In_{0.74}Ga_{0.26}As_{0.56}P_{0.44}$.

4-3 (a) If the radiative and nonradiative recombination lifetimes of the minority carriers in the active region of an LED are 3 ns and 100 ns, respectively, find the internal efficiency and the bulk recombination lifetime in the absence of self-absorption and recombination at the heterojunction.

(b) If the surface recombination velocity at the heterojunction interfaces is 5000 cm/s, what are the lifetime reductions for 1-μm- and 2-μm-thick active layers. Assume that the condition $L_D S/D \ll 1$ holds.

4-4 By using Eqs. (4-8) and (4-9) derive Eq. (4-10), which gives the density of excess electrons at the position x relative to the pn junction in the active layer of a light source.

4-5 Derive Eq. (4-12) by using Eqs. (4-10) and (4-11).

4-6 Show that Eq. (4-12) reduces to Eq. (4-13) under the conditions that $L_D S \ll D$ and $d \le L_D$.

4-7 Derive Eq. (4-16) from Eq. (4-14).

4-8 Calculate and plot the relative reduction in the internal quantum efficiency η_i^{dh}/η_0 for a 10^{18} cm^{-3} active-area doping concentration in a GaAs LED. Let the active-layer widths range from $d = 0.1L_D$ to $10L_D$ and assume the surface recombination velocity $S = 10^5$ cm/s. At a 10^{18} cm^{-3} doping level $\alpha_A = 4000$ cm^{-1}, $L_D = 5 \times 10^{-4}$ cm, and $D = 80$ cm^2/s for GaAs. Compare the results with the $\alpha_A L_D = 2.0$ curves for $S = 10^3$ cm/s and 10^6 cm/s shown in Fig. 4-8.

4-9 A particular InGaAsP LED emitting at 1.3 μm is found to have a radiative recombination coefficient of 3×10^{-10} cm^3/s, a carrier concentration of $n_0 + p_0 = 10^{17}$ cm^{-3} in its 1-μm-thick active region, and a surface recombination velocity of 10^4 cm/s. Using Eq. (4-21) for the effective carrier lifetime, plot the frequency response of the relative optical output intensity $I(\omega)/I_0$ of this LED for modulation frequencies ranging from 1 to 500 MHz at current densities of 0.5, 2.0, and 10 kA/cm^2. (*Note:* Use three-cycle semilog paper with the frequency assigned to the log scale.) What is the 3-dB modulation bandwidth at each of these current densities?

4-10 A practical surface-emitting LED has a 50-μm-diameter emitting area and operates at a peak modulation current of 100 mA. Assuming Eq. (4-26) holds, what is the bandwidth of a GaAlAs LED having a 2.0-μm active-area thickness? Take $B_r = 10^{-10}$ cm^3/s and $S = 10^4$ cm/s.

4-11 Derive Eq. (4-24). Show that this equation reduces to Eq. (4-25) when $\Delta n \ll n_0 + p_0$ and to Eq. (4-26) when $\Delta n \gg n_0 + p_0$.

4-12 An LED has a 500-pF space charge capacitance, a 1.0-pA saturation current, and a 5-ns minority carrier lifetime. Plot the half-current and 10 to 90 percent rise times as a function of current for drive current amplitudes ranging from 10 to 100 mA. (The fact that $t_{1/2}$ is greater than t_{10-90} in this plot for low drive currents in nonprebiased LEDs is a result of the time delay between the application of a current pulse and the onset of optical output power.)

Section 4-2

4-13 (*a*) A GaAlAs laser diode has a 500-μm cavity length which has an effective absorption coefficient of 10 cm^{-1}. For uncoated facets the reflectivities are 0.32 at each end. What is the optical gain at the lasing threshold?

(*b*) If one end of the laser is coated with a dielectric reflector so that its reflectivity is now 90 percent, what is the optical gain at the lasing threshold?

(*c*) If the internal quantum efficiency is 0.65, what is the external quantum efficiency in cases (*a*) and (*b*)?

4-14 Find the external quantum efficiency for a $Ga_{1-x}Al_xAs$ laser diode (with $x = 0.03$) which has the optical power versus drive current relationship shown in Fig. 4-14.

4-15 When a current pulse is applied to a laser diode, the injected carrier pair density Δn within the recombination region of width d changes with time according to the relationship

$$\frac{\partial(\Delta n)}{\partial t} = \frac{J}{qd} - \frac{\Delta n}{\tau}$$

(*a*) Assume τ is the average carrier lifetime in the recombination region when the injected carrier pair density is Δn_{th} near the threshold current density J_{th}. That is, in the steady state we have $\partial(\Delta n)/\partial t = 0$ so that

$$\Delta n_{th} = \frac{J_{th}\tau}{qd}$$

If a current pulse of amplitude I_p is applied to an unbiased laser diode, show that the time needed for the onset of stimulated emission is

$$t_d = \tau \ln \frac{I_p}{I_p - I_{th}}$$

Assume the drive current $I = JA$, where J is the current density and A is the area of the active region.

(*b*) If the laser is now prebiased to a current density $J_B = I_B/A$, so that the initial excess carrier pair density is $\Delta n_B = J_B\tau/qd$, then the current density in the active region during a current pulse I_p is $J = J_B + J_p$. Show that in this case Eq. (4-37) results.

Section 4-3

4-16 Consider the following Taylor series expansion of the optical power versus drive current relationship of an optical source about a given bias point

$$y(t) = a_1x(t) + a_2x^2(t) + a_3x^3(t) + a_4x^4(t)$$

Let the modulating signal $x(t)$ be the sum of two sinusoidal tones at frequencies ω_1 and ω_2 given by

$$x(t) = b_1 \cos \omega_1 t + b_2 \cos \omega_2 t$$

(*a*) Find the second-, third-, and fourth-order intermodulation distortion coefficients B_{mn} (where m and $n = \pm 1, \pm 2, \pm 3$, and ± 4) in terms of b_1, b_2, and the a_i.

(*b*) Find the second-, third-, and fourth-order harmonic distortion coefficients A_2, A_3, and A_4 in terms of b_1, b_2, and the a_i.

Section 4-5

4-17 An optical source is selected from a batch characterized as having lifetimes which follow a slow internal degradation mode. The -3-dB mean time to failure of these devices at room temperature is specified as 5×10^4 h. If the device emits 1 mW at room temperature, what is the expected optical output power after 1 month of operation? after 1 yr? after 5 yr?

4-18 A group of optical sources is found to have operating lifetimes of 4×10^4 h at 60°C and 6500 h at 90°C. What is the expected lifetime at 20°C if the device lifetime follows an Arrhenius-type relationship?

REFERENCES

1. H. Kressel and J. K. Butler, *Semiconductor Lasers and Heterojunction LEDs,* Academic, New York, 1977.
2. H. C. Casey, Jr. and M. B. Panish, *Heterostructure Lasers: Part A—Fundamental Principles; Part B—Materials and Operating Characteristics,* Academic, New York, 1978.
3. G. H. B. Thompson, *Physics of Semiconductor Laser Devices,* Wiley, New York, 1980.
4. H. Kressel, I. Ladany, M. Ettenberg, and H. Lockwood, "Light sources," *Physics Today,* **29,** 38–47, May 1976.
5. D. Botez and M. Ettenberg, "Comparison of surface and edge emitting LEDs for use in fiber optical communications," *IEEE Trans. Electron Devices,* **ED-26,** 1230–1238, Aug. 1979.
6. D. Marcuse, "LED fundamentals: Comparison of front- and edge-emitting diodes," *IEEE J. Quantum Electron.,* **QE-13,** 819–827, Oct. 1977.
7. M. B. Panish, "Heterostructure injection lasers," *Proc. IEEE,* **64,** 1512–1540, Oct. 1976.
8. H. Kressel, "Electroluminescent sources for fiber systems," in *Fundamentals of Optical Fiber Communications,* M. K. Barnoski (Ed.), Academic, New York, 1976, chap. 4.
9. C. A. Burrus and B. I. Miller, "Small-area double heterostructure AlGaAs electroluminescent diode sources for optical fiber transmission lines," *Opt. Commun.,* **4,** 307–309, Dec. 1971.
10. J. P. Wittke, M. Ettenberg, and H. Kressel, "High radiance LED for single fiber optical links," *RCA Rev.,* **37,** 159–183, June 1976.
11. Reference 1, chaps. 2 and 8.
12. T. P. Lee and A. J. Dentai, "Power and modulation bandwidth of GaAs-AlGaAs high radiance LEDs for optical communication systems," *IEEE J. Quantum Electron.,* **QE-14,** 150–159, Mar. 1978.
13. W. Harth, J. Heinen, and W. Huber, "Influence of active layer width on the performance of homojunction and single heterojunction GaAs light emitting diodes," *Electron. Lett.,* **11,** 23–24, Jan. 1975.
14. (a) H. Namizaki, M. Nagano, and S. Nakahara, "Frequency response of GaAlAs light emitting diodes," *IEEE Trans. Electron. Devices,* **ED-21,** 688-691, Nov. 1974.
    (b) Y. S. Lin and D. A. Smith, "The frequency response of an amplitude modulated GaAs luminescent diode," *Proc. IEEE,* **63,** 542-544, Mar. 1975.
15. T. P. Lee, "Effects of junction capacitance on the rise time of LEDs and the turn-on delay of injection lasers," *Bell Syst. Tech. J.,* **54,** 53–68, Jan. 1975.
16. I. Hino and K. Iwamoto, "LED pulse response analysis," *IEEE Trans. Electron. Devices,* **ED-26,** 1238–1242, Aug. 1979.
17. J. Zucker and R. B. Lauer, "Optimization and characterization of high radiance (Al, Ga)As double heterostructure LEDs for optical communication systems," *IEEE Trans. Electron. Devices,* **ED-25,** 193–198, Feb. 1978.
18. J. Zucker, "Closed-form calculation of the transient behavior of (Al, Ga)As double-heterojunction LEDs," *J. Appl. Phys.,* **49,** 2543–2545, Apr. 1978.
19. M. Ettenberg, "A new dielectric facet reflector for semiconductor lasers," *Appl. Phys. Lett.,* **32,** 724–725, June 1978.
20. M. Yamada and Y. Suematsu, "A condition of single mode longitudinal operation in injection laser diodes with index grading structure," *IEEE J. Quantum Electron.,* **QE-15,** 743–749, Aug. 1979.
21. (a) J. E. Ripper and J. C. Dyment, "Time delay and Q-switching in junction lasers," *IEEE J. Quantum Electron.,* **QE-5,** 396–403, Aug. 1979.

(b) J. C. Dyment, J. E. Ripper, and T. P. Lee, "Measurement and interpretation of long spontaneous lifetimes in double heterostructure lasers," *J. Appl. Phys.*, **43**, 452–457, Feb. 1972.

22. T. L. Paoli, "Near-threshold behavior of the intrinsic resonant frequency in a semiconductor laser," *IEEE J. Quantum Electron.*, **QE-15**, 807–811, Aug. 1979; "Modulation of diode lasers," *Laser Focus*, **13**, 54–57, Mar. 1977.

23. P. Torphammar, R. Tell, H. Eklund, and A. R. Johnston, "Minimizing pattern effects in semiconductor lasers at high rate pulse modulation," *IEEE J. Quantum Electron.*, **QE-15**, 1271–1276, Nov. 1979.

24. F. D. Nunes, N. B. Patel, and J. E. Ripper, "A theory on long time delays and Q-switching in GaAs junction lasers," *IEEE J. Quantum Electron.*, **QE-13**, 675–681, Aug. 1977.

25. "Quaternary compound semiconductor materials and devices—sources and detectors," Special Issue of *IEEE J. Quantum Electron.*, **QE-17**, Feb. 1981.

26. "Light sources and detectors," Special Issue of *IEEE Trans. Electron. Devices*, **ED-28**, Apr. 1981.

27. S. Wang, C. Y. Chen, A. S. H. Liao, and L. Figueroa, "Control of mode behavior in semiconductor lasers," *IEEE J. Quantum Electron.*, **QE-17**, 453–468, Apr. 1981.

28. M. Ettenberg, C. J. Neuse, and H. Kressel, "The temperature dependence of threshold current for double-heterojunction lasers," *J. Appl. Phys.*, **50**, 2949–2950, Apr. 1979.

29. R. Chin, N. Holonyak, B. A. Vojak, K. Hess, R. D. Dupuis, and P. D. Dapkus, "Temperature dependence of threshold current for quantum-well heterostructure laser diodes," *Appl. Phys. Lett.*, **36**, 19–21, Jan. 1980.

30. G. H. B. Thompson, "Temperature dependence of threshold current in GaInAsP DH lasers at 1.3 and 1.5 μm wavelength," *IEEE Proc.*, **128**, 37–43, Apr. 1981.

31. P. W. Shumate, Jr., F. S. Chen, and P. W. Dorman, "GaAlAs laser transmitter for lightwave transmission systems," *Bell Sys. Tech. J.*, **57**, 1823–1836, July–Aug. 1978.

32. F. S. Chen, "Simultaneous feedback control of bias and modulation currents for injection lasers," *Electron. Lett.*, **16**, 7–8, Jan. 1980.

33. D. W. Smith, "Laser level-control circuit for high-bit-rate systems using a slope detector," *Electron. Lett.*, **14**, 775–776, Nov. 1978.

34. J. F. Svacek, "Transmitter feedback techniques stabilize laser diode outputs," *EDN*, **25**, 107–111, Mar. 1980.

35. M. Ettenberg, D. R. Patterson, and E. J. Denlinger, "A temperature-compensated laser module for optical communications," *RCA Rev.*, **40**, 103–114, June 1979.

36. A. Albanese, "An automatic bias control circuit for injection lasers," *Bell Sys. Tech. J.*, **57**, 1533–1544, May 1978.

37. (a) A. B. Carlson, *Communication Systems*, McGraw-Hill, New York, 1975, chap. 4.
    (b) Tri T. Ha, *Solid-State Microwave Amplifier Design*, Wiley, New York, 1981, chap. 6.

38. K. Asatani and T. Kimura, "Analysis of LED nonlinear distortions," *IEEE Trans. Electron Devices*, **ED-25**, 199–207, Feb. 1978; "Linearization of LED nonlinearity by predistortions," *ibid.*, 207–212.

39. R. W. Dawson, "Frequency and bias dependence of video distortion in Burrus-type homostructure and heterostructure LEDs," *IEEE Trans. Electron Devices*, **ED-25**, 550–551, May 1978.

40. F. D. King, J. Straus, O. I. Szentesi, and A. J. Springthorpe, "High-radiance long-lived LEDs for analogue signalling," *Proc. IEE*, **123**, 619–622, June 1976.

41. T. Ozeki and E. H. Hara, "Measurement of nonlinear distortion in light emitting diodes," *Electron. Lett.*, **12**, 78–80, Feb. 1976.

42. K. Asatani, "Nonlinearity and its compensation of semiconductor laser diodes for analog intensity modulation systems," *IEEE Trans. Comm.* **COM-28**, 297–300, Feb. 1980.

43. J. Straus, "Linearized transmitters for analog fiber links," *Laser Focus*, **14**, 54–61, Oct. 1978.

44. J. Straus, A. J. Springthorpe, and O. I. Szentesi, "Phase shift modulation technique for the linearization of analog transmitters," *Electron. Lett.*, **13**, 149–151, Mar. 1977.

45. (a) R. E. Patterson, J. Straus, G. Blenman, and T. Witkowicz, "Linearization of multichannel analog optical transmitters by quasi-feedforward compensation technique," *IEEE Trans. Comm.*, **COM-27**, 582–588, Mar. 1979.
    (b) J. Straus and O. I. Szentesi, "Linearization of optical transmitters by a quasi-feedforward compensation technique," *Electron. Lett.*, **13**, 158–159, Mar. 1977.

46. D. Kato, "High-quality broadband optical communication by TDM-PAM: Nonlinearity in laser diodes," *IEEE J. Quantum Electron.*, **QE-14**, 343–346, May 1978.

47. R. E. Epworth, "The phenomenon of modal noise in analogue and digital optical fibre systems," *Proc. 4th European Conf. Opt. Commun.*, Geneva, Switzerland, 492–501, Sept. 1978.

48. Y. Okano, K. Nakagawa, and T. Ito, "Laser mode partition noise evaluation for optical fiber transmission," *IEEE Trans. Comm.*, **COM-28**, 238–243, Feb. 1980.

49. K. Sato and K. Asatani, "Speckle noise reduction in fiber optic analog video transmission using semiconductor laser diodes," *IEEE Trans. Comm.*, **COM-29**, 1017–1024, July 1981.

50. O. Hirota and Y. Suematsu, "Noise properties of injection lasers due to reflected waves," *IEEE J. Quantum Electron.*, **QE-15**, 142–149, Mar. 1979.

51. Y. C. Chen, "Noise characteristics of semiconductor laser diodes coupled to short optical fibers," *Appl. Phys. Lett.*, **37**, 587–589, Oct. 1980.

52. M. Ettenberg and H. Kressel, "The reliability of (AlGa)As CW laser diodes," *IEEE J. Quantum Electron.*, **QE-16**, 186–196, Feb. 1980.

53. S. Yamakoshi, O. Hasegawa, H. Hamaguchi, M. Abe, and T. Yamaoka, "Degradation of high-radiance $Ga_{1-x}Al_xAs$ LEDs," *Appl. Phys. Lett.*, **31**, 627–629, Nov. 1977.

54. L. R. Dawson, V. G. Keramidas, and C. L. Zipfel, "Reliable, high-speed LEDs for short-haul optical data links," *Bell Sys. Tech. J.*, **59**, 161–168, Feb. 1980.

55. S. Yamakoshi, M. Abe, O. Wada, S. Komiya, and T. Sakurai, "Reliability of high-radiance InGaAsP/InP LEDs operating in the 1.2–1.3 μm wavelength," *IEEE J. Quantum Electron.*, **QE-17**, 167-173, Feb. 1981.

56. O. Wada, S. Yamakoshi, H. Hamaguchi, T. Sanada, Y. Nishitani, and T. Sakurai, "Performance and reliability of high radiance InGaAsP/InP DH LEDs operating in the 1.15–1.5 μm wavelength region," *IEEE J. Quantum Electron.*, **QE-18**, 368-374, Mar. 1982.

57. R. E. Nahory, M. A. Pollack, W. D. Johnston, Jr., and R. L. Barns, "Band gap versus composition and demonstration of Vegard's law for InGaAsP lattice matched to InP," *Appl. Phys. Lett.*, **33**, 659–661, Oct. 1978.

FIVE

POWER LAUNCHING AND COUPLING

In implementing an optical fiber link two of the major system questions are how to launch optical power into a particular fiber from some type of luminescent source and how to couple optical power from one fiber into another. Launching optical power from a source into a fiber entails considerations such as the numerical aperture, core size, refractive-index profile, and core-cladding index difference of the fiber, plus the size, radiance, and angular power distribution of the optical source.

A measure of the amount of optical power emitted from a source that can be coupled into a fiber is usually given by the *coupling efficiency* η defined by

$$\eta = \frac{P_F}{P_S}$$

Here P_F is the power coupled into the fiber and P_S is the power emitted from the light source. The launching or coupling efficiency depends on the type of fiber that is attached to the source and on the coupling process, for example, whether or not lenses or other coupling improvement schemes are used.

In practice, many source suppliers offer devices with a short length of optical fiber (1 m or less) already attached in an optimum power-coupling configuration. This section of fiber is generally referred to as a *flylead* or a *pigtail*. The power-launching problem for these pigtailed sources thus reduces to a simpler one of coupling optical power from one fiber into another. The effects to be considered in this case include fiber misalignments, different core sizes, numerical apertures, and core refractive-index profiles, plus the need for clean and smooth fiber end faces that are perpendicular to the fiber axis.

Care must also be exercised when measuring the coupling efficiency between the fiber flylead and the cabled fiber, since the source can launch a significant amount of optical power into the cladding of the flylead. Although this power may be present at the end of the short flylead, it will not be coupled into the core of the following fiber. A true measure of the amount of power available from the flylead for coupling into a fiber can only be determined by stripping off the cladding modes before measuring the output optical power.

5-1 SOURCE-TO-FIBER POWER LAUNCHING

A convenient and useful measure of the optical output of a luminescent source is its radiance (or brightness) B at a given diode drive current. *Radiance* is the optical power radiated into a solid angle per unit emitting surface area and is generally specified in terms of watts per square centimeter per steradian. Since the optical power which can be coupled into a fiber depends on the radiance (that is, on the spatial distribution of the optical power) the radiance of an optical source rather than the total output power is the important parameter when considering source-to-fiber coupling efficiencies.

5-1-1 Source Output Pattern

To determine the optical power-accepting capability of a fiber, the spatial radiation pattern of the source must first be known. This pattern can be fairly complex. Consider Fig. 5-1 which shows a spherical coordinate system characterized by R, θ, and ϕ, with the normal to the emitting surface being the polar axis. The radiance may be a function of both θ and ϕ, and can also vary from point to point on the emitting surface. A reasonable assumption for simplicity of analysis is to take the emission to be uniform across the source area.

Surface-emitting LEDs are characterized by their lambertian output pattern, which means the source is equally bright when viewed from any direction. The power delivered at an angle θ measured relative to a normal to the emitting surface varies as cos θ because the projected area of the emitting surface varies as cos θ with

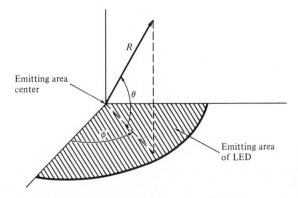

Figure 5-1 Spherical coordinate system for characterizing the emission pattern from an optical source.

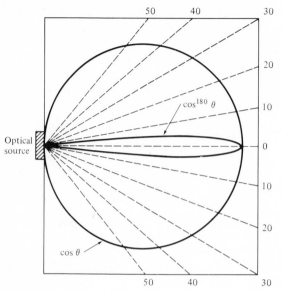

Figure 5-2 Radiance patterns for a lambertian source and the lateral output of a highly directional laser diode. Both sources have B_0 normalized to unity.

viewing direction. The emission pattern for a lambertian source thus follows the relationship

$$B(\theta, \phi) = B_0 \cos \theta \qquad (5\text{-}1)$$

where B_0 is the radiance along the normal to the radiating surface. The radiance pattern for this source is shown in Fig. 5-2.

Edge-emitting LEDs and laser diodes have a more complex emission pattern. These devices have different radiances $B(\theta,0°)$ and $B(\theta,90°)$ in the planes parallel and normal, respectively, to the emitting-junction plane of the device. These radiances can be approximated by the general form[1]

$$\frac{1}{B(\theta,\phi)} = \frac{\sin^2 \phi}{B_0 \cos^T\theta} + \frac{\cos^2 \phi}{B_0 \cos^L \theta} \qquad (5\text{-}2)$$

The integers T and L are the transverse and lateral power distribution coefficients, respectively. In general, for edge emitters, $L = 1$ (which is a lambertian distribution with a 120° half-power beam width) and T is significantly larger. For laser diodes, L can take on values over 100. As an example, we compare the lambertian pattern shown in Fig. 5-2 with a laser diode having a lateral ($\phi = 0°$) half-power beam width of $2\theta = 10°$, so that L is about 180. The much narrower output light beam from a laser diode allows significantly more light to be coupled into an optical fiber.

5-1-2 Power-Coupling Calculation

To calculate the maximum optical power coupled into a fiber, consider first the case shown in Fig. 5-3. Here the fiber end face is centered over the emitting surface of

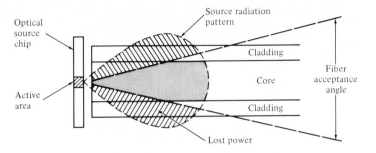

Figure 5-3 Schematic diagram of an optical source coupled to an optical fiber. Light outside of acceptance angle is lost.

the source and is positioned as close to it as possible. The coupled power can be found using the relationship

$$P = \int_0^{r_m} \int_0^{2\pi} \left[\int_0^{2\pi} \int_0^{\theta_{0,\max}} B(\theta,\phi) \sin\theta \; d\theta \; d\phi \right] d\theta_s r \; dr \qquad (5\text{-}3)$$

In this expression first the radiance $B(\theta,\phi)$ from an individual radiating point source on the emitting surface is integrated over the solid acceptance angle of the fiber. This is shown by the term in square brackets, where $\theta_{0,\max}$ is the maximum acceptance angle of the fiber, which is related to the numerical aperture NA through Eq. (2-15). The total coupled power is then determined by summing up the contributions from each individual emitting-point source of incremental area $d\theta_s r \; dr$, that is, integrating over the emitting area. For simplicity, here the emitting surface is taken as being circular. If the source radius r_s is less than the fiber core radius a, then the upper integration limit $r_m = r_s$; for source areas larger than the fiber core area, $r_m = a$.

As an example, assume a surface-emitting LED of radius r_s less than the fiber core radius a. Since this is a lambertian emitter, Eq. (5-1) applies and Eq. (5-3) becomes

$$P = \int_0^{r_s} \int_0^{2\pi} \left(2\pi B_0 \int_0^{\theta_{0,\max}} \cos\theta \sin\theta \; d\theta \right) d\theta_s r \; dr$$

$$= \pi B_0 \int_0^{r_s} \int_0^{2\pi} \sin^2\theta_{0,\max} \; d\theta_s r \; dr$$

$$= \pi B_0 \int_0^{r_s} \int_0^{2\pi} \mathrm{NA}^2 \; d\theta_s r \; dr \qquad (5\text{-}4)$$

where the numerical aperture NA is defined by Eq. (2-15). For step-index fibers the numerical aperture is independent of the positions θ_s and r on the fiber end face, so that Eq. (5-4) becomes (for $r_s < a$),

$$P_{\text{LED,step}} = \pi^2 r_s^2 \; B_0 (\mathrm{NA})^2 \simeq 2\pi^2 r_s^2 \; B_0 n_1^2 \; \Delta \qquad (5\text{-}5)$$

Consider now the total optical power P_s that is emitted from the source of area A_s into a hemisphere (2πsr). This is given by

$$P_s = A_s \int_0^{2\pi} \int_0^{\pi/2} B(\theta,\phi) \sin\theta \, d\theta \, d\phi$$

$$= \pi r_s^2 \, 2\pi B_0 \int_0^{\pi/2} \cos\theta \sin\theta \, d\theta$$

$$= \pi^2 r_s^2 \, B_0 \tag{5-6}$$

Equation (5-5) can, therefore, be expressed in terms of P_s:

$$P_{\text{LED,step}} = P_s \, (\text{NA})^2 \qquad \text{for } r_s \le a \tag{5-7}$$

When the radius of the emitting area is larger than the radius a of the fiber core area, Eq. (5-7) becomes

$$P_{\text{LED,step}} = \left(\frac{a}{r_s}\right)^2 P_s(\text{NA})^2 \qquad \text{for } r_s > a \tag{5-8}$$

In the case of a graded-index fiber, the numerical aperture depends on the distance r from the fiber axis through the relationship defined by Eq. (2-62). Thus using Eqs. (2-62) and (2-63) the power coupled from a surface-emitting LED into a graded-index fiber becomes (for $r_s < a$)

$$P_{\text{LED,graded}} = 2\pi^2 B_0 \int_0^{r_s} [n^2(r) - n_2^2] r \, dr$$

$$= 2\pi^2 r_s^2 \, B_0 n_1^2 \, \Delta \left[1 - \frac{2}{\alpha + 2} \left(\frac{r_s}{a}\right)^\alpha \right]$$

$$= 2P_s n_1^2 \, \Delta \left[1 - \frac{2}{\alpha + 2} \left(\frac{r_s}{a}\right)^\alpha \right] \tag{5-9}$$

where the last expression was obtained from Eq. (5-6).

The foregoing analyses assumed perfect coupling conditions between the source and the fiber. This can only be achieved if the refractive index of the medium separating the source and the fiber end matches the index n_1 of the fiber core. If the refractive index n of this medium is different from n_1, then the power coupled into the fiber reduces by the factor

$$R = \left(\frac{n_1 - n}{n_1 + n}\right)^2 \tag{5-10}$$

where R is the Fresnel reflection coefficient at the fiber core end face.

The calculation of power coupling for nonlambertian emitters following a cylindrical $\cos^m \theta$ distribution is left as an exercise. The power launched into a fiber from an edge-emitting LED having a noncylindrical distribution is rather complex. An example of this has been given by Marcuse[2] to which the reader is referred for details.

5-1-3 Power Launching Versus Wavelength

It is of interest to note that the optical power launched into a fiber does not depend on the wavelength of the source but only on its brightness, that is, its radiance. Let us explore this a little further. We saw in Eq. (2-79) that the number of modes which can propagate in a graded-index fiber of core size a and index profile α is

$$M = \frac{\alpha}{\alpha + 2} \left(\frac{2\pi a n_1}{\lambda} \right)^2 \Delta \qquad (5\text{-}11)$$

Thus, for example, twice as many modes propagate in a given fiber at 900 nm than at 1300 nm.

The radiated power per mode P_s/M from a source at a particular wavelength is given by the radiance multiplied by the square of the nominal source wavelength[3]

$$\frac{P_s}{M} = B_0 \lambda^2 \qquad (5\text{-}12)$$

Thus twice as much power is launched into a given mode at 1300 nm than at 900 nm. Hence two identically sized sources operating at different wavelengths but having identical radiances will launch equal amounts of optical power into the same fiber.

5-1-4 Equilibrium Numerical Aperture

As we noted earlier, a light source is often supplied with a short (0.1- to 1-m) fiber flylead attached to it in order to facilitate coupling the source to a system fiber. To achieve a low coupling loss, this flylead should be connected to a system fiber having a nominally identical NA and core diameter. A certain amount of optical power (ranging from 0.1 to 1 dB) is lost at this junction, the exact loss depending on the connecting mechanism; this is discussed in Sec. 5-3. In addition to the coupling loss, an excess power loss will occur in the first few tens of meters of the system fiber. This excess loss is a result of nonpropagating modes scattering out of the fiber as the launched modes come to an equilibrium condition (see Sec. 3-6). This is of particular importance for surface-emitting LEDs, which tend to launch power into all modes of the fiber. Fiber-coupled lasers are less prone to this effect since they tend to excite fewer nonpropagating fiber modes.

The excess power loss must be analyzed carefully in any system design since it can be significantly higher for some types of fibers than for others.[4,5] An example of the excess power loss is shown in Fig. 5-4 in terms of the fiber numerical aperture. At the input end of the fiber the light acceptance is described in terms of the launch numerical aperture NA_{in}. If the light-emitting area of the LED is less than the cross-sectional area of the fiber core, then at this point the power coupled into the fiber is given by Eq. (5-7), where $NA = NA_{in}$.

However, when the optical power is measured in long fiber lengths after the launched modes have come to equilibrium (which is often taken to be 50 m), the

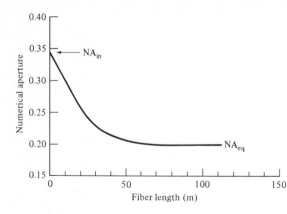

Figure 5-4 Example of the change in numerical aperture as a function of fiber length.

effect of the equilibrium numerical aperture NA_{eq} becomes apparent. At this point the optical power in the fiber scales as

$$P_{eq} = P_{50} \left(\frac{NA_{eq}}{NA_{in}} \right)^2 \tag{5-13}$$

where P_{50} is the power expected in the fiber at the 50-m point based on the launch NA. The degree of mode coupling occurring in a fiber is primarily a function of the core-cladding index difference. It can thus vary significantly among different fiber types. Since most optical fibers attain 80 to 90 percent of their equilibrium NA after about 50 m, it is the value of NA_{eq} that is important when calculating launched optical power in telecommunication systems.

5-2 LENSING SCHEMES FOR COUPLING IMPROVEMENT

The optical power-launching analysis given in Sec. 5-1 is based on centering a flat fiber end face directly over the light source and as close to it as possible. If the source-emitting area is larger than the fiber core area, then the resulting optical power coupled into the fiber is the maximum that can be achieved. This is a result of fundamental energy and radiance conservation principles[6] (also known as the *law of brightness*). However, if the emitting area of the source is smaller than the core area, a miniature lens may be placed between the source and the fiber to improve the power-coupling efficiency.[7-13]

The function of the microlens is to magnify the emitting area of the source to match exactly the core area of the fiber end face. If the emitting area is increased by a magnification factor M, the solid angle within which optical power is coupled to the fiber from the LED is increased by the same factor.

Several possible lensing schemes[1,7-13] are shown in Fig. 5-5. These include a rounded-end fiber, a small glass sphere (nonimaging microsphere) in contact with both the fiber and the source, a larger spherical lens used to image the source on the core area of the fiber end, a cylindrical lens generally formed from a short

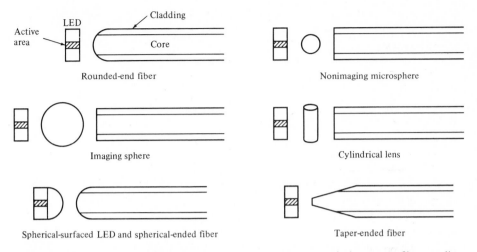

Figure 5-5 Examples of possible lensing schemes used to improve optical source-to-fiber coupling efficiency.

section of fiber, a system consisting of a spherical-surfaced LED and a spherical-ended fiber, and a taper-ended fiber.

Although these techniques can improve the source-to-fiber coupling efficiency, they also create additional complexities. One problem is that the lens size is similar to the source and fiber core dimensions, which introduces fabrication and handling difficulties. In the case of the taper-ended fiber the mechanical alignment must be done with greater precision since the coupling efficiency becomes a more sharply peaked function of the spatial alignment. However, alignment tolerances are increased for other types of lensing systems.

5-2-1 Nonimaging Microsphere

One of the most efficient lensing methods is the use of a nonimaging microsphere.[9-12] Let us first examine its use for a surface emitter, as shown in Fig. 5-6. We first make the following practical assumptions: the spherical lens has a refractive index of about 2.0, the outside medium is air ($n = 1.0$), and the emitting area

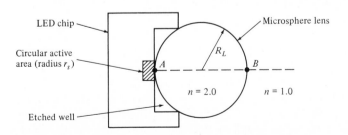

Figure 5-6 Schematic of an LED emitter with a microsphere lens.

is circular. To collimate the output from the LED, the emitting surface should be located at the focal point of the lens. The focal point can be found from the gaussian lens formula[14]

$$\frac{n}{s} + \frac{n'}{q} = \frac{n' - n}{r} \tag{5-14}$$

where s and q are the object and image distances, respectively, as measured from the lens surface, n is the refractive index of the lens, n' is the refractive index of the outside medium, and r is the radius of curvature of the lens surface.

The following sign conventions are used with Eq. (5-14):

1. Light travels from left to right.
2. Object distances are measured as positive to the left of a vertex and negative to the right.
3. Image distances are measured as positive to the right of a vertex and negative to the left.
4. All convex surfaces encountered by the light have a positive radius of curvature, and concave surfaces have a negative radius.

With the use of these conventions we shall now find the focal point for the right-hand surface of the lens shown in Fig. 5-6. To find the focal point, we set $q = \infty$ and solve for s in Eq. (5-14), where s is measured from point B. With $n = 2.0$, $n' = 1.0$, $q = \infty$, and $r = -R_L$, Eq. (5-14) yields

$$s = f = 2R_L$$

Thus the focal point is located on the lens surface at point A. (This, of course, changes if the refractive index of the sphere is not equal to 2.0.)

Placing the LED close to the lens surface thus results in a magnification M of the emitting area. This is given by the ratio of the cross-sectional area of the lens to that of the emitting area:

$$M = \frac{\pi R_L^2}{\pi r_s^2} = \left(\frac{R_L}{r_s}\right)^2 \tag{5-15}$$

Using Eq. (5-4) we can show that, with the lens, the optical power P_L that can be coupled into a full aperture angle 2θ is given by

$$P_L = P_s \left(\frac{R_L}{r_s}\right)^2 \sin^2 \theta \tag{5-16}$$

where P_s is the total output power from the LED without the lens.

The theoretical coupling efficiency that can be achieved is based on energy and radiance conservation principles.[15] This efficiency is usually determined by the size of the fiber. For a fiber of radius a and numerical aperture NA, the maximum

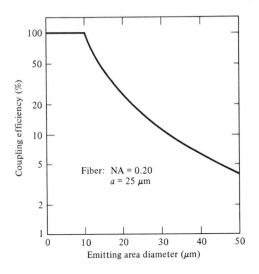

Figure 5-7 Theoretical coupling efficiency for a surface-emitting LED as a function of the emitting diameter. Coupling is to a fiber with NA = 0.20 and radius a = 25 μm.

coupling efficiency η_{max} is given by

$$\eta_{max} = \left(\frac{a}{r_s}\right)^2 (NA)^2 \qquad \text{for } \frac{r_s}{a} > NA$$

$$\eta_{max} = 1 \qquad \text{for } \frac{r_s}{a} \leq NA$$

(5-17)

Thus when the radius of the emitting area is larger than the fiber radius, no improvement in coupling efficiency is possible with a lens. In this case the best coupling efficiency is achieved by a direct-butt method.

Based on Eq. (5-17) the theoretical coupling efficiency as a function of the emitting diameter is shown in Fig. 5-7 for a fiber with a numerical aperture of 0.20 and a 50-μm core diameter.

5-2-2 Laser Diode-to-Fiber Coupling

As we noted in Chap. 4 laser diodes have an emission pattern which nominally has a full-width half-maximum (FWHM) of 30 to 50° in the plane perpendicular to the active-area junction and an FWHM of 5 to 10° in the plane parallel to the junction. Since the angular output distribution of the laser is greater than the fiber acceptance angle and since the laser emitting area is much smaller than the fiber core, spherical or cylindrical lenses can also be used to improve the coupling efficiency between laser diodes and optical fibers.[12, 16, 17]

The use of homogeneous glass microsphere lenses has been tested in a series of several hundred laser diode assemblies by Khoe and Kuyt.[16] Spherical glass lenses with a refractive index of 1.9 and diameters ranging between 50 and 60 μm were epoxied to the ends of 50-μm core-diameter graded-index fibers having a numerical aperture of 0.2. The measured FWHM values of the laser output beams were:

1. Between 3 and 9 μm for the near field parallel to the junction
2. Between 30 and 60° for the field perpendicular to the junction
3. Between 15 and 55° for the field parallel to the junction

Coupling efficiencies in these experiments ranged between 50 and 80 percent.

5-3 FIBER-TO-FIBER JOINTS

A significant factor in any fiber optic system installation is the interconnection of fibers in a low-loss manner. These interconnections occur at the optical source, at the photodetector, at intermediate points within a cable where two fibers are joined, and at intermediate points in a link where two cables are connected. The particular technique selected for joining the fibers depends on whether a permanent bond or an easily demountable connection is desired. A permanent bond is generally referred to as a *splice,* whereas a demountable joint is known as a *connector.*

Every jointing technique is subject to certain conditions which can cause various amounts of optical power loss at the joint. These losses depend on parameters such as the input power distribution to the joint, the length of the fiber between the optical source and the joint, the geometrical and waveguide characteristics of the two fibers being coupled, the various types of mechanical misalignments between the two fiber ends at the joint, and the fiber end face qualities.

The optical power that can be coupled from one fiber to another is limited by the number of modes that can propagate in each fiber. For example if a fiber in which 500 modes can propagate is connected to a fiber in which only 400 modes can propagate, then at most 80 percent of the optical power from the first fiber can be coupled into the second fiber (if we assume that all modes are equally excited). The total number of modes in a fiber can be found from Eq. (2-76). Letting the maximum value of the parameter $\beta = kn_2$, where $k = 2\pi/\lambda$, we have, for a fiber of radius a,

$$M = k^2 \int_0^a [n^2(r) - n_2^2] r \, dr \tag{5-18}$$

where $n(r)$ defines the variation in the refractive-index profile of the core. This can be related to a general local numerical aperture NA(r) through Eq. (2-62) to yield

$$M = k^2 \int_0^a \text{NA}^2(r) r \, dr$$

$$= k^2 \text{NA}^2(0) \int_0^a \left[1 - \left(\frac{r}{a} \right)^\alpha \right] r \, dr \tag{5-19}$$

In general, any two fibers that are to be joined will have varying degrees of differences in their radii a, axial numerical apertures NA(0), and index profiles α.

Thus the fraction of energy coupled from one fiber to another is proportional to the common mode volume M_{comm} (if a uniform distribution of energy over the modes is assumed). The fiber-to-fiber coupling efficiency η_F is given by

$$\eta_F = \frac{M_{comm}}{M_E} \qquad (5\text{-}20)$$

where M_E is the number of modes in the *emitting fiber* (the fiber which launches power into the next fiber).

The fiber-to-fiber coupling loss L_F is given in terms of η_F as

$$L_F = -10 \log \eta_F \qquad (5\text{-}21)$$

An analytical estimate of the optical power loss at a joint between multimode fibers is difficult to make, since the loss depends on the power distribution among the modes in the fiber.[18] For example, consider first the case where all modes in a fiber are equally excited, as shown in Fig. 5-8a. The emerging optical beam thus fills the entire exit numerical aperture of this emitting fiber. Suppose now that a second identical fiber, which we shall call the *receiving fiber*, is to be joined to the emitting fiber. For the receiving fiber to accept all the optical power emitted by the first fiber, there must be perfect mechanical alignment between the two optical waveguides, and their geometric and waveguide characteristics must match precisely.

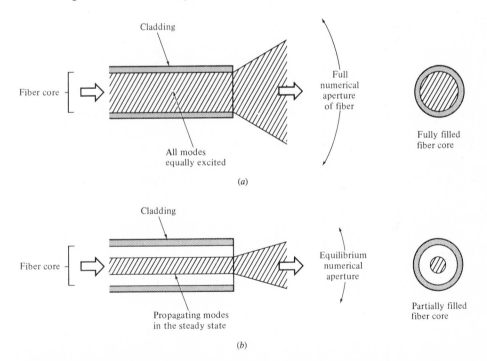

Figure 5-8 Different modal distributions of the optical beam emerging from a fiber lead to different degrees of coupling loss. (*a*) When all modes are equally excited, the output beam fills the entire output NA; (*b*) for a steady-state modal distribution only the equilibrium NA is filled by the output beam.

On the other hand, if steady-state modal equilibrium has been established in the emitting fiber, most of the energy is concentrated in the lower-order fiber modes. This means that the optical power is concentrated near the center of the fiber core, as shown in Fig. 5-8b. The optical power emerging from the fiber then fills only the equilibrium numerical aperture (see Fig. 5-4). In this case since the input NA of the receiving fiber is larger than the equilibrium NA of the emitting fiber, slight mechanical misalignments of the two joined fibers and small variations in their geometric characteristics do not contribute significantly to joint loss.

Steady-state modal equilibrium is generally established in long fiber lengths (see Chap. 3). Thus when estimating joint losses between long fibers, calculations based on a uniform modal power distribution tend to lead to results which may be too pessimistic. However, if a steady-state equilibrium modal distribution is assumed, the estimate may be too optimistic, since mechanical misalignments and variations in fiber-to-fiber characteristics cause a redistribution of power among the modes in the second fiber. As the power propagates along the second fiber, an additional loss will thus occur when a steady-state distribution is again established.

An exact calculation of coupling loss between different optical fibers, which takes into account nonuniform distribution of power among the modes and propagation effects in the second fiber, is lengthy and involved.[19] Here we shall, therefore, make the assumption that all modes in the fiber are equally excited. Although this gives a somewhat pessimistic prediction of joint loss, it will allow an estimate of the relative effects of losses resulting from mechanical misalignments, geometrical mismatches, and variations in the waveguide properties between two joined fibers.

5-3-1 Mechanical Misalignment

Mechanical alignment is a major problem when joining two fibers because of their microscopic size.[20-24] A standard graded-index fiber core is 50 μm in diameter, which is roughly the thickness of a human hair. Radiation losses result from mechanical misalignments because the radiation cone of the emitting fiber does not match the acceptance cone of the receiving fiber. The magnitude of the radiation loss depends on the degree of misalignment. The three fundamental types of misalignments between fibers are shown in Fig. 5-9.

Longitudinal separation occurs when the fibers have the same axis but have a gap s between their end faces. *Angular misalignment* results when the two axes form an angle so that the fiber end faces are no longer parallel. *Axial displacement*

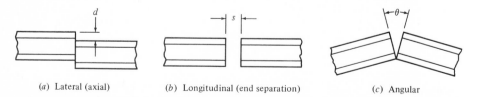

(a) Lateral (axial) (b) Longitudinal (end separation) (c) Angular

Figure 5-9 Three types of mechanical misalignments that can occur between two joined fibers.

(which is also often called *lateral displacement*) results when the axes of the two fibers are separated by a distance d.

The most common misalignment occurring in practice, which also causes the greatest power loss, is axial displacement. This axial offset reduces the overlap area of the two fiber core end faces, as illustrated in Fig. 5-10, and consequently reduces the amount of optical power that can be coupled from one fiber into the other.

To illustrate the effects of axial misalignment, let us first consider the simple case of two identical step-index fibers of radii a. Suppose that their axes are offset by a separation d at their common junction, as is shown in Fig. 5-10, and assume there is a uniform modal power distribution in the emitting fiber. Since the numerical aperture is constant across the end faces of the two fibers, the optical power coupled from one fiber to another is simply proportional to the common area A_{comm} of the two fiber cores. It is straightforward to show that this is (see Prob. 5-9)

$$A_{comm} = 2a^2 \arccos \frac{d}{2a} - d\left(a^2 - \frac{d^2}{4}\right)^{1/2} \tag{5-22}$$

For the step-index fiber the coupling efficiency is simply the ratio of the common core area to the core end face area

$$\eta_{F,step} = \frac{A_{comm}}{\pi a^2} = \frac{2}{\pi} \arccos \frac{d}{2a} - \frac{d}{\pi a}\left[1 - \left(\frac{d}{2a}\right)^2\right]^{1/2} \tag{5-23}$$

The calculation of power coupled from one graded-index fiber into another identical one is more complex since the numerical aperture varies across the fiber end face. Because of this, the total power coupled into the receiving fiber at a given point in the common core area is limited by the numerical aperture of the transmitting or receiving fiber, depending on which is smaller at that point.

If the end face of a graded-index fiber is uniformly illuminated, the optical power accepted by the core will be that power that falls within the numerical aperture of the fiber. The optical power density $p(r)$ at a point r on the fiber end is proportional to the square of the local numerical aperture $NA(r)$ at that point:[25]

$$p(r) = p(0) \frac{NA^2(r)}{NA^2(0)} \tag{5-24}$$

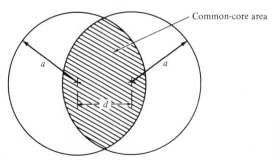
Common-core area

Figure 5-10 Axial offset reduces the common core area of the two fiber end faces.

where NA(r) and NA(0) are defined by Eqs. (2-62) and (2-63), respectively. The parameter $p(0)$ is the power density at the core axis which is related to the total power P in the fiber by

$$P = \int_0^{2\pi} \int_0^a p(r)r \, dr \, d\theta \qquad (5\text{-}25)$$

For an arbitrary index profile the double integral in Eq. (5-25) must be evaluated numerically. However, an analytic expression can be found by using a fiber with a parabolic index profile ($\alpha = 2.0$). Using Eq. (2-62) the power density expression at a point r given by Eq. (5-24) becomes

$$p(r) = p(0)\left[1 - \left(\frac{r}{a}\right)^2\right] \qquad (5\text{-}26)$$

Using Eqs. (5-25) and (5-26) the relationship between the axial power density $p(0)$ and the total power P in the emitting fiber is

$$P = \frac{\pi a^2}{2}p(0) \qquad (5\text{-}27)$$

Let us now calculate the power transmitted across the butt joint of the two parabolic graded-index fibers with an axial offset d, as shown in Fig. 5-11. The overlap region must be considered separately for the areas A_1 and A_2. In area A_1 the numerical aperture is limited by that of the emitting fiber, whereas in area A_2 the numerical aperture of the receiving fiber is smaller than that of the emitting fiber. The vertical dashed line separating the two areas is the locus of points where the numerical apertures are equal.

To determine the power coupled into the receiving fiber, the power density given by Eq. (5-26) is integrated separately over areas A_1 and A_2. Since the numerical aperture of the emitting fiber is smaller than that of the receiving fiber in area A_1, all of the power emitted in this region will be accepted by the receiving

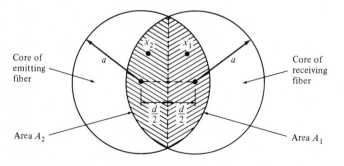

Figure 5-11 Core overlap region for two identical parabolic graded-index fibers with an axial separation d. Points x_1 and x_2 are arbitrary points of symmetry in areas A_1 and A_2.

fiber. The received power P_1 in area A_1 is, thus,

$$P_1 = 2 \int_0^{\theta_1} \int_{r_1}^a p(r) r \, dr \, d\theta$$

$$= 2p(0) \int_0^{\theta_1} \int_{r_1}^a \left[1 - \left(\frac{r}{a} \right)^2 \right] r \, dr \, d\theta \qquad (5\text{-}28)$$

where the limits of integration shown in Fig. 5-12 are

$$r_1 = \frac{d}{2 \cos \theta}$$

and

$$\theta_1 = \arccos \frac{d}{2a}$$

Carrying out the integration yields

$$P_1 = \frac{a^2}{2} p(0) \left\{ \arccos \frac{d}{2a} - \left[1 - \left(\frac{d}{2a} \right)^2 \right]^{1/2} \frac{d}{6a} \left(5 - \frac{d^2}{2a^2} \right) \right\} \qquad (5\text{-}29)$$

where $p(0)$ is given by Eq. (5-27). The derivation of Eq. (5-29) is left as an exercise.

In area A_2 the emitting fiber has a larger numerical aperture than the receiving fiber. This means that the receiving fiber will accept only that fraction of the emitted optical power that falls within its own numerical aperture. This power can be found easily from symmetry considerations.[26] The numerical aperture of the receiving fiber at a point x_2 in area A_2 is the same as the numerical aperture of the emitting fiber at the symmetrical point x_1 in area A_1. Thus the optical power accepted by the receiving fiber at any point x_2 in area A_2 is equal to that emitted from the symmetrical point x_1 in area A_1. The total power P_2 coupled across area A_2 is thus equal to the power P_1 coupled across area A_1. Combining these results, we have that the

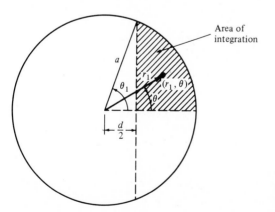

Area of integration

Figure 5-12 Area and limits of integration for the common core area of two parabolic graded-index fibers.

total power P_T accepted by the receiving fiber is

$$P_T = 2P_1$$

$$= \frac{2}{\pi} P \left\{ \arccos \frac{d}{2a} - \left[1 - \left(\frac{d}{2a} \right)^2 \right]^{1/2} \frac{d}{6a} \left(5 - \frac{d^2}{2a^2} \right) \right\} \qquad (5\text{-}30)$$

When the axial misalignment d is small compared to the core radius a, Eq. (5-30) can be approximated by[26]

$$P_T \simeq P \left(1 - \frac{8d}{3\pi a} \right) \qquad (5\text{-}31)$$

This is accurate to within 1 percent for $d/a < 0.4$. The coupling loss for the offsets given by Eqs. (5-30) and (5-31) is

$$L_F = -10 \log \eta_F = -10 \log \frac{P_T}{P} \qquad (5\text{-}32)$$

The effect of separating the two fiber ends longitudinally by a gap s is shown in Fig. 5-13. All the higher-mode optical power emitted in the ring of width x will not be intercepted by the receiving fiber. It is straightforward to show that, for a step-index fiber, the loss occurring in this case is

$$L_F = -10 \log \left(\frac{a}{a + s \tan \theta_c} \right)^2 \qquad (5\text{-}33)$$

where θ_c is the critical acceptance angle of the fiber.

When the axes of two joined fibers are angularly misaligned at the joint, the optical power that leaves the emitting fiber outside of the solid acceptance angle of the receiving fiber will be lost. For two step-index fibers having an angular misalignment θ, the optical power loss at the joint has been shown to be[27, 28]

$$L_F = -10 \log \left(\cos \theta \left\{ \frac{1}{2} - \frac{1}{\pi} p (1 - p^2)^{1/2} - \frac{1}{\pi} \arcsin p \right. \right.$$

$$\left. \left. - q \left[\frac{1}{\pi} y (1 - y^2)^{1/2} + \frac{1}{\pi} \arcsin y + \frac{1}{2} \right] \right\} \right) \qquad (5\text{-}34)$$

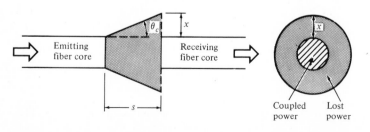

Figure 5-13 Loss effect when the fiber ends are separated longitudinally by a gap s.

where

$$p = \frac{\cos \theta_c (1 - \cos \theta)}{\sin \theta_c \sin \theta}$$

$$q = \frac{\cos^3 \theta_c}{(\cos^2 \theta_c - \sin^2 \theta)^{3/2}}$$

$$y = \frac{\cos^2 \theta_c (1 - \cos \theta) - \sin^2 \theta}{\sin \theta_c \cos \theta_c \sin \theta}$$

Here n is the refractive index of the material between the fibers ($n = 1.0$ for air). The derivation of Eq. (5-34) again assumes that all modes are uniformly excited.

An experimental comparison[22] of the losses induced by the three types of mechanical misalignments is shown in Fig. 5-14. The measurements were based on two independent experiments using LED sources and graded-index fibers. The core diameters were 50 and 55 μm for the first and second experiments, respectively. A 1.83-m-long fiber was used in the first test and a 20-m length in the second. In either case the power output from the fibers was first optimized. The fibers were then cut at the center, so that the mechanical misalignment loss measurements were done on identical fibers. The axial offset and longitudinal separation losses are plotted as functions of misalignments normalized to the core radius. A normalized angular misalignment of 0.1 corresponds to a 1° angular offset.

Figure 5-14 shows that of the three mechanical misalignments the dominant loss arises from lateral displacement. In practice, angular misalignments of less

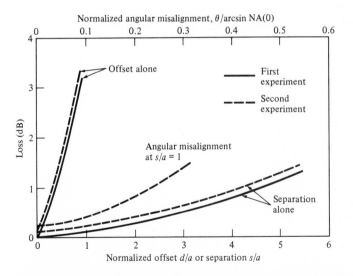

Figure 5-14 Experimental comparison of loss (in dB) as a function of mechanical misalignments. *(Reproduced with permission from Chu and McCormick,*[22] *copyright 1978, American Telephone and Telegraph Company.)*

than 1° are readily achievable in splices and connectors. From the experimental data shown in Fig. 5-14, these misalignments result in losses of less than 0.5 dB.

For splices the separation losses are normally negligible since the fibers should be in relatively close contact. In most connectors the fiber ends are intentionally separated by a small gap. This prevents them from rubbing against each other and becoming damaged during connector engagement. Typical gaps in these applications range from 0.025 to 0.10 mm, which results in losses of less than 0.8 dB for a 50-μm-diameter fiber.

5-3-2 Fiber-Related Losses

In addition to mechanical misalignments, differences in the geometrical and waveguide characteristics of any two waveguides being joined can have a profound effect on fiber-to-fiber coupling loss. These include variations in core diameter, core area ellipticity, numerical aperture, refractive-index profile, and core-cladding concentricity of each fiber. Since these are manufacturer-related variations, the user generally has little control over them. Theoretical and experimental studies[19, 29–31] of the effects of these variations have shown that, for a given percentage mismatch, differences in core radii and numerical apertures have a significantly larger effect on joint loss than mismatches in the refractive-index profile or core ellipticity.

The joint losses resulting from core diameter, numerical aperture, and core refractive-index-profile mismatches can easily be found from Eqs. (5-19) and (5-20). For simplicity, let the subscripts E and R refer to the emitting and receiving fibers, respectively. If the radii a_E and a_R are not equal but the axial numerical apertures and the index profiles are equal [$NA_E(0) = NA_R(0)$ and $\alpha_E = \alpha_R$], then the coupling loss is

$$L_F(a) = -10 \log \left(\frac{a_R}{a_E}\right)^2 \qquad \text{for } a_R < a_E$$

$$L_F(a) = 0 \qquad \text{for } a_R \geq a_E$$

(5-35)

If the radii and the index profiles of the two coupled fibers are identical but their axial numerical apertures are different, then

$$L_F(NA) = -10 \log \left[\frac{NA_R(0)}{NA_E(0)}\right]^2 \qquad \text{for } NA_R(0) < NA_E(0)$$

$$L_F(NA) = 0 \qquad \text{for } NA_R(0) \geq NA_E(0)$$

(5-36)

Finally if the radii and the axial numerical apertures are the same but the core refractive-index profiles differ in two joined fibers, then the coupling loss is

$$L_F(\alpha) = -10 \log \frac{\alpha_R(\alpha_E + 2)}{\alpha_E(\alpha_R + 2)} \qquad \text{for } \alpha_R < \alpha_E$$

$$L_F(\alpha) = 0 \qquad \text{for } \alpha_R \geq \alpha_E$$

(5-37)

This results because for $\alpha_R < \alpha_E$ the number of modes that can be supported by the receiving fiber is less than the number of modes in the emitting fiber. If $\alpha_R > \alpha_E$, then all modes in the emitting fiber can be captured by the receiving fiber. The derivations of Eqs. (5-35) to (5-37) are left as an exercise (see Probs. 5-14 through 5-16).

5-3-3 Fiber End Face Preparation

One of the first steps that must be followed before fibers are connected or spliced to each other is to properly prepare the fiber end faces. In order not to have light deflected or scattered at the joint, the fiber ends must be flat, perpendicular to the fiber axis, and smooth. End-preparation techniques that have been extensively used include sawing, grinding and polishing, and controlled fracture.[32–34]

Conventional grinding and polishing techniques can produce a very smooth surface that is perpendicular to the fiber axis. However, this method is quite time consuming and requires a fair amount of operator skill. Although it is often implemented in a controlled environment such as a laboratory or a factory, it is not readily adaptable for field use. The procedure employed in the grinding and polishing technique is to use successively finer abrasives to polish the fiber end face. The end face is polished with each successive abrasive until the scratches created by the previous abrasive material are replaced by the finer scratches of the present abrasive. The number of abrasives used depends on the degree of smoothness that is desired.

Controlled-fracture techniques are based on score-and-break methods for cleaving fibers. In this operation the fiber to be cleaved is first scratched to create a stress concentration at the surface. The fiber is then bent over a curved form while tension is simultaneously applied, as shown in Fig. 5-15. This action produces a stress distribution across the fiber. The maximum stress occurs at the scratch point so that a crack starts to propagate through the fiber.

A highly smooth and perpendicular end face can be produced this way. However, this method requires careful control of the curvature of the fiber and of the amount of tension applied. If the stress distribution across the fiber is not properly controlled, the fracture propagating across the fiber can fork into several cracks. This forking produces either a lip or a hackled portion on the fiber end, as shown

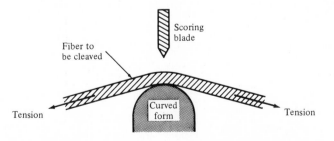

Figure 5-15 Controlled-fracture procedure for fiber end preparation.

Figure 5-16 Improperly cleaved fiber end faces.

in Fig. 5-16. A number of different tools based on the controlled-fracture technique have been developed[34–39] and are being used both in the field and factory environments.

5-4 SPLICING TECHNIQUES

Many different fiber-splicing techniques have arisen during the evolution of optical fiber technology. Three popular ones are the fusion splice,[40–43] the V-groove butt splice,[44–46] and the elastic-tube splice.[47]

Fusion splices are made by thermally bonding together prepared fiber ends, as pictured in Fig. 5-17. In this method the fiber ends are first prealigned and butted together. This is done either in a grooved fiber holder or under a microscope with micromanipulators. The butt joint is then heated with an electric arc or a laser pulse so that the fiber ends are momentarily melted and, hence, bonded together. This technique can produce very low splice losses (0.1 to 0.2 dB for identical fibers). However, care must be exercised in this technique, since surface damage due to handling, surface defect growth created during heating, and residual stresses induced near the joint as a result of changes in chemical composition arising from the material melting can produce a weak splice.

In the V-groove splice technique the prepared fiber ends are first butted together in a V-shaped groove, as shown in Fig. 5-18. They are then bonded together with an adhesive or are held in place by means of a cover plate. The V-shaped channel could be either a grooved silicon, plastic, ceramic, or metal substrate. The splice loss in this method depends strongly on the fiber size (outside dimensions and core diameter variations) and eccentricity (the position of the core relative to the center of the fiber).

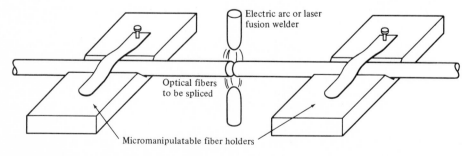

Figure 5-17 Fusion splicing of optical fibers.

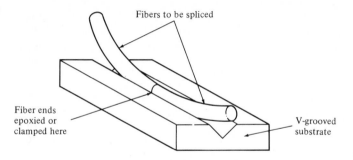

Figure 5-18 V-groove optical fiber-splicing technique.

The elastic-tube splice shown cross-sectionally in Fig. 5-19 is a unique device that automatically performs lateral, longitudinal, and angular alignment.[47] It splices multimode fibers with losses in the same range as commercial fusion splices, but much less equipment and skill are needed. The splice mechanism is basically a tube made of an elastic material. The central hole diameter is slightly smaller than that of the fiber to be spliced and is tapered on each end for easy fiber insertion. When a fiber is inserted it expands the hole diameter so that the elastic material exerts a symmetrical force on the fiber. This symmetry feature allows an accurate and automatic alignment of the axes of the two joined fibers. A wide range of fiber diameters can be inserted into the elastic tube. Thus the fibers to be spliced do not have to be equal in diameter since each fiber moves into position independently relative to the tube axis. This feature is a great advantage over the V-groove splice in which fibers of unequal diameter are automatically misaligned.

5-5 OPTICAL FIBER CONNECTORS

A wide variety of optical fiber connectors based on different principles of operation has evolved during the exploratory years of fiber optics. Among these concepts are:

1. Watch jewel ferrule connectors[48]
2. Groove- or channel-based connectors[27, 45, 49–51]
3. Concentric sleeve connectors[48, 52]

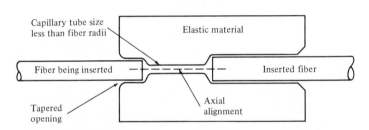

Figure 5-19 Elastic-tube splice.

4. Molded connectors[53, 54]
5. Expanded beam (lensed) connectors[55–57]

Some of the principal goals of a connector design are to have:

1. Low coupling losses even after numerous connects and disconnects
2. Interchangeability with connectors of the same type
3. Ease of connection
4. Simple and low-cost construction
5. Reliability of connection
6. Low sensitivity to environmental conditions such as temperature, dust, and moisture

Losses of less than 1 dB between matched 50-μm core-diameter graded-index fibers can be achieved with the molded connector and the expanded beam connector.

5-6 SUMMARY

In this chapter we have addressed the problem of launching optical power from a light source into a fiber and the factors involved in coupling light from one fiber into another. The coupling of optical power from a luminescent source into a fiber is influenced by the following:

1. The numerical aperture of the fiber, which defines the light acceptance cone of the fiber.
2. The cross-sectional area of the fiber core compared to the source-emitting area. If the emitting area is smaller than the fiber core, then lensing schemes can be used to improve the coupling efficiency.
3. The radiance (or brightness) of the light source, which is the optical power radiated into a solid angle per unit emitting area (measured in watts per square centimeter per steradian).
4. The spatial radiation pattern of the source; the incompatibility between the wide beam divergence of LEDs and the narrow acceptance cone of the fiber is a major contributor to coupling loss. This holds to a lesser extent for laser diodes.

In practice, many suppliers offer optical sources which have a short length of optical fiber (1 m or less) already attached in an optimum power coupling configuration. This fiber, which is referred to as a flylead or a pigtail, makes it easier for the user to couple the source to a system fiber. The power-launching problem now becomes a simpler one of coupling optical power from one fiber into another. To achieve a low coupling loss, the fiber flylead should be connected to a system fiber having a nominally identical numerical aperture and core diameter.

Fiber-to-fiber joints can exist between the source flylead and the system fiber, at the photodetector, at intermediate points within a cable where two fibers are joined, and at intermediate points in a link where two cable sections are connected. The two principal types of joints are splices, which are permanent bonds between the two fibers, and connectors which are used when an easily demountable connection between two fibers is desired.

Each jointing technique is subject to certain conditions which can cause varying degrees of optical power loss at the joint. These losses depend on parameters such as:

1. *The geometrical characteristics of the fibers.* For example, optical power will be lost because of area mismatches if an emitting fiber has a larger core diameter than the receiving fiber.
2. *The waveguide characteristics of the fibers.* For example, if an emitting fiber has a larger numerical aperture than the receiving fiber, all optical power falling outside of the acceptance cone of the receiving fiber is lost.
3. *The various mechanical misalignments between the two fiber ends at the joint.* These include longitudinal separation, angular misalignment, and axial (or lateral) displacement. The most common misalignment occurring in practice, which also causes the greatest power loss, is axial misalignment.
4. *The input power distribution to the joint.* If all the modes of an emitting fiber are equally excited, there must be perfect mechanical alignment between the two optical waveguides, and their geometric and waveguide characteristics must match precisely in order for no optical power loss to occur at the joint. On the other hand, if steady-state modal equilibrium has been established in the emitting fiber (which happens in long fiber lengths), most of the energy is concentrated in the lower-order fiber modes. In this case, slight mechanical misalignments of the two joined fibers and small variations in their geometric and waveguide characteristics do not contribute significantly to joint loss.
5. *The fiber end face quality.* One criterion for low-loss joints is that the fiber end faces be clean, smooth, and perpendicular to the fiber axis. End preparation techniques include sawing, grinding and polishing, and controlled fracture.

PROBLEMS

Section 5-1

5-1 Use polar graph paper to compare the emission patterns from a lambertian source and a source with an emission pattern given by $B(\theta) = B_0 \cos^3 \theta$. Assume B_0 is normalized to unity in both cases.

5-2 A laser diode has lateral ($\phi = 0°$) and transverse ($\phi = 90°$) half-power beam widths of $2\theta = 60$ and $30°$, respectively. What are the transverse and lateral power distribution coefficients for this device?

5-3 An LED with a circular emitting area of radius 20 μm has a lambertian emission pattern with a 100 W/(cm^2 · sr) axial radiance at 100-mA drive current. How much optical power can be coupled into a step-index fiber having a 100-μm core diameter and NA = 0.22? How much optical power can be coupled from this source into a 50-μm core-diameter graded-index fiber having $\alpha = 2.0$, $n_1 = 1.48$, and $\Delta = 0.01$?

5-4 Derive the right-hand side of Eq. (5-9).

5-5 When the refractive index of the medium between an optical source and a fiber end is different from that of the fiber core, part of the optical energy incident on the fiber end face is lost through Fresnel reflection. What is this loss in percent and in decibels when the intervening medium is air ($n = 1.0$) and the fiber core has a refractive index $n_1 = 1.48$?

5-6 Use Eq. (5-3) to derive an expression for the power coupled into a step-index fiber from an LED having a radiant distribution given by

$$B(\theta) = B_0 \cos^m \theta$$

Section 5-2

5-7 Use Eq. (5-4) to verify Eq. (5-16).

5-8 On the same graph plot the maximum coupling efficiencies as a function of the source radius r_s for the following fibers:
   (a) core radius of 25 μm and NA = 0.16
   (b) core radius of 50 μm and NA = 0.20
Let r_s range from 0 to 50 μm. In what regions can a lens improve the coupling efficiency?

Section 5-3

5-9 Verify that Eq. (5-22) gives the common core area of the two axially misaligned step-index fibers shown in Fig. 5-10. If $d = 0.1a$, what is the coupling efficiency in decibels?

5-10 Derive Eq. (5-27) by using Eqs. (5-25) and (5-26).

5-11 Derive Eq. (5-29) for the received power in area A_1 of Fig. 5-11.

5-12 Show that, when the axial misalignment d is small compared to the core radius a, Eq. (5-30) can be approximated by Eq. (5-31). Plot Eqs. (5-30) and (5-31) in terms of P_T/P as a function of d/a for $0 \le d/a \le 0.4$.

5-13 Verify that Eq. (5-33) gives the coupling loss between two step-index fibers separated by a gap s, as shown in Fig. 5-13.

5-14 Using Eqs. (5-19) and (5-20) show that Eq. (5-35) gives the coupling loss for two fibers with unequal core radii. Make a plot of the coupling loss as a function of a_R/a_E for $0.5 \le a_R/a_E \le 1.0$.

5-15 Using Eqs. (5-19) and (5-20) show that Eq. (5-36) gives the coupling loss for two fibers with unequal axial numerical apertures. Plot this coupling loss as a function of $NA_R(0)/NA_E(0)$ for $0.5 \le NA_R(0)/NA_E(0) \le 1.0$.

5-16 Show that Eq. (5-37) gives the coupling loss for two fibers having different core refractive-index profiles. Plot this coupling loss as a function of α_R/α_E for the range $0.75 \le \alpha_R/\alpha_E \le 1.0$. Take $\alpha_E = 2.0$.

REFERENCES

1. (a) Y. Uematsu, T. Ozeki, and Y. Unno, "Efficient power coupling between an MH LED and a taper-ended multimode fiber," *IEEE J. Quantum Electron.*, **QE-15**, 86–92, Feb. 1979.
   (b) H. Kuwahara, M. Sasaki, and N. Tokoyo, "Efficient coupling from semiconductor lasers into single-mode fibers with tapered hemispherical ends," *Appl. Opt.*, **19**, 2578–2583, Aug. 1980.
2. D. Marcuse, "Excitation of parabolic-index fibers with incoherent sources," *Bell Sys. Tech. J.*, **54**, 1507-1530, Nov. 1975; "LED fundamentals: comparison of front and edge-emitting diodes," *IEEE J. Quantum Electron.*, **QE-13**, 819–827, Oct. 1977.
3. A. Yariv, *Quantum Electronics*, Wiley, New York, 1967.
4. S. Zemon and D. Fellows, "Characterization of the approach to steady state and the steady state

properties of multimode optical fibers using LED excitation," *Optics Commun.*, **13**, 198–202, Feb. 1975.

5. R. B. Lauer and J. Schlafer, "LEDs or DLs: Which light source shines brightest in fiber optic telecomm systems?," *Electron. Design*, **8**, 131–135, Apr. 12, 1980.

6. M. Born and E. Wolf, *Principles of Optics*, Pergamon, Oxford, 1965, p. 189.

7. K. H. Yang and J. D. Kingsley, "Calculation of coupling losses between light emitting diodes and low loss optical fibers," *Appl. Opt.*, **14**, 288–293, Feb. 1975.

8. B. S. Kawasaki and D. C. Johnson, "Bulb-ended fiber coupling to LED sources," *Opt. Quantum Electron.*, **7**, 281–288, 1975.

9. J. G. Ackenhusen, "Microlenses to improve LED-to-optical fiber coupling and alignment tolerance," *Appl. Opt.*, **18**, 3694–3699, Nov. 1979.

10. S. Horiuchi, K. Ikeda, T. Tanaka, and W. Susaki, "A new LED structure with a self-aligned sphere lens for efficient coupling to optical fibers," *IEEE Trans. Electron. Devices*, **ED-24**, 986–990, July 1977.

11. R. A. Abram, R. W. Allen, and R. C. Goodfellow, "The coupling of light-emitting diodes to optical fibers using sphere lenses," *J. Appl. Phys.*, **46**, 3468–3474, Aug. 1975.

12. M. L. Dakss and B. Kim, "Effects of fiber propagation loss on diode laser-fiber coupling," *Electron. Lett.*, **16**, 421–422, May 1980; "Simple self-centering technique for mounting microsphere coupling lens on a fiber," *ibid.*, **16**, 463–464, June 1980.

13. O. Hasegawa, R. Namazu, M. Abe, and Y. Toyoma, "Coupling of spherical-surfaced LED and spherical-ended fiber," *J. Appl. Phys.*, **51**, 30–36, Jan. 1980.

14. F. A. Jenkins and H. E. White, *Fundamentals of Optics*, McGraw-Hill, New York, 1976.

15. M. C. Hudson, "Calculation of the maximum optical coupling efficiency into multimode optical waveguides," *Appl. Opt.*, **13**, 1029–1033, May 1974.

16. G. K. Khoe and G. Kuyt, "Realistic efficiency of coupling light from GaAs laser diodes into parabolic-index optical fibers," *Electron. Lett.*, **14**, 667–669, Sept. 28, 1978.

17. L. d'Auria, Y. Combemale, and C. Moronvalle, "High index microlenses for GaA1As laser-fibre coupling," *Electron. Lett.*, **16**, 322–324, Apr. 24, 1980.

18. (*a*) R. B. Kummer, "Lightguide splice loss—effects of launch beam numerical aperture," *Bell Sys. Tech. J.*, **59**, 441–447, Mar. 1980.
    (*b*) A. H. Cherin and P. J. Rich, "Measurement of loss and output numerical aperture of optical fiber splices," *Appl. Opt.*, **17**, 642–645, Feb. 1978.
    (*c*) Y. Daido, E. Miyauchi, and T. Iwama, "Measuring fiber connection loss using steady-state power distribution: a method," *ibid.*, **20**, 451–456, Feb. 1981.

19. P. DiVita and U. Rossi, "Realistic evaluation of coupling loss between different optical fibers," *J. Opt. Commun.*, **1**, 26–32, Sept. 1980; "Evaluation of splice losses induced by mismatch in fibre parameters," *Opt. Quantum Electron.*, **13**, 91–94, Jan. 1981.

20. M. J. Adams, D. N. Payne, and F. M. E. Sladen, "Splicing tolerances in graded index fibers," *Appl. Phys. Lett.*, **28**, 524–526, May 1976.

21. D. Gloge, "Offset and tilt loss in optical fiber splices," *Bell Sys. Tech. J.*, **55**, 905–916, Sept. 1976.

22. T. C. Chu and A. R. McCormick, "Measurement of loss due to offset, end separation, and angular misalignment in graded index fibers excited by an incoherent source," *Bell Sys. Tech. J.*, **57**, 595–602, Mar. 1978.

23. P. DiVita and U. Rossi, "Theory of power coupling between multimode optical fibers," *Opt. Quantum Electron.*, **10**, 107–117, Jan. 1978.

24. C. M. Miller, "Transmission vs. transverse offset for parabolic-profile fiber splices with unequal core diameters," *Bell Sys. Tech. J.*, **55**, 917–927, Sept. 1976.

25. D. Gloge and E. A. J. Marcatili, "Multimode theory of graded-core fibers," *Bell Sys. Tech. J.*, **52**, 1563–1578, Nov. 1973.

26. H. G. Unger, *Planar Optical Waveguides and Fibres*, Clarendon, Oxford, 1977.

27. F. L. Thiel and R. M. Hawk, "Optical waveguide cable connection," *Appl. Opt.*, **15**, 2785–2791, Nov. 1976.

28. F. L. Thiel and D. H. Davis, "Contributions of optical-waveguide manufacturing variations to joint loss," *Electron. Lett.*, **12**, 340–341, June 1976.

29. S. C. Mettler, "A general characterization of splice loss for multimode optical fibers," *Bell Sys. Tech. J.*, **58**, 2163–2182, Dec. 1979.

30. S. E. Miller and A. G. Chynoweth, *Optical Fiber Telecommunications*, Academic, New York, 1979, chap. 3.

31. D. J. Bond and P. Hensel, "The effects of joint losses of tolerances in some geometrical parameters of optical fibres," *Opt. Quantum Electron.*, **13**, 11–18, Jan. 1981.

32. D. Gloge, P. W. Smith, D. L. Bisbee, and E. L. Chinnock, "Optical fiber end preparation for low-loss splices," *Bell Sys. Tech. J.*, **52**, 1579–1588, Nov. 1973.

33. Ref. 30, chap. 14.

34. G. D. Khoe, G. Kuyt, and J. A. Luijendijk, "Optical fiber end preparation: a new method for producing perpendicular fractures in glass fibers, coated-glass fibers, and plastic-clad fibers," *Appl. Opt.*, **20**, 707–714, Feb. 1981.

35. J. E. Fulenwider and M. L. Dakss, "Hand-held tool for optical fibre end preparation," *Electron. Lett.*, **13**, 578–580, Sept. 1977.

36. D. Gloge, P. W. Smith, and E. L. Chinnock, "Apparatus for breaking brittle rods or fibers," U.S. Patent 4,027,817, 1977.

37. A. Albanese and L. Maggi, "New fiber breaking tool," *Appl. Opt.*, **16**, 2604–2605, Oct. 1977.

38. C. Belmonte, M. L. Dakss, and J. E. Fulenwider, "Hand-held tool for optical fiber waveguide end preparation," U.S. Patent 4,159,793, July 1979.

39. P. C. Hensel, "Simplified optical-fibre breaking machine," *Electron. Lett.*, **11**, 581–582, Nov. 1975; "Spark-induced fracture of optical fibres," *ibid.*, **13**, 603–604, Sept. 1977.

40. I. Hatakeyama and H. Tsuchiya, "Fusion splices for single mode optical fibers," *IEEE J. Quantum Electron.*, **QE-14**, 614–619, Aug. 1978.

41. D. L. Bisbee, "Splicing silica fibers with an electric arc," *Appl. Opt.*, **15**, 796–798, Mar. 1976.

42. M. Hirai and N. Uchida, "Melt splice of multimode optical fiber with an electric arc," *Electron. Lett.*, **13**, 123–125, Mar. 1977.

43. J. T. Krause, C. R. Kurkjian, and U. C. Paek, "Strength of fusion splices for fiber lightguides," *Electron. Lett.*, **17**, 232–233, Mar. 1981.

44. C. M. Miller, "Losse tube splices for optical fibers," *Bell Sys. Tech. J.*, **54**, 1215–1225, Sept. 1975.

45. C. M. Miller, "Fiber-optic array splicing with etched silicon chips," *Bell Sys. Tech. J.*, **57**, 75–90, Jan. 1978.

46. E. E. Basch, R. A. Beaudette, and H. A. Carnes, "Optical transmission for interoffice trunks," *IEEE Trans. Commun.*, **COM-26**, 1007–1014, July 1978.

47. (*a*) M. L. Dakss, W. J. Carlsen, and J. E. Benasutti, "Field installable connectors and splices for glass optical fiber communication systems," *12th Annual Connector Symp.*, 17–18, Oct. 1979, Cherry Hill, N.J.
    (*b*) W. J. Carlsen, "An elastic-tube fiber splice," *Laser Focus*, **16**, 58–62, Apr. 1980.

48. Technical Staff of CSELT, *Optical Fibre Communication*, McGraw-Hill, New York, 1980.

49. P. Hensel, "Triple ball connector for optical fibers," *Electron. Lett.*, **13**, 734–735, Nov. 1977.

50. F. Auracher and K. H. Zeitler, "Multiple fiber connectors for multimode fibers," *Opt. Commun.*, **18**, 556–558, Sept. 1976.

51. K. Höllerl, J. Thürnbeck, and L. Reiter, "Retention grooves for optical fiber connectors," *Electron. Lett.*, **13**, 74–76, Feb. 1977.

52. M. L. Dakss and A. Bridger, "Plug-in fibre-to-fibre coupler," *Electron. Lett.*, **10**, 280–281, July 1974.

53. P. K. Runge and S. S. Cheng, "Demountable single-fiber optic connectors and their measurement on location," *Bell Sys. Tech. J.*, **57**, 1771–1790, July–Aug. 1978.

54. Ref. 30, chap. 15.

55. A. Nicia, "Practical low-loss lens connector for optical fibres," *Electron. Lett.*, **14**, 511–512, Aug. 1978.

56. W. J. Carlsen and P. Melman, "Connectors that stretch," *Opt. Spectra*, **14**, 41–42, Oct. 1980.

57. J. C. Baker and D. N. Payne, "Expanded beam connector design study," *Appl. Opt.*, **20**, 2861–2867, Aug. 1981.

PHOTODETECTORS

At the output end of an optical transmission line there must be a receiving device which interprets the information contained in the optical signal. The first element of this receiver is a photodetector. The photodetector senses the luminescent power falling upon it and converts the variation of this optical power into a correspondingly varying electric current. Since the optical signal is generally weakened and distorted when it emerges from the end of the fiber, the photodetector must meet very high performance requirements. Among the foremost of these requirements are a high response or sensitivity in the emission wavelength range of the optical source being used, a minimum addition of noise to the system, and a fast response speed or sufficient bandwidth to handle the desired data rate. The photodetector should also be insensitive to variations in temperature, be compatible with the physical dimensions of the optical fiber, have a reasonable cost in relation to the other components of the system, and have a long operating life.

Several different types of photodetectors are in existence. Among these are photomultipliers,[1,2] pyroelectric detectors,[3] and semiconductor-based photoconductors, phototransistors, and photodiodes.[4] However, many of these detectors do not meet one or more of the foregoing requirements. Photomultipliers consisting of a photocathode and an electron multiplier packaged in a vacuum tube are capable of very high gain and very low noise. Unfortunately, their large size and high voltage requirements make them unsuitable for optical fiber systems. Pyroelectric photodetectors involve the conversion of photons to heat. Photon absorption results in a temperature change of the detector material. This gives rise to a variation in the dielectric constant which is usually measured as a capacitance change. The response of this detector is quite flat over a broad spectral band, but its speed is limited by the detector cooling rate after it has been excited. Its principal use is for detecting high-speed laser pulses, and it is not well suited for optical fiber systems.

Of the semiconductor-based photodetectors, the photodiode is used almost exclusively for fiber optic systems because of its small size, suitable material, high sensitivity, and fast response time. The two types of photodiodes used are the *pin* photodetector and the avalanche photodiode (APD). Detailed reviews of these photodiodes have been presented in the literature by Melchior,[1,2] Sze,[5] Stillman and Wolfe,[6] Lee and Li,[8] Webb, McIntyre, and Conradi,[7] and Carni.[9] We shall discuss the fundamental characteristics of these two device types in the following sections. In describing these devices we shall make use of some elementary principles of semiconductor device physics. A review of these principles is given in App. E.

6-1 PHYSICAL PRINCIPLES OF PHOTODIODES

6-1-1 The *pin* Photodetector

The most common semiconductor photodetector is the *pin* photodiode shown schematically in Fig. 6-1. The device structure consists of p and n regions separated by a very lightly n-doped intrinsic (i) region. In normal operation a sufficiently large reverse-bias voltage is applied across the device so that the intrinsic region is fully depleted of carriers. That is, the intrinsic n and p carrier concentrations are negligibly small in comparison to the impurity concentration in this region.

When an incident photon has an energy greater than or equal to the band gap energy of the semiconductor material, the photon can give up its energy and excite an electron from the valence band to the conduction band. This process generates free electron-hole pairs which are known as *photocarriers*, since they are photon-generated charge carriers, as is shown in Fig. 6-2. The photodetector is normally designed so that these carriers are generated mainly in the depletion region (the depleted intrinsic region) where most of the incident light is absorbed. The high electric field present in the depletion region causes the carriers to separate and be collected across the reverse-biased junction. This gives rise to a current flow in an external circuit, with one electron flowing for every carrier pair generated. This current flow is known as the *photocurrent*.

As the charge carriers flow through the material some electron-hole pairs will recombine and, hence, disappear. On the average, the charge carriers move a distance L_n or L_p for electrons and holes, respectively. This distance is known as the *diffusion length*. The time it takes for an electron or hole to recombine is known as the *carrier lifetime* and is represented by τ_n and τ_p, respectively. The lifetimes

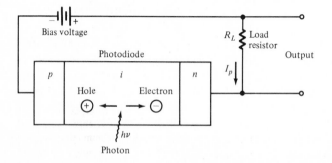

Figure 6-1 Schematic representation of a *pin* photodiode circuit with an applied reverse bias.

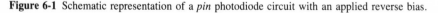

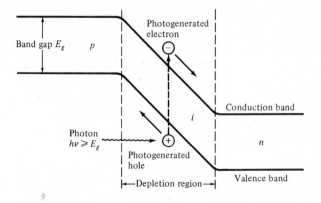

Figure 6-2 Simple energy band diagram for a *pin* photodiode. Photons with an energy greater than or equal to the band gap energy E_g can generate free electron-hole pairs which act as photocurrent carriers.

and the diffusion lengths are related by the expressions

$$L_n = (D_n \tau_n)^{1/2} \quad \text{and} \quad L_p = (D_p \tau_p)^{1/2}$$

where D_n and D_p are the electron and hole diffusion coefficients (or constants), respectively, which are expressed in units of centimeters squared per second.

Optical radiation is absorbed in the semiconductor material according to the exponential law

$$P(x) = P_0(1 - e^{-\alpha_s(\lambda)x}) \tag{6-1}$$

Here $\alpha_s(\lambda)$ is the *absorption coefficient* at a wavelength λ, P_0 is the incident optical power level, and $P(x)$ is the optical power absorbed in a distance x.

The dependence of the optical absorption coefficient on wavelength is shown in Fig. 6-3 for several photodiode materials.[10] As the curves clearly show, α_s

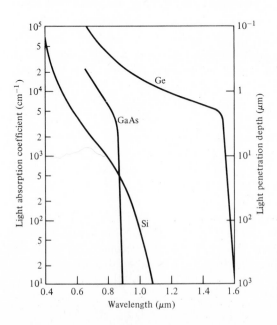

Figure 6-3 Optical absorption coefficient as a function of wavelength for silicon, germanium, and gallium arsenide. (*Reproduced with permission from Miller, Marcatili, and Li,*[10] *copyright 1973, IEEE.*)

depends strongly on the wavelength. Thus a particular semiconductor material can only be used over a limited wavelength range. The upper wavelength cutoff λ_c is determined by the band gap energy E_g of the material. If E_g is expressed in units of electron volts (eV), then λ_c is given in units of micrometers (μm) by

$$\lambda_c \ (\mu\text{m}) = \frac{hc}{E_g} = \frac{1.24}{E_g \ (\text{eV})} \tag{6-2}$$

The cutoff wavelength is about 1.06 μm for Si and 1.6 μm for Ge. For longer wavelengths the photon energy is not sufficient to excite an electron from the valence to the conduction band.

At the lower wavelength end, the photoresponse cuts off as a result of the very large values of α_s at the shorter wavelengths. In this case the photons are absorbed very close to the photodetector surface where the recombination time of the generated electron-hole pairs is very short. The generated carriers thus recombine before they can be collected by the photodetector circuitry.

If the depletion region has a width w, then, from Eq. (6-1), the total power absorbed in the distance w is

$$P(w) = P_0(1 - e^{-\alpha_s w}) \tag{6-3}$$

If we take into account a reflectivity R_f at the entrance face of the photodiode, then the primary photocurrent I_p resulting from the power absorption of Eq. (6-3) is given by

$$I_p = \frac{q}{h\nu}P_0(1 - e^{-\alpha_s w})(1 - R_f) \tag{6-4}$$

where P_0 is the optical power incident on the photodetector, q is the electron charge, and $h\nu$ is the photon energy.

Two important characteristics of a photodetector are its quantum efficiency and its response speed. These parameters depend on the material band gap, the operating wavelength, and the doping and thickness of the p, i, and n regions of the device. The *quantum efficiency* η is the number of electron-hole carrier pairs generated per incident photon of energy $h\nu$ and is given by

$$\eta = \frac{\text{Number of electron-hole pairs generated}}{\text{Number of incident photons}} = \frac{I_p/q}{P_0/h\nu} \tag{6-5}$$

Here I_p is the average photocurrent generated by a steady-state average optical power P_0 incident on the photodetector.

In a practical photodiode, 100 photons will create between 30 and 95 electron-hole pairs, thus giving a detector quantum efficiency ranging from 30 to 95 percent. To achieve a high quantum efficiency, the depletion layer must be thick enough to permit a large fraction of the incident light to be absorbed. However, the thicker the depletion layer, the longer it takes for the photogenerated carriers to drift across the reverse-biased junction. Since the carrier drift time determines the response speed of the photodiode, a compromise has to be made between response speed and quantum efficiency. We shall discuss this further in Sec. 6-3.

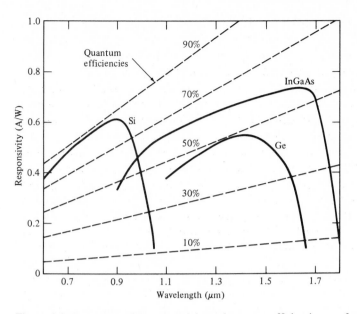

Figure 6-4 Comparison of the responsivity and quantum efficiencies as a function of wavelength for *pin* photodiodes constructed of different materials.

The performance of a photodiode is often characterized by the *responsivity* $\mathcal{R}$. This is related to the quantum efficiency by

$$\mathcal{R} = \frac{I_p}{P_0} = \frac{\eta q}{h\nu} \tag{6-6}$$

This parameter is quite useful since it specifies the photocurrent generated per unit optical power. Typical *pin* photodiode responsivities as a function of wavelength are shown in Fig. 6-4. Representative values are 0.65 μA/μW for silicon at 800 nm, 0.45 μA/μW for germanium at 1.3 μm, and 0.6 μA/μW for InGaAs at 1.3 μm.

In most photodiodes the quantum efficiency is independent of the power level falling on the detector at a given photon energy. Thus the responsivity is a linear function of the optical power. That is, the photocurrent I_p is directly proportional to the optical power P_0 incident upon the photodetector, so that the responsivity $\mathcal{R}$ is constant at a given wavelength (a given value of $h\nu$). Note, however, that the quantum efficiency is not a constant at all wavelengths since it varies according to the photon energy. Consequently, the responsivity is a function of the wavelength and of the photodiode material (since different materials have different band gap energies). For a given material, as the wavelength of the incident photon becomes longer, the photon energy becomes less than that required to excite an electron from the valence band to the conduction band. The responsivity thus falls off rapidly beyond the cutoff wavelength, as can be seen in Fig. 6-4.

6-1-2 Avalanche Photodiodes

Avalanche photodiodes (APDs) internally multiply the primary signal photocurrent before it enters the input circuitry of the following amplifier. This increases receiver sensitivity since the photocurrent is multiplied before encountering the thermal noise associated with the receiver circuit. In order for carrier multiplication to take place, the photogenerated carriers must traverse a region where a very high electric field is present. In this high-field region a photogenerated electron or hole can gain enough energy so that it ionizes bound electrons in the valence band upon colliding with them. This carrier multiplication mechanism is known as *impact ionization*. The newly created carriers are also accelerated by the high electric field, thus gaining enough energy to cause further impact ionization. This phenomenon is the *avalanche effect*. Below the diode breakdown voltage a finite total number of carriers are created, whereas above breakdown the number can be infinite.

A commonly used structure for achieving carrier multiplication with very little excess noise is the *reach-through* construction[7,10-15] shown in Fig. 6-5. The reach-through avalanche photodiode (RAPD) is composed of a high-resistivity p-type material deposited as an epitaxial layer on a p^+ (heavily doped p-type) substrate. A p-type diffusion or ion implant is then made in the high-resistivity material followed by the construction of an n^+ (heavily doped n-type) layer. For silicon the dopants used to form these layers are normally boron and phosphorous, respectively. This configuration is referred to as a $p^+\pi p n^+$ *reach-through* structure. The π layer is basically an intrinsic material that inadvertently has some p doping because of imperfect purification.

The term "reach-through" arises from the photodiode operation. When a low reverse-bias voltage is applied, most of the potential drop is across the pn^+ junction. The depletion layer widens with increasing bias until a certain voltage is reached at which the peak electric field at the pn^+ junction is about 5 to 10 percent below that needed to cause avalanche breakdown. At this point the depletion layer just "reaches through" to the nearly intrinsic π region.

In normal usage the RAPD is operated in the fully depleted mode. Light enters

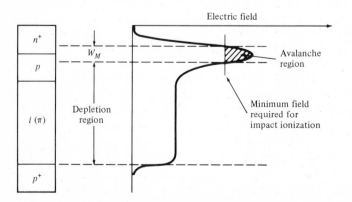

Figure 6-5 Reach-through avalanche photodiode structure and the electric fields in the depletion and multiplication regions.

the device through the p^+ region and is absorbed in the π material, which acts as the collection region for the photogenerated carriers. Upon being absorbed the photon gives up its energy, thereby creating electron-hole pairs, which are then separated by the electric field in the π region. The photogenerated electrons drift through the π region to the pn^+ junction where a high electric field exists. It is in this high-field region that carrier multiplication takes place.

The average number of electron-hole pairs created by a carrier per unit distance traveled is called the *ionization rate*. Most materials exhibit different *electron ionization rates* α and *hole ionization rates* β. Experimentally obtained values of α and β for five different semiconductor materials are shown in Fig. 6-6. The ratio $k = \beta/\alpha$ of the two ionization rates is a measure of the photodetector performance. As we shall see in Sec. 6-4, avalanche photodiodes constructed of materials in which one type of carrier largely dominates impact ionization exhibit low noise and

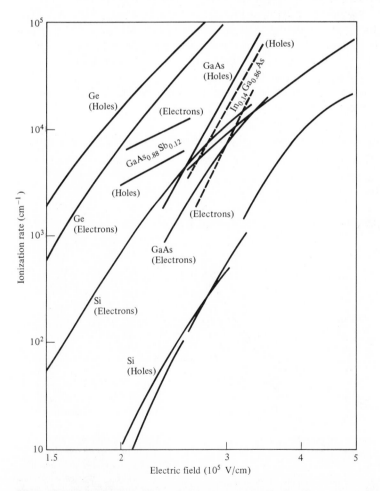

Figure 6-6 Carrier ionization rates obtained experimentally for silicon, germanium, gallium arsenide, gallium arsenide antimonide, and indium gallium arsenide. *(Reproduced with permission from Melchior[2].)*

large gain-bandwidth products. As shown in Fig. 6-6, of all the materials examined so far,[16-34] only silicon has a significant difference between electron and hole ionization rates.

The multiplication M for all carriers generated in the photodiode is defined by

$$M = \frac{I_M}{I_p} \tag{6-7}$$

where I_M is the average value of the total multiplied output current and I_p is the primary unmultiplied photocurrent defined in Eq. (6-4). In practice, the avalanche mechanism is a statistical process since not every carrier pair generated in the diode experiences the same multiplication. Thus the measured value of M is expressed as an average quantity.

Typical current gains for different wavelengths[15] as a function of bias voltage for a silicon reach-through avalanche photodiode are shown in Fig. 6-7. The dependence of the gain on the excitation wavelength is attributable to mixed initiation of the avalanche process by electrons and holes when most of the light is absorbed in the n^+p region close to the detector surface. This is especially noticeable at short wavelengths, where a larger portion of the optical power is absorbed close to the surface, as compared to longer wavelengths. Since the ionization coefficient for holes is smaller than that for electrons in silicon, the total current gain is reduced at the short wavelengths.

Analogous to the *pin* photodiode, the performance of an APD is characterized

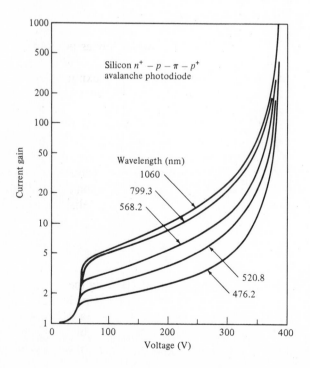

Figure 6-7 Typical room temperature current gains of a silicon reach-through avalanche photodiode for different wavelengths as a function of bias voltage. *(Reproduced with permission from Melchior, Hartman, Schinke, and Seidel,[15] copyright 1978, the American Telephone and Telegraph Co.)*

by its responsivity $\mathcal{R}_{APD}$ which is given by

$$\mathcal{R}_{APD} = \frac{\eta q}{h\nu}M = \mathcal{R}_0 M \tag{6-8}$$

where $\mathcal{R}_0$ is the unity gain responsivity.

6-2 PHOTODETECTOR NOISE

In fiber optic communication systems the photodiode is generally required to detect very weak optical signals. Detection of the weakest possible optical signals requires that the photodetector and its following amplification circuitry be optimized so that a given signal-to-noise ratio is maintained. The power signal-to-noise ratio S/N at the output of an optical receiver is defined by

$$\frac{S}{N} = \frac{\text{Signal power from photocurrent}}{\text{Photodetector noise power} + \text{amplifier noise power}} \tag{6-9}$$

The noise sources in the receiver arise from the photodetector noises resulting from the statistical nature of the photon-to-electron conversion process and the thermal noises associated with the amplifier circuitry.

To achieve a high signal-to-noise ratio, the following conditions should be met:

1. The photodetector must have a high quantum efficiency to generate a large signal power.
2. The photodetector and amplifier noises should be kept as low as possible.

For most applications it is the noise currents which determine the maximum optical power level that can be detected, since the photodiode quantum efficiency is normally close to its maximum possible value.

The sensitivity of a photodetector in an optical fiber communication system is describable in terms of the *minimum detectable optical power*. This is the optical power necessary to produce a photocurrent of the same magnitude as the root mean square of the total noise current or, equivalently, a signal-to-noise ratio of one. A thorough understanding of the source, characteristics, and interrelationships of the various noises in a photodetector is, therefore, necessary to make a reliable design and to evaluate optical receivers.

6-2-1 Noise Sources

To see the interrelationship of the different types of noises affecting the signal-to-noise ratio, let us examine the simple receiver model and its equivalent circuit shown in Fig. 6-8. The photodiode has a small series resistance R_s, a total capacitance C_d consisting of junction and packaging capacitances, and a bias (or load) resistor R_L. The amplifier following the photodiode has an input capacitance C_a and

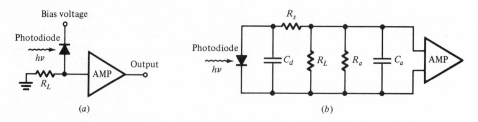

Figure 6-8 (a) Simple model of a photodetector receiver, and (b) its equivalent circuit.

a resistance R_a. For practical purposes, R_s is much smaller than the load resistance R_L and can be neglected.

If a modulated signal of optical power $P(t)$ falls on the detector, the primary photocurrent $i_{ph}(t)$ generated is[8-10]

$$i_{ph}(t) = \frac{\eta q}{h\nu} P(t) \tag{6-10}$$

This primary current consists of a dc value I_p, which is the average photocurrent due to the signal power, and a signal component $i_p(t)$. For *pin* photodiodes the mean square signal current $\langle i_s^2 \rangle$ is

$$\langle i_s^2 \rangle = \langle i_p^2(t) \rangle \tag{6-11a}$$

whereas for avalanche photodetectors,

$$\langle i_s^2 \rangle = \langle i_p^2(t) \rangle M^2 \tag{6-11b}$$

where M is the average of the statistically varying avalanche gain as defined in Eq. (6-7). For a sinusoidally varying input signal of modulation index m, the signal component $\langle i_p^2 \rangle$ is of the form (see Prob. 6-5)

$$\langle i_p^2(t) \rangle = \frac{m^2}{2} I_p^2 \tag{6-12}$$

where m is defined in Eq. (4-40).

The principal noises associated with photodetectors having no internal gain are quantum noise, dark-current noise generated in the bulk material of the photodiode, and surface leakage current noise. The *quantum* or *shot noise* arises from the statistical nature of the production and collection of photoelectrons when an optical signal is incident on a photodetector. It has been demonstrated[35] that these statistics follow a Poisson process. Since the fluctuations in the number of photocarriers created from the photoelectric effect are a fundamental property of the photo-detection process, they set the lower limit on the receiver sensitivity when all other conditions are optimized. The quantum noise current has a mean square value in a bandwidth B which is proportional to the average value of the photocurrent I_p

$$\langle i_Q^2 \rangle = 2q I_p B M^2 F(M) \tag{6-13}$$

where $F(M)$ is a noise figure associated with the random nature of the avalanche process. From experimental results it has been found that to a reasonable approximation $F(M) \simeq M^x$, where x (with $0 \le x \le 1.0$) depends on the material. This is discussed in more detail in Sec. 6-4. For *pin* photodiodes M and $F(M)$ are unity.

The photodiode dark current is the current that continues to flow through the bias circuit of the device when no light is incident on the photodiode. This is a combination of bulk and surface currents. The *bulk dark current* i_{DB} arises from electrons and/or holes which are thermally generated in the *pn* junction of the photodiode. In an APD these liberated carriers also get accelerated by the high electric field present at the *pn* junction, and are, therefore, multiplied by the avalanche gain mechanism. The mean square value of this current is given by

$$\langle i_{DB}^2 \rangle = 2qI_D M^2 F(M)B \qquad (6\text{-}14)$$

where I_D is the primary (unmultiplied) detector bulk dark current.

The *surface dark current* is also referred to as a *surface leakage current* or simply the leakage current. It is dependent on surface defects, cleanliness, bias voltage, and surface area. An effective way of reducing surface dark current is through the use of a guard ring structure which shunts surface leakage currents away from the load resistor. The mean square value of the surface dark current is given by

$$\langle i_{DS}^2 \rangle = 2qI_L B \qquad (6\text{-}15)$$

where I_L is the surface leakage current. Note that since avalanche multiplication is a bulk effect, the surface dark current is not affected by the avalanche gain.

A comparison[30] of typical dark currents for Si, Ge, GaAs, and In$_x$Ga$_{1-x}$As photodiodes is given in Fig. 6-9 as a function of applied voltage normalized to the

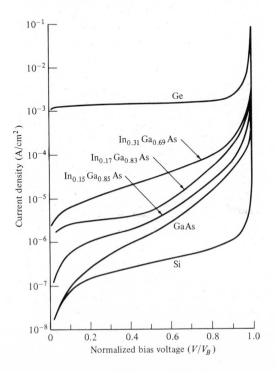

Figure 6-9 A comparison of typical dark currents for Si, Ge, GaAs, and InGaAs photodiodes as a function of normalized bias voltage. *(Reproduced with permission from Susa, Yamauchi, and Kanbe,[30] copyright 1980, IEEE.)*

breakdown voltage V_B. Note that for $In_xGa_{1-x}As$ photodiodes the dark current increases with the composition x. Under a reverse bias, both dark currents also increase with area. The surface dark current increases in proportion to the square root of the active area and the bulk dark current is directly proportional to the area.

Since the dark currents and the signal current are uncorrelated, the total mean square photodetector noise current $\langle i_N^2 \rangle$ can be written as

$$\langle i_N^2 \rangle = \langle i_Q^2 \rangle + \langle i_{DB}^2 \rangle + \langle i_{DS}^2 \rangle$$

$$= 2q(I_p + I_D)M^2F(M)B + 2qI_LB \qquad (6\text{-}16)$$

To simplify the analysis of the receiver circuitry, we shall assume here that the amplifier input impedance is much greater than the load resistance so that its thermal noise is much smaller than that of R_L. The photodetector load resistor contributes a mean square thermal (Johnson) noise current

$$\langle i_T^2 \rangle = \frac{4k_BT}{R_L}B \qquad (6\text{-}17)$$

where k_B is Boltzmann's constant and T is the absolute temperature. This noise can be reduced by using a load resistor which is large but still consistent with the receiver bandwidth requirements. Further details on this are given in Chap. 7 along with a detailed discussion of the amplifier noise current i_{amp}.

6-2-2 Signal-to-Noise Ratio

Substituting Eqs. (6-11), (6-16), and (6-17) into Eq. (6-9) for the signal-to-noise ratio at the input of the amplifier, we have

$$\frac{S}{N} = \frac{\langle i_p^2 \rangle M^2}{2q(I_p + I_D)M^2F(M)B + 2qI_LB + 4k_BTB/R_L} \qquad (6\text{-}18)$$

In general, when *pin* photodiodes are used, the dominating noise currents are those of the detector load resistor (the thermal current i_T) and the active elements of the amplifier circuitry (i_{amp}). For avalanche photodiodes the thermal noise is of lesser importance, and the photodetector noises usually dominate.[36]

From Eq. (6-18) it can be seen that the signal power is multiplied by M^2 and the quantum noise plus bulk dark current are multiplied by $M^2F(M)$. The surface leakage current is not altered by the avalanche gain mechanism. Since the noise figure $F(M)$ increases with M, there always exists an optimum value of M that maximizes the signal-to-noise ratio. The optimum gain at the maximum signal-to-noise ratio can be found by differentiating Eq. (6-18) with respect to M, setting the result equal to zero, and solving for M. Doing so for a sinusoidally modulated signal, with $m = 1$ and $F(M)$ approximated by M^x, yields

$$M_{opt}^{x+2} = \frac{2qI_L + 4k_BT/R_L}{xq(I_p + I_D)} \qquad (6\text{-}19)$$

6-3 DETECTOR RESPONSE TIME

6-3-1 Depletion Layer Photocurrent

To understand the frequency response of photodiodes, let us first consider the schematic representation of a reverse-biased *pin* photodiode shown in Fig. 6-10. Light enters the device through the *p* layer and produces electron-hole pairs as it is absorbed in the semiconductor material. Those electron-hole pairs that are generated in the depletion region or within a diffusion length of it will be separated by the reverse-bias voltage-induced electric field, thereby leading to a current flow in the external circuit as the carriers drift across the depletion layer.

Under steady-state conditions the total current density J_{tot} flowing through the reverse-biased depletion layer is[37]

$$J_{tot} = J_{dr} + J_{diff} \tag{6-20}$$

Here J_{dr} is the drift current density resulting from carriers generated inside the depletion region and J_{diff} is the diffusion current density arising from the carriers that are produced outside of the depletion layer in the bulk of the semiconductor (that is, in the *n* and *p* regions) and diffuse into the reverse-biased junction. The drift current density can be found from Eq. (6-4),

$$J_{dr} = \frac{I_p}{A} = q\Phi_0(1 - e^{-\alpha_s w}) \tag{6-21}$$

where A is the photodiode area and Φ_0 is the incident photon flux per unit area given by

$$\Phi_0 = \frac{P_0(1 - R_f)}{Ah\nu} \tag{6-22}$$

The surface *p* layer of a *pin* photodiode is normally very thin. The diffusion current is thus principally determined by hole diffusion from the bulk *n* region. The hole diffusion in this material can be determined by the one-dimensional diffusion equation[5]

$$D_p \frac{\partial^2 p_n}{\partial x^2} - \frac{p_n - p_{n0}}{\tau_p} + G(x) = 0 \tag{6-23}$$

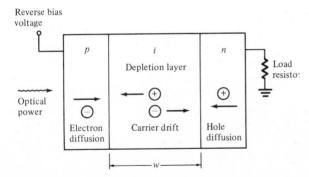

Figure 6-10 Schematic representation of a reverse-biased *pin* photodiode.

where D_p is the hole diffusion coefficient, p_n is the hole concentration in the n-type material, τ_p is the excess hole lifetime, p_{n0} is the equilibrium hole density, and $G(x)$ is the electron-hole generation rate given by

$$G(x) = \Phi_0 \alpha_s e^{-\alpha_s x} \qquad (6\text{-}24)$$

From Eq. (6-23) the diffusion current density is found to be (see Prob. 6-9)

$$J_{\text{diff}} = q\Phi_0 \frac{\alpha_s L_p}{1 + \alpha_s L_p} e^{-\alpha_s w} + q p_{n0} \frac{D_p}{L_p} \qquad (6\text{-}25)$$

Substituting Eqs. (6-21) and (6-25) into Eq. (6-20) we have that the total current density through the reverse-biased depletion layer is

$$J_{\text{tot}} = q\Phi_0 \left(1 - \frac{e^{-\alpha_s w}}{1 + \alpha_s L_p}\right) + q p_{n0} \frac{D_p}{L_p} \qquad (6\text{-}26)$$

The term involving p_{n0} is normally small so that the total photogenerated current is proportional to the photon flux Φ_0.

6-3-2 Response Time

The response time of a photodiode together with its output circuit (see Fig. 6-8) depends mainly on the following three factors:

1. The transit time of the photocarriers in the depletion region
2. The diffusion time of the photocarriers generated outside the depletion region
3. The RC time constant of the photodiode and its associated circuit

The photodiode parameters responsible for these three factors are the absorption coefficient α_s, the depletion region width w, the photodiode junction and package capacitances, the amplifier capacitance, the detector load resistance, the amplifier input resistance, and the photodiode series resistance. The photodiode series resistance is generally only a few ohms and can be neglected in comparison with the large load resistance and the amplifier input resistance.

Let us first look at the transit time of the photocarriers in the depletion region. The response speed of a photodiode is fundamentally limited by the time it takes photogenerated carriers to travel across the depletion region. This transit time t_d depends on the carrier drift velocity v_d and the depletion layer width w, and is given by

$$t_d = \frac{w}{v_d} \qquad (6\text{-}27)$$

In general, the electric field in the depletion region is large enough so that the carriers have reached their scattering-limited velocity. For silicon the maximum velocities for electrons and holes are 8.4×10^6 and 4.4×10^6 cm/s, respectively, when the field strength is on the order of 2×10^4 V/cm. A typical high-speed silicon photodiode with a 10-μm depletion layer width thus has a response time limit of about 0.1 ns.

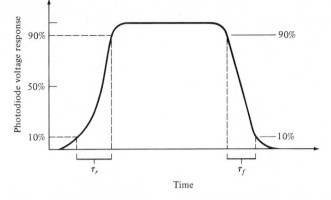

Figure 6-11 Photodiode response to an optical input pulse showing the 10 to 90 percent rise time and the 90 to 10 percent fall time.

The diffusion processes are slow compared to the drift of carriers in the high-field region. Therefore, to have a high-speed photodiode, the photocarriers should be generated in the depletion region or so close to it that the diffusion times are less than or equal to the carrier drift times. The effect of long diffusion times can be seen by considering the photodiode response time. This response time is described by the rise time and fall time of the detector output when the detector is illuminated by a step input of optical radiation. The rise time τ_r is typically measured from the 10 to the 90 percent points of the leading edge of the output pulse, as is shown in Fig. 6-11. For fully depleted photodiodes the rise time τ_r and fall time τ_f are generally the same. However, they can be different at low bias levels where the photodiode is not fully depleted, since the photon collection time then starts to become a significant contributor to the rise time. In this case, charge carriers produced in the depletion region are separated and collected quickly. On the other hand, electron-hole pairs generated in the n and p regions must slowly diffuse to the depletion region before they can be separated and collected. A typical response time of a partially depleted photodiode is shown in Fig. 6-12. The fast carriers allow the device output to rise to 50 percent of its maximum value in

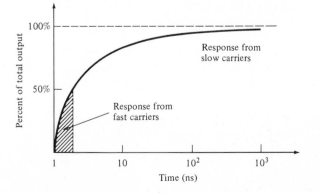

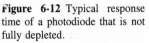

Figure 6-12 Typical response time of a photodiode that is not fully depleted.

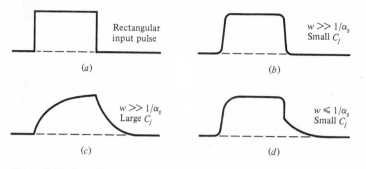

Figure 6-13 Photodiode pulse responses under various detector parameters.

approximately 1 ns, but the slow carriers cause a relatively long delay before the output reaches its maximum value.

To achieve a high quantum efficiency the depletion layer width must be much larger than $1/\alpha_s$ (the inverse of the absorption coefficient) so that most of the light will be absorbed. The response to a rectangular input pulse of a low-capacitance photodiode having $w \gg 1/\alpha_s$ is shown in Fig. 6-13b. The rise and fall times of the photodiode follow the input pulse quite well. If the photodiode capacitance is larger, the response time becomes limited by the RC time constant of the load resistor R_L and the photodiode capacitance. The photodetector response then begins to appear as that shown in Fig. 6-13c.

If the depletion layer is too narrow, any carriers created in the undepleted material would have to diffuse back into the depletion region before they could be collected. Devices with very thin depletion regions thus tend to show distinct slow- and fast-response components, as shown in Fig. 6-13d. The fast component is a result of absorption in the depletion region, whereas the slow component, which appears as the slowly decaying tails shown in Fig. 6-13d, arises from the carriers created in the undepleted material. Also, if w is too thin, the junction capacitance will become excessive. The junction capacitance C_j is

$$C_j = \frac{\epsilon_s A}{w} \tag{6-28}$$

where ϵ_s = the permittivity of the semiconductor material = $\epsilon_0 K_s$

K_s = the semiconductor dielectric constant

ϵ_0 = 8.8542 × 10^{-12} F/m is the free-space permittivity

A = the diffusion layer area.

This excessiveness will then give rise to a large RC time constant which limits the detector response time. A reasonable compromise between high-frequency response and high quantum efficiency is found for absorption region thicknesses between $1/\alpha_s$ and $2/\alpha_s$.

If R_T is the combination of the load and amplifier input resistances and C_T is the sum of the photodiode and amplifier capacitances, as shown in Fig. 6-8, the detector behaves approximately like a simple RC low-pass filter with a passband given by

$$B = \frac{1}{2\pi R_T C_T} \tag{6-29}$$

For example, if the photodiode capacitance is 3 pF, the amplifier capacitance is 4 pF, the load resistor is 1 kΩ, and the amplifier input resistance is 1 MΩ then $C_T = 7$ pF and $R_T \simeq 1$ kΩ, so that the circuit bandwidth is

$$B = \frac{1}{2\pi R_T C_T} = 23 \text{ MHz} \qquad (6\text{-}30)$$

6-4 AVALANCHE MULTIPLICATION NOISE

As we noted earlier, the avalanche process is statistical in nature since not every photogenerated carrier pair undergoes the same multiplication.[38-41] The probability distribution of possible gains that any particular electron-hole pair might experience is sufficiently wide so that the mean square gain is greater than the average gain squared. That is, if m denotes the statistically varying gain then

$$\langle m^2 \rangle > \langle m \rangle^2 = M^2 \qquad (6\text{-}31)$$

where the symbols $\langle \ \rangle$ denote an ensemble average and $\langle m \rangle = M$ is the average carrier gain defined in Eq. (6-7). Since the noise created by the avalanche process depends on the mean square gain $\langle m^2 \rangle$, the noise in an avalanche photodiode can be relatively high. From experimental observations it has been found that, in general, $\langle m^2 \rangle$ can be approximated by

$$\langle m^2 \rangle \simeq M^{2+x} \qquad (6\text{-}32)$$

where the exponent x varies between 0 and 1.0 depending on the photodiode material and structure.

The ratio of the actual noise generated in an avalanche photodiode to the noise that would exist if all carrier pairs were multiplied by exactly M is denoted by the *excess noise factor F* defined by

$$F = \frac{\langle m^2 \rangle}{\langle m \rangle^2} = \frac{\langle m^2 \rangle}{M^2} \qquad (6\text{-}33)$$

This excess noise factor is a measure of the increase in detector noise resulting from the randomness of the multiplication process. It depends on the ratio of the electron and hole ionization rates and on the carrier multiplication.

The derivation of an expression for F is complex since the electric field in the avalanche region (of width W_M, as shown in Fig. 6-5) is not uniform, and both holes and electrons produce impact ionization. McIntyre[40] has shown that, for injected electrons and holes, the excess noise factors are

$$F_e = \frac{k_2 - k_1^2}{1 - k_2} M_e + 2\left[1 - \frac{k_1(1 - k_1)}{1 - k_2}\right] - \frac{(1 - k_1)^2}{M_e(1 - k_2)} \qquad (6\text{-}34)$$

$$F_h = \frac{k_2 - k_1^2}{k_1^2(1 - k_2)} M_h - 2\left[\frac{k_2(1 - k_1)}{k_1^2(1 - k_2)} - 1\right] + \frac{(1 - k_1)^2 k_2}{k_1^2(1 - k_2)M_h} \qquad (6\text{-}35)$$

where the subscripts e and h refer to electrons and holes, respectively. The weighted ionization rate ratios k_1 and k_2 take into account the nonuniformity of the gain and the carrier ionization rates in the avalanche region. They are given by

$$k_1 = \frac{\displaystyle\int_0^{W_M} \beta(x)M(x)\ dx}{\displaystyle\int_0^{W_M} \alpha(x)M(x)\ dx} \tag{6-36}$$

$$k_2 = \frac{\displaystyle\int_0^{W_M} \beta(x)M^2(x)\ dx}{\displaystyle\int_0^{W_M} \alpha(x)M^2(x)\ dx} \tag{6-37}$$

Normally to a first approximation k_1 and k_2 do not change much with variations in gain and can be considered as constant and equal. Thus Eqs. (6-34) and (6-35) can be simplified as[7]

$$F_e = M_e\left[1 - (1 - k_{\text{eff}})\left(1 - \frac{1}{M_e}\right)^2\right]$$

$$= k_{\text{eff}}M_e + \left(2 - \frac{1}{M_e}\right)(1 - k_{\text{eff}}) \tag{6-38}$$

for electron injection, and

$$F_h = M_h\left[1 - \left(1 - \frac{1}{k'_{\text{eff}}}\right)\left(1 - \frac{1}{M_h}\right)^2\right]$$

$$= k'_{\text{eff}}M_h - \left(2 - \frac{1}{M_h}\right)(k'_{\text{eff}} - 1) \tag{6-39}$$

for hole injection, where the effective ionization rate ratios are

$$k_{\text{eff}} = \frac{k_2 - k_1^2}{1 - k_2} \simeq k_2$$

$$k'_{\text{eff}} = \frac{k_{\text{eff}}}{k_1^2} \simeq \frac{k_2}{k_1^2} \tag{6-40}$$

Figure 6-14 shows F_e as a function of the average electron gain M_e for various values of the effective ionization ratio k_{eff}. If the ionization rates are equal, the excess noise is at its maximum so that F is at its upper limit of M_e. As the ratio β/α decreases from unity, the electron ionization rate starts to be the dominant contributor to impact ionization, and the excess noise factor becomes smaller. If only electrons cause ionization, $\beta = 0$ and F reaches its lower limit of 2.

This shows that, to keep the excess noise factor at a minimum, it is desirable to have small values of k_{eff}. Referring back to Fig. 6-6, we thus see the superiority of silicon over other materials for making avalanche photodiodes. For silicon

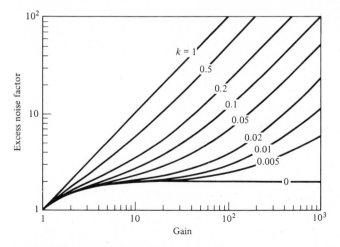

Figure 6-14 Variation of the electron excess noise factor F_e as a function of the electron gain for various values of the effective ionization rate ratio k_{eff}. *(Reproduced with permission from Webb, McIntyre, and Conradi.[7])*

photodiodes k_{eff} ranges from 0.015 to 0.035. In germanium[17,42] k_{eff} varies between 0.6 and 1.0.

From the empirical relationship for the mean square gain given by Eq. (6-32), the excess noise factor can be approximated by

$$F = M^x \tag{6-41}$$

The parameter x takes on values of 0.5 for silicon avalanche photodiodes and 0.85 to 1.0 for germanium.

6-5 TEMPERATURE EFFECT ON AVALANCHE GAIN

The gain mechanism of an avalanche photodiode is very temperature sensitive because of the temperature dependence of the electron and hole ionization rates.[43,44] This temperature dependence is particularly critical at high bias voltages, where small changes in temperature can cause large variations in gain. An example of this is shown in Fig. 6-15 for a silicon avalanche photodiode. For example, if the operating temperature decreases and the applied bias voltage is kept constant, the ionization rates for electrons and holes will increase and so will the avalanche gain.

To maintain a constant gain as the temperature changes, the electric field in the multiplying region of the *pn* junction must also be changed. This requires that the receiver incorporate a compensation circuit which adjusts the applied bias voltage on the photodetector when the temperature changes.

The dependence of gain on temperature has been studied in detail by Conradi.[44] In that work the gain curves were described by using the explicit temperature dependence of the ionization rates α and β together with a detailed knowledge of

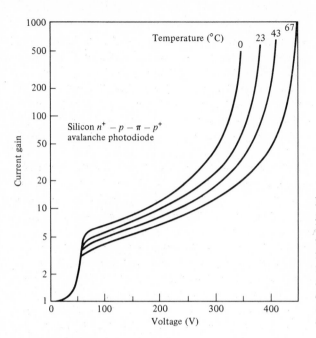

Figure 6-15 Example of how the gain mechanism of a silicon avalanche photodiode depends on temperature. The measurements for this device were done at 825 nm. *(Reproduced with permission from Melchior, Hartman, Schinke, and Seidel,*[15] *copyright 1978, the American Telephone and Telegraph Co.)*

the device structure. Although excellent agreement was found between the theoretically computed and the experimentally measured gains, the calculations are rather involved. However, a simple temperature-dependent expression can be obtained from the empirical relationship[45]

$$M = \frac{1}{1 - (V/V_B)^n} \tag{6-42}$$

where V_B is the breakdown voltage at which M goes to infinity, the parameter n varies between 2.5 and 7, depending on the material, and $V = V_a - I_M R_M$, with V_a being the reverse-bias voltage applied to the detector, I_M is the multiplied photocurrent, and R_M accounts for the photodiode series resistance and the detector load resistance. Since the breakdown voltage is known to vary with temperature as[46,47]

$$V_B(T) = V_B(T_0)[1 + a(T - T_0)] \tag{6-43}$$

the temperature dependence of the avalanche gain can be approximated by substituting Eq. (6-43) into Eq. (6-42) together with the expression

$$n(T) = n(T_0)[1 + b(T - T_0)] \tag{6-44}$$

The constants a and b are positive for reach-through avalanche photodiodes and can be determined from experimental curves of gain versus temperature.

6-6 PHOTODIODE MATERIALS

The responsivity of a photodetector is principally determined by the construction of the detector and the type of material used. The absorption coefficient α_s of semiconductor materials varies greatly with wavelength, as is shown in Fig. 6-3. For a practical photodiode, the best responsivity and the highest quantum efficiency are obtained in a material having a band gap energy slightly less than the energy of the photons at the longest wavelength of interest. In addition to ensuring good detector quantum efficiency and response speed, this condition simultaneously keeps the dark current low.

Any of a number of different materials, including Si, Ge, GaAs, InGaAs, and InGaAsP, could be used for photodiode operation in the 800- to 900-nm spectral region. However, silicon is used almost exclusively because, in addition to its highly developed technology, it exhibits the lowest avalanche carrier multiplication noise, thus permitting high receiver sensitivity.

For operation at wavelengths above 1.0 μm the responsivity of Si is too low for it to be used as a photodiode, since photons at these wavelengths do not have enough energy to excite an electron across the 1.17-eV silicon band gap. Various high-sensitivity photodetectors have been developed for the 1.0- to 1.65-μm range. The materials examined have included Ge,[16,17] InP,[18] InGaAsP,[19-22] GaSb,[23,24] GaAlSb,[25] GaAlAsSb,[26] HgCdTe,[27] and InGaAs.[28-32] Germanium has a large absorption coefficient of approximately 10^4 cm^{-1} over the wavelength range of 1.0 to 1.55 μm, which should make it an ideal photodetector for long-wavelength applications. A number of Ge photodetectors with reasonable sensitivity and fast response times have been fabricated, but the material exhibits a number of shortcomings. For example, Ge has a high excess noise factor for avalanche multiplication owing to a carrier ionization rate ratio of only 2. Furthermore, since the band gap of Ge is narrower than that of Si, the bulk dark current is much higher, thus limiting the usable avalanche gain. Despite these limitations, Ge avalanche photodiodes have been successfully used in high-data transmission experiments,[33] one example being an 800-Mb/s link operating at 1.3 μm over an 11-km distance.

In addition to Ge a variety of III-V semiconductor alloys such as InGaAsP, GaAlSb, InGaAs, GaSb, and GaAsSb have been investigated for long-wavelength applications.[34] There are several reasons for examining these materials. First, since the band gaps of these alloys depend on their molecular composition, the absorption edge can be selected to be just above the longest wavelength of operation by varying the molecular concentrations of the constituent elements of the alloys. This results in detectors with high quantum efficiency, fast response speed, and low dark current. Another reason for studying these alloys is to search for a material having a large difference in the electron and hole ionization rates. Unfortunately, the ionization rate ratios in all III-V materials measured to date are inferior to silicon. This tends to limit operation of these devices to moderate avalanche gains of between 10 and 30.

6-7 SUMMARY

Semiconductor *pin* and avalanche photodiodes are the principal devices used as photon detectors in optical fiber links because of their size compatibility with fibers, their high sensitivities at the desired optical wavelengths, and their fast response times. In addition to these two photodiode structures, considerable attention is being given to the heterojunction phototransistor,[48–50] which is capable of satisfying many of the detector requirements of a fiber communication system.

When light having photon energies greater than or equal to the band gap energy of the semiconductor material is incident on a photodetector, the photons can give up their energy and excite electrons from the valence band to the conduction band. This process generates free electron-hole pairs, which are known as photocarriers. When a reverse-bias voltage is applied across the photodetector, the resultant electric field in the device causes the carriers to separate. This gives rise to a current flow in an external circuit, which is known as the photocurrent.

The quantum efficiency η is an important photodetector performance parameter. This is defined as the number of electron-hole carrier pairs generated per incident photon of energy $h\nu$. In practice, quantum efficiencies range from 30 to 95 percent. Another important factor is the responsivity. This is related to the quantum efficiency by

$$\mathfrak{R} = \frac{\eta q}{h\nu}$$

This parameter is quite useful since it specifies the photocurrent generated per unit optical power. Representative responsivities for *pin* photodiodes are 0.65 $\mu A/\mu W$ for Si at 800 nm, 0.45 $\mu A/\mu W$ for Ge at 1300 nm, and 0.6 $\mu A/\mu W$ for InGaAs at 1300 nm.

Avalanche photodiodes (APDs) internally multiply the primary signal photocurrent. This increases receiver sensitivity since the photocurrent is multiplied before encountering the thermal noise associated with the receiver circuitry. The carrier multiplication M is a result of impact ionization. Since the avalanche mechanism is a statistical process, not every carrier pair generated in the photodiode experiences the same multiplication. Thus the measured value of M is expressed as an average quantity. Analogous to the *pin* photodiode, the performance of an APD is characterized by its responsivity

$$\mathfrak{R}_{APD} = \frac{\eta q}{h\nu} M = \mathfrak{R}_0 M$$

where $\mathfrak{R}_0$ is the unity gain responsivity.

The sensitivity of a photodetector and its associated receiver is essentially determined by the photodetector noises resulting from the statistical nature of the photon-to-electron conversion process and the thermal noises in the amplifier circuitry. The main noise currents of photodetectors are:

1. Quantum or shot noise current arising from the statistical nature of the production and collection of photoelectrons
2. Bulk dark current arising from electrons and/or holes which are thermally generated in the *pn* junction of the photodiode
3. Surface dark current (or leakage current) which depends on surface defects, cleanliness, bias voltage, and surface area

generated consists of a dc (average) component I_p and a signal current i_p given by

$$\langle i_s^2 \rangle = I_p^2 + \langle i_p^2 \rangle = (\mathcal{R}_0 P_0)^2 + \tfrac{1}{2}(m\mathcal{R}_0 P_0)^2$$

where the responsivity $\mathcal{R}_0$ is given by Eq. (6-6).

6-6 Consider an avalanche photodiode receiver having the following parameters: dark current $I_D = 1$ nA, leakage current $I_L = 1$ nA, quantum efficiency $\eta = 0.85$, gain $M = 100$, excess noise factor $F = M^{1/2}$, load resistor $R_L = 10^4 \; \Omega$, and bandwidth $B = 10$ kHz. Suppose a sinusoidally varying 850-nm signal having a modulation index $m = 0.85$ falls on the photodiode which is at room temperature $(T = 300 \text{ K})$. To compare the contributions from the various noise terms to the signal-to-noise ratio for this particular set of parameters, plot the following terms in decibels [that is, $10 \log (S/N)$] as a function of the average received optical power P_0. Let P_0 range from -70 to 0 dBm, that is, from 0.1 nW to 1.0 mW:

(a) $\left(\dfrac{S}{N}\right)_Q = \dfrac{\langle i_s^2 \rangle}{\langle i_Q^2 \rangle}$

(b) $\left(\dfrac{S}{N}\right)_{DB} = \dfrac{\langle i_s^2 \rangle}{\langle i_{DB}^2 \rangle}$

(c) $\left(\dfrac{S}{N}\right)_{DS} = \dfrac{\langle i_s^2 \rangle}{\langle i_{DS}^2 \rangle}$

(d) $\left(\dfrac{S}{N}\right)_T = \dfrac{\langle i_s^2 \rangle}{\langle i_T^2 \rangle}$

What happens to these curves if either the load resistor, the gain, the dark current, or the bandwidth is changed?

6-7 Suppose an avalanche photodiode has the following parameters: $I_L = 1$ nA, $I_D = 1$ nA, $\eta = 0.85$, $F = M^{1/2}$, $R_L = 10^3 \; \Omega$, and $B = 1$ kHz. Consider a sinusoidally varying 850-nm signal, which has a modulation index $m = 0.85$ and an average power level $P_0 = -50$ dBm, to fall on the detector at room temperature. Plot the signal-to-noise ratio as a function of M for gains ranging from 20 to 100. At what value of M does the maximum signal-to-noise ratio occur?

6-8 Derive Eq. (6-19).

Section 6-3

6-9 (a) Show that under the boundary conditions

$$p_n = p_{n0} \qquad \text{for } x = \infty$$

and

$$p_n = 0 \qquad \text{for } x = w$$

the solution to Eq. (6-23) is given by

$$p_n = p_{n0} - (p_{n0} + Be^{-\alpha_s w})e^{(w-x)/L_p} + Be^{-\alpha_s x}$$

where $L_p = (D_p \tau_p)^{1/2}$ is the diffusion length and

$$B = \left(\frac{\Phi_0}{D_p}\right)\frac{\alpha_s L_p^2}{1 - \alpha_s^2 L_p^2}$$

(b) Derive Eq. (6-25) using the relationship

$$J_{\text{diff}} = qD_p\left(\frac{\partial p_n}{\partial x}\right)_{x=w}$$

(c) Verify that J_{tot} is given by Eq. (6-26).

6-10 Consider a modulated photon flux density

$$\Phi = \Phi_0 e^{j\omega t} \qquad \text{photons}/(\text{s} \cdot \text{cm}^2)$$

to fall on a photodetector, where ω is the modulation frequency. The total current through the depletion region generated by this photon flux can be shown to be[37]

$$J_{\text{tot}} = \left(\frac{j\omega \epsilon_s V}{w} + q\Phi_0 \frac{1 - e^{-j\omega t_d}}{j\omega t_d} \right) e^{j\omega t}$$

where ϵ_s is the material permittivity, V is the voltage across the depletion layer, and t_d is the transit time of carriers through the depletion region.

    (*a*) From the short-circuit current density ($V = 0$), find the value of ωt_d at which the photocurrent amplitude is reduced by $\sqrt{2}$.

    (*b*) If the depletion region thickness is assumed to be $1/\alpha_s$, what is the 3-dB modulation frequency in terms of α_s and v_d (the drift velocity)?

6-11 Suppose we have a silicon *pin* photodiode which has a depletion layer width $w = 20$ μm, an area $A = 0.05$ mm^2, and a dielectric constant $K_s = 11.7$. If the photodiode is to operate with a 10-kΩ load resistor at 800 nm, where the absorption coefficient $\alpha_s = 10^3$ cm^{-1}, compare the *RC* time constant and the carrier drift time of this device. Is carrier diffusion time of importance in this photodiode?

Section 6-4

6-12 Verify that, when the weighted ionization rate ratios k_1 and k_2 are assumed to be approximately equal, Eqs. (6-34) and (6-35) can be simplified to yield Eqs. (6-38) and (6-39).

6-13 Derive the limits of F_e given by Eq. (6-38) when (*a*) only electrons cause ionization; (*b*) the ionization rates α and β are equal.

REFERENCES

1. H. Melchior, "Sensitive high speed photodetectors for the demodulation of visible and near infra-red light," *J. Luminescence,* **7,** 390–414, 1973.
2. H. Melchior, "Detectors for lightwave communications," *Phys. Today,* **30,** 32–39, Nov. 1977.
3. E. H. Putley, "The pyro-electric detector," in *Semiconductors and Semimetals,* R. K. Willardson and A. C. Beer (Eds.), vol. 5, Academic, New York, 1970; vol. 12, Academic, New York, 1977.
4. L. A. Murray, "Selecting a photodetector," *Electro-Opt. Sys. Design,* **10,** 38–42, July 1978.
5. S. M. Sze, *Physics of Semiconductor Devices,* 2d ed., chap 13, Wiley, New York, 1981.
6. G. E. Stillman and C. M. Wolfe, "Avalanche photodiodes," in *Semiconductors and Semimetals,* R. K. Willardson and A. C. Beer (Eds.), vol. 12, Academic, New York, 1977.
7. P. P. Webb, R. J. McIntyre, and J. Conradi, "Properties of avalanche photodiodes," *RCA Rev.,* **35,** 234–278, June 1974.
8. T. P. Lee and T. Li, "Photodetectors," in *Optical Fiber Communications,* S. E. Miller and A. C. Chynoweth (Eds.), Academic, New York, 1979.
9. P. L. Carni, "Photodetectors," in *Optical Fibre Communication,* Tech. Staff of CSELT, McGraw-Hill, New York, 1980, pt. II, chap. 2.
10. S. E. Miller, E. A. J. Marcatili, and T. Li, "Research toward optical-fiber transmission systems," *Proc. IEEE,* **61,** 1703–1751, Dec. 1973.
11. H. W. Ruegg, "An optimized avalanche photodiode," *IEEE Trans. Electron Devices,* **ED-14,** 239–251, May 1967.
12. K. Berchtold, O. Krumpholz, and J. Suri, "Avalanche photodiodes with a gain-bandwidth product of more than 200 GHz," *Appl. Phys. Lett.,* **26,** 585–587, May 1975.
13. H. Kanbe, T. Kimura, Y. Mizushima, and K. Kajiyama, "Silicon avalanche photodiodes with low multiplication noise and high-speed response," *IEEE Trans. Electron Devices,* **ED-23,** 1337–1343, Dec. 1976.

14. T. Kaneda, H. Matsumoto, and T. Yamaoka, "A model for reach-through avalanche photo-diodes," *J. Appl. Phys.*, **47**, 3135–3139, July 1976.

15. H. Melchior, A. R. Hartman, D. P. Schinke, and T. E. Seidel, "Planar epitaxial silicon avalanche photodiode," *Bell Sys. Tech. J.*, **57**, 1791–1807, July–Aug. 1978.

16. H. Melchior and W. T. Lynch, "Signal and noise response of high speed germanium avalanche photodiodes," *IEEE Trans. Electron Devices*, **ED-13**, 829–838, Dec. 1966.

17. H. Ando, H. Kanbe, T. Kimura, T. Yamaoka, and T. Kaneda, "Characteristics of germanium avalanche photodiodes in the wavelength region of 1–1.6 μm," *IEEE J. Quantum Electron.*, **QE-14**, 804–809, Nov. 1978.

18. T. P. Lee, C. A. Burrus, A. G. Dentai, A. A. Ballman, and W. A. Bonner, "High avalanche gain in small-area InP photodiodes," *Appl. Phys. Lett.*, **35**, 511-513, Oct. 1979.

19. R. Yeats and S. H. Chiao, "Long-wavelength InGaAsP avalanche photodiodes," *Appl. Phys. Lett.*, **34**, 581–583, May 1979; "Leakage current in InGaAsP avalanche photodiodes," *ibid.*, **36**, 167–170, Jan. 1980.

20. T. Shirai, S. Yamasaki, F. Osaka, K. Nakajima, and T. Kaneda, "Multiplication noise in planar InP/InGaAsP heterostructure avalanche photodiodes," *Appl. Phys. Lett.*, **40**, 532–533, Mar. 1982.

21. M. A. Washington, R. E. Nahory, M. A. Pollack, and E. D. Beebe, "High efficiency $In_{1-x}Ga_xAs_yP_{1-y}$/InP photodetectors with selective wavelength response between 0.9 and 1.7 μm," *Appl. Phys. Lett.*, **33**, 854–856, Nov. 1978.

22. C. E. Hurwitz and J. J. Hsieh, "GaInAsP/InP avalanche photodiodes," *Appl. Phys. Lett.*, **32**, 487–489, Apr. 1978.

23. F. Capasso, M. B. Panish, S. Sumski, and P. W. Foy, "Very high quantum efficiency GaSb mesa photodetectors between 1.3 and 1.6 μm," *Appl. Phys. Lett.*, **36**, 165–167, Jan. 1980.

24. Y. Nagao, T. Hariu, and Y. Shibata, "GaSb Schottky diodes for infrared detectors," *IEEE Trans. Electron Devices*, **ED-28**, 407–411, Apr. 1981.

25. H. D. Law, L. R. Tomasetta, K. Nakano, and J. S. Harris, "1.0–1.4 μm high speed avalanche photodiodes," *Appl. Phys. Lett.*, **33**, 416–417, Sept. 1978.

26. L. R. Tomasetta, H. D. Law, R. C. Eden, I. Deyhimy, and K. Nakano, "High sensitivity optical receivers for 1.0–1.4 μm fiber optic systems," *IEEE J. Quantum Electron.*, **QE-14**, 800–804, Nov. 1978.

27. M. Chu, S. H. Shin, H. D. Law, and D. T. Cheung, "1.33 μm HgCdTe/CdTe photodiodes," *Appl. Phys. Lett.*, **37**, 318–320, Aug. 1980.

28. T. P. Pearsall, "$Ga_{0.47}In_{0.53}As$: A ternary semiconductor for photodiode application," *IEEE J. Quantum Electron.*, **QE-16**, 709–720, July 1980.

29. S. R. Forrest, R. F. Leheny, R. E. Nahory, and M. A. Pollack, "$In_{0.53}Ga_{0.47}As$ photodiodes with dark current limited by generation-recombination and tunneling," *Appl. Phys. Lett.* **37**, 322–325, Aug. 1980.

30. N. Susa, Y. Yamauchi, and H. Kanbe, "Vapor phase epitaxially grown InGaAs photodiodes," *IEEE Trans. Electron Devices*, **ED-27**, 92–98, Jan. 1980.

31. O. K. Kim, S. R. Forrest, W. A. Bonner, and R. G. Smith, "A high gain $In_{0.53}Ga_{0.47}As$/InP avalanche photodiode with no tunneling leakage current," *Appl. Phys. Lett.*, **39**, 402–404, Sept. 1981.

32. A series of papers on photodetectors made of InGaAs and other materials can be found in "Special issue on quaternary compound semiconductor materials and devices—sources and detectors," *IEEE J. Quantum Electron.*, **QE-17**, 210–288, Feb. 1981.

33. (*a*) J. Yamada, M. Saruwatari, K. Asatani, H. Tsuchiya, A. Kawana, K. Sugiyama, and T. Kimura, "High-speed optical pulse transmission at 1.29 μm wavelength using low-loss single-mode fibers," *IEEE J. Quantum Electron.*, **QE-14**, 791–800, Nov. 1978.
    (*b*) M. C. Brain, "Comparison of available detectors for digital optical fiber systems for the 1.2–1.55 μm wavelength range," *ibid.*, **QE-18**, 219–224, Feb. 1982.

34. H. D. Law, K. Nakano, and L. R. Tomasetta, "III-V alloy heterostructure high speed avalanche photodiodes," *IEEE J. Quantum Electron.*, **QE-15**, 549–558, July 1979.

35. B. M. Oliver, "Thermal and quantum noise," *Proc. IEEE*, **53**, 436–454, May 1965.

36. W. M. Hubbard, "Utilization of optical-frequency carriers for low and moderate bandwidth channels," *Bell Sys. Tech. J.*, **52**, 731–765, May–June 1973.

37. W. W. Gaertnar, "Depletion-layer photoeffects in semiconductors," *Phys. Rev.*, **116**, 84–87, Oct. 1959.
38. S. D. Personick, "New results on avalanche multiplication statistics with applications to optical detection," *Bell Sys. Tech. J.*, **50**, 167–189, Jan. 1971.
39. S. D. Personick, "Statistics of a general class of avalanche detectors with applications to optical communications," *Bell Sys. Tech. J.*, **50**, 3075–3096, Dec. 1971.
40. R. J. McIntyre, "The distribution of gains in uniformly multiplying avalanche photodiodes: Theory," *IEEE Trans. Electron Devices*, **ED-19**, 703–713, June 1972.
41. J. Conradi, "The distribution of gains in uniformly multiplying avalanche photodiodes: Experimental," *IEEE Trans. Electron Devices*, **ED-19**, 713–718, June 1972.
42. T. Mikawa, S. Kagawa, T. Kaneda, Y. Toyama, and O. Mikami, "Crystal orientation dependence of ionization rates in germanium," *Appl. Phys. Lett.*, **37**, 387–389, Aug. 1980.
43. C. R. Crowell and S. M. Sze, "Temperature dependence of avalanche multiplication in semiconductors," *Appl. Phys. Lett.*, **9**, 242–244, Sept. 1966.
44. J. Conradi, "Temperature effects in silicon avalanche photodiodes," *Solid State Electron.*, **17**, 99–106, Jan. 1974.
45. S. L. Miller, "Avalanche breakdown in germanium," *Phys. Rev.*, **99**, 1234–1241, Aug. 1955.
46. M. S. Tyagi, "Zener and avalanche breakdown in silicon alloyed p-n junction," *Solid State Electron.*, **11**, 99–115, Feb. 1968.
47. N. Susa, H. Nakagome, H. Ando, and H. Kanbe, "Characteristics in InGaAs/InP avalanche photodiodes with separated absorption and multiplication regions," *IEEE J. Quantum Electron.*, **QE-17**, 243–250, Feb. 1981.
48. J. C. Campbell, A. G. Dentai, C. A. Burrus, Jr., and J. F. Ferguson, "InP/InGaAs heterojunction phototransistors," *IEEE J. Quantum Electron.*, **QE-17**, 264–269, Feb. 1981.
49. K. Tabatabaie-Alavi and C. G. Fonstad, Jr., "Performance comparison of heterojunction phototransistors, *pin* FETs, and APD-FETs for optical fiber communication systems," *IEEE J. Quantum Electron.*, **QE-17**, 2259–2261, Feb. 1981.
50. R. A. Milano, P. D. Dapkus, and G. E. Stillman, "An analysis of the performance of heterojunction phototransistors for fiber optic communications," *IEEE Trans. Electron Devices*, **ED-29**, 266–274, Feb. 1982.

SEVEN

OPTICAL RECEIVER OPERATION

Having discussed the characteristics and operation of photodetectors in the previous chapter, we now turn our attention to the optical receiver. An optical receiver consists of a photodetector, an amplifier, and signal-processing circuitry. It has the task of first converting the optical energy emerging from the end of a fiber into an electric signal, and then amplifying this signal to a large enough level so that it can be processed by the electronics following the receiver amplifier.

In these processes various noises and distortions will unavoidably be introduced which can lead to errors in the interpretation of the received signal. As we saw in the previous chapter, the current generated by the photodetector is generally very weak and is adversely affected by the random noises associated with the photodetection process. When this electric signal output from the photodiode is amplified, additional noises arising from the amplifier electronics will further corrupt the signal. Noise considerations are thus important in the design of optical receivers, since the noise sources operating in the receiver generally set a lower limit for the signals that can be processed.

In designing a receiver it is desirable to predict its performance based on mathematical models of the various receiver stages. These models must take into account the noises and distortions added to the signal by the components in each stage, and they must show the designer which components to choose so that the desired performance criteria of the receiver are met.

The most meaningful criterion for measuring the performance of a digital communication system is the average error probability. In an analog system the fidelity criterion is usually specified in terms of a peak signal-to-rms-noise ratio. The calculation of the error probability for a digital optical communication receiver differs from that of conventional electric systems. This is because of the discrete quantum nature of the optical signal and also because of the probabilistic character of the gain process when an avalanche photodiode is used. Various authors[1-15] have used different numerical methods to derive approximate predictions for receiver performance. In carrying out these approximations a tradeoff generally results between simplicity of the analysis and accuracy of the approximation.

In this chapter we first present an overview of the fundamental operational characteristics of the various stages of an optical receiver. This consists of tracing the path of a digital signal through the receiver and showing what happens at each step along the way. This is followed in Sec. 7-2 by mathematical models for predicting the performance of a digital receiver under various noise and distortion conditions. Practical receiver design examples and their performance predictions based on these models are discussed next. The chapter concludes with an analysis of analog receiver systems.

7-1 FUNDAMENTAL RECEIVER OPERATION

The design of an optical receiver is much more complicated than that of an optical transmitter because the receiver must first detect weak, distorted signals and then make decisions on what type of data was sent based on an amplified version of this distorted signal. To get an appreciation of the function of the optical receiver, we first examine what happens to a signal as it is sent through the optical data link shown in Fig. 7-1. Since most fiber optic systems use a two-level binary digital signal, we shall analyze receiver performance by using this signal form first. Analog systems are discussed in Sec. 7-4.

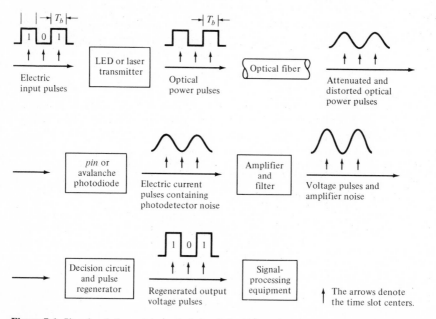

Figure 7-1 Signal path through an optical data link. *(Adapted with permission from Personick et al.,[5] copyright 1977, IEEE.)*

7-1-1 Digital Signal Transmission

A typical digital fiber transmission system is shown in Fig. 7-1. The transmitted signal is a two-level binary data stream consisting of either a 0 or a 1 in a time slot of duration T_b. This time slot is referred to as a *bit period*. Electrically there are many ways of sending a given digital message.[16] One of the simplest (but not necessarily the most efficient) techniques for sending binary data is amplitude-shift keying, wherein a voltage level is switched between two values which are usually *on* and *off*. The resultant signal wave thus consists of a voltage pulse of amplitude V relative to the zero voltage level when a binary 1 occurs and a zero voltage level space when a binary 0 occurs. Depending on the coding scheme to be used, a 1 may or may not fill the time slot T_b. For simplicity here we assume that when a 1 is sent, a voltage pulse of duration T_b occurs, whereas for a 0 the voltage remains at its zero level. A discussion of more efficient transmission codes is given in Chap. 8.

The function of the optical transmitter is to convert the electric signal to an optical signal. As we saw in Chap. 4, an electric current $i(t)$ can be used to modulate directly an optical source (either an LED or a laser diode) to produce an optical output power $P(t)$. Thus, in the optical signal emerging from the transmitter, a 1 is represented by a pulse of optical power (light) of duration T_b, whereas a 0 is the absence of any light.

The optical signal that gets coupled from the light source to the fiber becomes attenuated and distorted as it propagates along the fiber waveguide. Upon reaching the receiver either a *pin* or an avalanche photodiode converts the optical signal back to an electrical format. After the electric signal produced by the photodetector is amplified and filtered, a decision circuit compares the signal in each time slot to a certain reference voltage known as the *threshold level*. If the received signal level is greater than the threshold level, a 1 is said to have been received. If the voltage is below the threshold level, a 0 is assumed to have been received.

7-1-2 Error Sources

Errors in the detection mechanism can arise from various noises and disturbances associated with the signal detection system, as shown in Fig. 7-2. The term *noise* is used customarily to describe unwanted components of an electric signal that tend

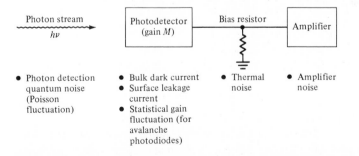

Figure 7-2 Noise sources and disturbances in the optical pulse detection mechanism.

to disturb the transmission and processing of the signal in a physical system, and over which we have incomplete control. The noise sources can be either external to the system (for example, atmospheric noise, equipment-generated noise) or internal to the system. Here we shall be concerned mainly with internal noise, which is present in every communication system and represents a basic limitation on the transmission or detection of signals. This noise is caused by the spontaneous fluctuations of current or voltage in electric circuits. The two most common examples of these spontaneous fluctuations are shot noise and thermal noise. Shot noise arises in electronic devices because of the discrete nature of current flow in the device. Thermal noise arises from the random motion of electrons in a conductor. Detailed treatments of electric noise may be found in Ref. 17.

As we saw in Chap. 6, the random arrival rate of signal photons produces a quantum (or shot) noise at the photodetector. Since this noise depends on the signal level, it is of particular importance for *pin* receivers having large optical input levels and for avalanche photodiode receivers. When using an avalanche photodiode an additional shot noise arises from the statistical nature of the multiplication process. This noise level increases with increasing avalanche gain M. Additional photodetector noises come from the dark current and leakage current. These are independent of the photodiode illumination and can generally be made very small in relation to other noise currents by a judicious choice of components.

Thermal noises arising from the detector load resistor and from the amplifier electronics tend to dominate in applications with low signal-to-noise ratio when a *pin* photodiode is used. When an avalanche photodiode is used in low optical signal level applications, the optimum avalanche gain is determined by a design tradeoff between the thermal noise and the gain-dependent quantum noise.

Since the thermal noises are of a gaussian nature, they can be readily treated by standard techniques. This is shown in Sec. 7-2. The analysis of the noises and the resulting error probabilities associated with the primary photocurrent generation and the avalanche multiplication are complicated since neither of these processes are gaussian. The primary photocurrent generated by the photodiode is a time-varying Poisson process resulting from the random arrival of photons at the detector. If the detector is illuminated by an optical signal $P(t)$, then the average number of electron-hole pairs $\overline{N}$ generated in a time τ is

$$\overline{N} = \frac{\eta}{h\nu} \int_0^\tau P(t) \, dt = \frac{\eta E}{h\nu} \tag{7-1}$$

where η is the detector quantum efficiency, $h\nu$ is the photon energy, and E is the energy received in a time interval τ. The actual number of electron-hole pairs n that are generated fluctuates from the average according to the Poisson distribution

$$P_r(n) = \overline{N}^n \frac{e^{-\overline{N}}}{n!} \tag{7-2}$$

where $P_r(n)$ is the probability that n electrons are emitted in an interval τ. The reason that it is not possible to predict exactly how many electron-hole pairs are generated by a known optical power incident on the detector is the origin of the type

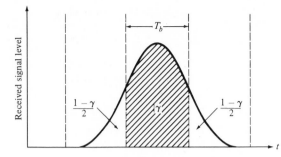

Figure 7-3 Pulse spreading in an optical signal which leads to intersymbol interference.

of shot noise called *quantum noise*. The random nature of the avalanche multiplication process gives rise to another type of shot noise. Recall from Chap. 6 that, for a detector with a mean avalanche gain M and an ionization ratio k, the excess noise factor $F(M)$ for electron injection is

$$F(M) = kM + \left(2 - \frac{1}{M}\right)(1 - k)$$

This equation is often approximated by the empirical expression

$$F(M) \simeq M^x \tag{7-3}$$

where the factor x ranges between 0 and 1.0 depending on the photodiode material.

A further error source is attributed to *intersymbol interference* (ISI), which results from pulse spreading in the optical fiber. When a pulse is transmitted in a given time slot, most of the pulse energy will arrive in the corresponding time slot at the receiver, as shown in Fig. 7-3. However, because of pulse spreading induced by the fiber, some of the transmitted energy will progressively spread into neighboring time slots as the pulse propagates along the fiber. The presence of this energy in adjacent time slots results in an interfering signal, hence the term *intersymbol interference*. In Fig. 7-3 the fraction of energy remaining in the appropriate time slot is designated by γ, so that $1 - \gamma$ is the fraction of energy that has spread into adjacent time slots.

7-1-3 Receiver Configuration

A schematic diagram of a typical optical receiver is shown in Fig. 7-4. The three basic stages of the receiver are a photodetector, an amplifier, and an equalizer. The photodetector can be either an avalanche photodiode with a mean gain M or a *pin* photodiode for which $M = 1$. The photodiode has a quantum efficiency η and a capacitance C_d. The detector bias resistor has a resistance R_b which generates a thermal noise current $i_b(t)$.

The amplifier has an input impedance represented by the parallel combination of a resistance R_a and a shunt capacitance C_a. Voltages appearing across this impedance cause current to flow in the amplifier output. This amplifying function is represented by the voltage-controlled current source which is characterized by a

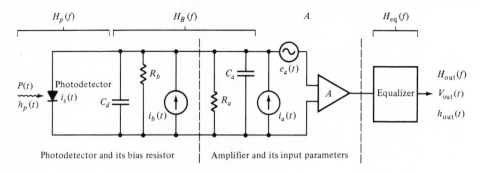

Figure 7-4 Schematic diagram of a typical optical receiver.

transconductance g_m (given in amperes/volt or siemens). There are two amplifier noise sources: The input noise current source $i_a(t)$ arises from the thermal noise of the amplifier input resistance R_a, whereas the noise voltage source $e_a(t)$ represents the thermal noise of the amplifier channel. These noise sources are assumed gaussian in statistics, flat in spectrum (which characterizes *white* noise), and uncorrelated (statistically independent). They are thus completely described by their noise spectral densities[16] S_I and S_E (see App. F).

The equalizer that follows the amplifier is normally a linear frequency-shaping filter that is used to mitigate the effects of signal distortion and intersymbol interference. Ideally[18] it accepts the combined frequency response of the transmitter, the transmission medium, and the receiver, and transforms it into a signal response that is suitable for the following signal-processing electronics. In some cases, the equalizer may be used only to correct for the electric frequency response of the photodetector and the amplifier.

To account for the fact that the rectangular digital pulses that were sent out by the transmitter arrive rounded and distorted at the receiver, the binary digital-pulse-train incident on the photodetector can be described by

$$P(t) = \sum_{n=-\infty}^{\infty} b_n h_p(t - nT_b) \tag{7-4}$$

Here $P(t)$ is the received optical power, T_b is the bit period, b_n is an amplitude parameter representing the nth message digit, and $h_p(t)$ is the received pulse shape which is positive for all t. For binary data the parameter b_n can take on the two values b_{on} and b_{off} corresponding to a binary 1 and 0, respectively. If we let the nonnegative photodiode input pulse $h_p(t)$ be normalized to have unit area

$$\int_{-\infty}^{\infty} h_p(t) \, dt = 1 \tag{7-5}$$

then b_n represents the energy in the nth pulse.

The mean output current from the photodiode at time t resulting from the pulse train given in Eq. (7-4) is (neglecting dc components arising from dark noises)

$$\langle i(t) \rangle = \frac{\eta q}{h\nu} MP(t) = \mathcal{R}_0 M \sum_{n=-\infty}^{\infty} b_n h_p(t - nT_b) \qquad (7\text{-}6)$$

where $\mathcal{R}_0 = \eta q / h\nu$ is the photodiode responsivity as given in Eq. (6-6). This current is then amplified and filtered to produce a mean voltage at the output of the equalizer given by the convolution of the current with the amplifier impulse response (see App. F):

$$\langle v_{\text{out}}(t) \rangle = A\mathcal{R}_0 MP(t) * h_B(t) * h_{\text{eq}}(t)$$

$$= \mathcal{R}_0 GP(t) * h_B(t) * h_{\text{eq}}(t) \qquad (7\text{-}7)$$

Here A is the amplifier gain, we define $G = AM$ for brevity, $h_B(t)$ is the impulse response of the bias circuit, $h_{\text{eq}}(t)$ is the equalizer impulse response, and $*$ denotes convolution.

From Fig. 7-4 $h_B(t)$ is given by the inverse Fourier transform of the bias circuit transfer function $H_B(f)$:

$$h_B(t) = F^{-1}[H_B(f)] = \int_{-\infty}^{\infty} H_B(f) e^{j2\pi ft} \, df \qquad (7\text{-}8)$$

where F denotes the Fourier transform operation. The bias circuit transfer function $H_B(f)$ is simply the impedance of the parallel combination of R_b, R_a, C_d, and C_a:

$$H_B(f) = \frac{1}{1/R + j2\pi fC} \qquad (7\text{-}9)$$

where

$$\frac{1}{R} = \frac{1}{R_a} + \frac{1}{R_b} \qquad (7\text{-}10)$$

and

$$C = C_a + C_d \qquad (7\text{-}11)$$

Analogous to Eq. (7-4) the mean voltage output from the equalizer can be written in the form

$$\langle v_{\text{out}}(t) \rangle = \sum_{n=-\infty}^{\infty} b_n h_{\text{out}}(t - nT_b) \qquad (7\text{-}12)$$

where

$$h_{\text{out}}(t) = \mathcal{R}_0 G h_p(t) * h_B(t) * h_{\text{eq}}(t) \qquad (7\text{-}13)$$

is the shape of an isolated amplified and filtered pulse. The Fourier transform of Eq. (7-13) can be written as[16] (see App. F)

$$H_{\text{out}}(f) = \int_{-\infty}^{\infty} h_{\text{out}}(t) e^{-j2\pi ft} \, dt = \mathcal{R}_0 G H_p(f) H_B(f) H_{\text{eq}}(f) \qquad (7\text{-}14)$$

Here $H_p(f)$ is the Fourier transform of the received pulse shape $h_p(t)$ and $H_{eq}(f)$ is the transfer function of the equalizer.

7-2 DIGITAL RECEIVER PERFORMANCE CALCULATION

In a digital receiver the amplified and filtered signal emerging from the equalizer is compared to a threshold level once per time slot to determine whether or not a pulse is present at the photodetector in that time slot. Ideally the output signal $v_{out}(t)$ would always exceed the threshold voltage when a 1 is present and would be less than the threshold when no pulse (a 0) was sent. In actual systems, deviations from the average value of $v_{out}(t)$ are caused by various noises, interference from adjacent pulses, and conditions wherein the light source is not completely extinguished during a zero pulse.

In practice there are several standard ways of measuring the rate of error occurrences in a digital data stream.[19] One common approach is to divide the number of errors N_e occurring over a certain time interval t by the number of pulses (ones and zeros) N_t transmitted during this interval. This is called either the *error rate* or the *bit error rate,* which is commonly abbreviated BER. Thus we have

$$\text{BER} = \frac{N_e}{N_t} = \frac{N_e}{Bt} \tag{7-15}$$

where $B = 1/T_b$ is the bit rate (that is, the pulse transmission rate). The error rate is expressed by a number such as 10^{-6}, for example, which states that on the average one error occurs for every million pulses sent. Typical error rates for optical fiber telecommunication systems range from 10^{-6} to 10^{-10}. This error rate depends on the signal-to-noise ratio at the receiver (the ratio of signal power to noise power). The system error rate requirements and the receiver noise levels thus set a lower limit on the optical signal power level that is required at the photodetector.

To compute the bit error rate at the receiver, we have to know the probability distribution[20] of the signal at the equalizer output. Knowing the signal probability distribution at this point is important because it is here that the decision is made as to whether a 0 or a 1 was sent. The shapes of two signal probability distributions are shown in Fig. 7-5. These are

$$P_1(v) = \int_{-\infty}^{v} p(y \mid 1) \, dy \tag{7-16}$$

which is the probability that the equalizer output voltage is less than v when a 1 pulse was sent, and

$$P_0(v) = \int_{v}^{\infty} p(y \mid 0) \, dy \tag{7-17}$$

which is the probability that the output voltage exceeds v when a 0 was transmitted. The functions $p(y \mid 1)$ and $p(y \mid 0)$ are the conditional probability distribution

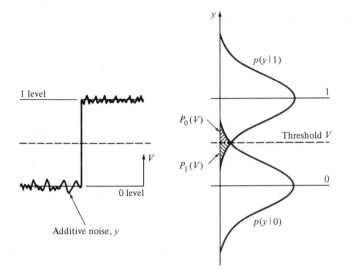

Figure 7-5 Probability distributions for 0 and 1 signal levels.

functions;[6, 20] that is, $p(y \mid x)$ is the probability that the output voltage is y, given that an x was transmitted.

If the threshold voltage is v_{th} then the error probability P_e is defined as

$$P_e = aP_1(v_{th}) + bP_0(v_{th}) \tag{7-18}$$

The weighting factors a and b are determined by the *a priori* distribution of the data. That is, a and b are the probabilities that either a 1 or 0 occurs, respectively. For unbiased data with equal probability of 1 and 0 occurrences, $a = b = 0.5$. The problem to be solved now is to select the decision threshold at that point where P_e is minimum.

To calculate the error probability we require a knowledge of the mean square noise voltage $\langle v_N^2 \rangle$ which is superimposed on the signal voltage at the decision time. The statistics of the output voltage at the sampling time are very complicated, so that an exact calculation is rather tedious to perform. A number of different approximations[1-15] have therefore been used to calculate the performance of a binary optical fiber receiver. In applying these approximations, we have to make a tradeoff between computational simplicity and accuracy of the results. The simplest method is based on a gaussian approximation. In this method it is assumed that, when the sequence of optical input pulses is known, the equalizer output voltage $v_{out}(t)$ is a gaussian random variable. Thus, to calculate the error probability, we only need to know the mean and standard deviation of $v_{out}(t)$. Other approximations which have been investigated are more involved[5, 12, 14, 15] and will not be discussed here.

For simplicity we shall first assume that the optical power is completely extinguished in those time slots where a 0 occurs. In practice there usually is some optical signal in the 0 time slot of a baseband binary signal, since the optical source is often biased slightly on at all times. As we saw in Chap. 4, this is done to increase the response speed of the light source. The effect of biasing the light source slightly on during a 0 time slot results in a nonzero *extinction ratio* ϵ. This is defined as the ratio of the optical power in a 0 pulse to the power in a 1 pulse. Its effect on receiver performance is discussed in Sec. 7-2-5.

7-2-1 Receiver Noises

We now turn our attention to calculating the noise voltages. If $v_N(t)$ is the noise voltage causing $v_{\text{out}}(t)$ to deviate from its average value, then the actual equalizer output voltage is of the form

$$v_{\text{out}}(t) = \langle v_{\text{out}}(t) \rangle + v_N(t) \tag{7-19}$$

The noise voltage at the equalizer output for the receiver shown in Fig. 7-4 can be represented by

$$v_N^2(t) = v_s^2(t) + v_R^2(t) + v_I^2(t) + v_E^2(t) \tag{7-20}$$

where

$v_s(t)$ is the quantum (or shot) noise resulting from the random multiplied Poisson nature of the photocurrent $i_s(t)$ produced by the photodetector.
$v_R(t)$ is the thermal (or Johnson) noise associated with the bias resistor R_b.
$v_I(t)$ results from the amplifier input noise current source $i_a(t)$.
$v_E(t)$ results from the amplifier input voltage noise source $e_a(t)$.

The amplifier noise sources will be assumed as independent of each other and gaussian in their statistics.

Here we are interested in the mean square noise voltage $\langle v_N^2 \rangle$, which is given by

$$\langle v_N^2 \rangle = \langle [v_{\text{out}}(t) - \langle v_{\text{out}}(t) \rangle]^2 \rangle$$
$$= \langle v_{\text{out}}^2(t) \rangle - \langle v_{\text{out}}(t) \rangle^2$$
$$= \langle v_s^2(t) \rangle + \langle v_R^2(t) \rangle + \langle v_I^2(t) \rangle + \langle v_E^2(t) \rangle \tag{7-21}$$

We shall first evaluate the last three thermal noise terms of Eq. (7-21) at the output of the equalizer. The thermal noise of the load resistor R_b is[16]

$$\langle v_R^2(t) \rangle = \frac{4k_B T}{R_b} B_{bae} R^2 A^2 \tag{7-22}$$

Here $k_B T$ is Boltzmann's constant times the absolute temperature, R is given by Eq. (7-10), A is the amplifier gain, and B_{bae} is the noise equivalent bandwidth of the bias circuit, amplifier, and equalizer defined for positive frequencies only:[16]

$$2B_{bae} = \frac{1}{|H_B(0)H_{eq}(0)|^2} \int_{-\infty}^{\infty} |H_B(f)H_{eq}(f)|^2 \, df$$

$$= \frac{1}{|H_{out}(0)/H_p(0)|^2} \int_{-\infty}^{\infty} \left| \frac{H_{out}(f)}{H_p(f)} \right|^2 \, df \qquad (7\text{-}23)$$

where we have used Eq. (7-14) for the last equality.

Since the thermal noise contributions from the amplifier input noise current source $i_a(t)$ and from the amplifier input noise voltage source $e_a(t)$ are assumed to be gaussian and independent, they are completely characterized by their noise spectral densities.[16] Thus,

$$\langle v_I^2(t) \rangle = 2S_I B_{bae} R^2 A^2 \qquad (7\text{-}24)$$

and

$$\langle v_E^2(t) \rangle = 2S_E B_e A^2 \qquad (7\text{-}25)$$

where S_I is the spectral density of the amplifier input noise current source (measured in amperes squared per hertz), S_E is the spectral density of the amplifier noise voltage source (measured in volts squared per hertz), and

$$2B_e = \frac{1}{|H_{eq}(0)|^2} \int_{-\infty}^{\infty} |H_{eq}(f)|^2 \, df$$

$$= \frac{R^2}{|H_{out}(0)/H_p(0)|^2} \int_{-\infty}^{\infty} \left| \frac{H_{out}(f)}{H_p(f)} \left(\frac{1}{R} + j2\pi fC \right) \right|^2 \, df \qquad (7\text{-}26)$$

is the noise equivalent bandwidth of the equalizer. The last equality comes from Eq. (7-14). The noise spectral densities are described further in Sec. 7-3.

7-2-2 Shot Noise

The nongaussian nature of the photodetection process and the avalanche multiplication noise makes the evaluation of the shot noise term $\langle v_s^2(t) \rangle$ more difficult than that of the thermal noise. Personick[1] carried out a detailed analysis that evaluated the shot noise as a function of time within the bit slot. This results in an accurate estimate of the shot noise contribution to the equalizer output noise voltage, but at the expense of computational difficulty.

Smith and Garrett[8] subsequently proposed a simplification of Personick's expressions by relating the mean square shot noise voltage $\langle v_s^2(t) \rangle$ at the decision time to the average unity gain photocurrent $\langle i_0 \rangle$ over the bit time T_b through the standard shot noise expression[17]

$$\langle v_s^2(t) \rangle = 2q \langle i_0 \rangle \langle m^2 \rangle B_{bae} R^2 A^2 \qquad (7\text{-}27)$$

Here $\langle m^2 \rangle$ is the mean square avalanche gain [Eq. (6-32)], which we shall assume takes the form M^{2+x} with $0 < x \leq 1.0$. The other terms are as defined in Eq. (7-22).

We now calculate $\langle i_0 \rangle$ at the decision time within a particular bit slot. For this we must take into account not only the shot noise contribution from a pulse within this particular time slot but also the shot noises resulting from all other pulses that overlap into this bit period. The shot noise within a time slot will thus depend on the shape of the received pulse (that is, how much of it has spread into adjacent time slots as shown in Fig. 7-3) and on the data sequence (the distribution of 1 and 0 pulses in the data stream). The worst case of shot noise in any particular time slot occurs when all neighboring pulses are 1, since this causes the greatest amount of intersymbol interference. For this case the mean unity gain photocurrent over a bit time T_b for a 1 pulse is

$$\langle i_0 \rangle_1 = \sum_{n=-\infty}^{\infty} \frac{\eta q}{h\nu} b_{on} \frac{1}{T_b} \int_{-T_b/2}^{T_b/2} h_p(t - nT_b)\, dt$$

$$= \frac{\eta q}{h\nu} \frac{b_{on}}{T_b} \int_{-\infty}^{\infty} h_p(t)\, dt = \frac{\eta q}{h\nu} \frac{b_{on}}{T_b} \tag{7-28}$$

where we have made use of Eq. (7-5).

For a 0 pulse (with all adjacent pulses being 1), we assume $b_{off} = 0$, so that

$$\langle i_0 \rangle_0 = \sum_{n \neq 0} \frac{\eta q}{h\nu} b_{on} \frac{1}{T_b} \int_{-T_b/2}^{T_b/2} h_p(t - nT_b)\, dt$$

$$= \frac{\eta q}{h\nu} \frac{b_{on}}{T_b} \left[\sum_{n=-\infty}^{\infty} \int_{-T_b/2}^{T_b/2} h_p(t - nT_b)\, dt - \int_{-T_b/2}^{T_b/2} h_p(t)\, dt \right]$$

$$= \frac{\eta q}{h\nu} \frac{b_{on}}{T_b} (1 - \gamma) \tag{7-29}$$

The parameter

$$\gamma = \int_{-T_b/2}^{T_b/2} h_p(t)\, dt \tag{7-30}$$

is the fractional energy of a 1 pulse that is contained within its bit period, as shown by the shaded area in Fig. 7-3. The factor $1 - \gamma$ is thus the fractional energy of a pulse that has spread outside of its bit period as it traveled through the optical fiber.

Equations (7-28) and (7-29) can now be substituted back into Eq. (7-27) to find the worst-case shot noise for a 1 and 0 pulse, respectively.

7-2-3 Receiver Sensitivity Calculation

To calculate the sensitivity of an optical receiver, we first simplify the noise voltage expressions by using the notation of Personick.[1] We begin by assuming that the equalized pulse stream has no intersymbol interference at the sampling times nT_b,

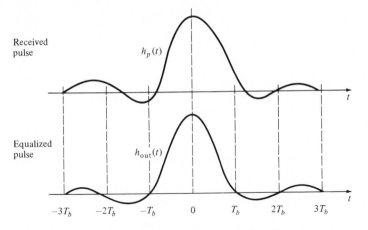

Figure 7-6 Equalized output pulse with no intersymbol interference at the decision time.

as shown in Fig. 7-6, and that the maximum value of $h_{\text{out}}(t)$ at $t = 0$ is unity. This means that

$$h_{\text{out}}(t = 0) = 1$$
$$h_{\text{out}}(t = nT_b) = 0 \qquad \text{for } n \neq 0$$

(7-31)

Substituting this into Eq. (7-12), we then have from Eq. (7-19) that the actual equalizer output voltage at the sampling times $t = nT_b$ is

$$v_{\text{out}} = b_n h_{\text{out}}(0) + v_N(nT_b)$$

(7-32)

This shows that the noise $v_N(t)$ depends on all the b_n values and on the time t.

Furthermore, we introduce the dimensionless time and frequency variables $\tau = t/T_b$ and $\phi = fT_b$ to make the bandwidth integrals of Eqs. (7-23) and (7-26) independent of the bit period T_b. This means that the numerical value will depend only on the shapes of the received and equalized pulses and not on their scale. Using these values it can be shown by considering the Fourier transforms of $h_p(\tau)$ and $h_{\text{out}}(\tau)$ that the normalized transforms, denoted by $H'_p(\phi)$ and $H'_{\text{out}}(\phi)$, are related to $H_p(f)$ and $H_{\text{out}}(f)$ by

$$H'_p(\phi) = H_p(f)$$
$$H'_{\text{out}}(\phi) = \frac{1}{T_b} H_{\text{out}}(f)$$

(7-33)

From these it follows that

$$I_2 = \frac{1}{T_b} \int_{-\infty}^{\infty} \left| \frac{H_{\text{out}}(f)}{H_p(f)} \right|^2 df = \int_{-\infty}^{\infty} \left| \frac{H'_{\text{out}}(\phi)}{H'_p(\phi)} \right|^2 d\phi$$

(7-34)

and

$$I_3 = T_b \int_{-\infty}^{\infty} \left| \frac{H_{\text{out}}(f)}{H_p(f)} \right|^2 f^2 \, df$$

$$= \int_{-\infty}^{\infty} \left| \frac{H'_{\text{out}}(\phi)}{H'_p(\phi)} \right|^2 \phi^2 \, d\phi \tag{7-35}$$

where we have used the notation of Personick[1] for the normalized dimensionless bandwidth integrals I_2 and I_3.

Thus using Eqs. (7-34) and (7-35) the bandwidth integrals in Eqs. (7-23) and (7-26) become

$$2B_{bae} = \frac{I_2}{T_b} \tag{7-36}$$

and

$$2B_e = \frac{I_2}{T_b} + \frac{(2\pi RC)^2}{T_b^3} I_3 \tag{7-37}$$

In evaluating Eqs. (7-23) and (7-26) we used the normalization conditions on $h_p(t)$ and $h_{\text{out}}(t)$ given by Eqs. (7-5) and (7-31), respectively, to derive the relationships

$$H_p(0) = 1 = H'_p(0) \qquad \text{and} \qquad H_{\text{out}}(0) = T_b$$

so that $H'_{\text{out}}(0) = 1$.

Using these expressions for B_{bae} and B_e and from Eqs. (7-22), (7-24), (7-25), and (7-27), the total mean square noise voltage in Eq. (7-21) becomes

$$\langle v_N^2 \rangle = \frac{R^2 A^2}{T_b} \left(q\langle i_0 \rangle M^{2+x} + \frac{2k_B T}{R_b} + S_I + \frac{S_E}{R^2} \right) I_2 + \frac{(2\pi RC)^2 A^2}{T_b^3} S_E I_3$$

$$= \left(\frac{qRA}{T_b} \right)^2 \left(\frac{\langle i_0 \rangle}{q} M^{2+x} T_b I_2 + W \right) \tag{7-38}$$

where

$$W = \frac{T_b}{q^2} \left(S_I + \frac{2k_B T}{R_b} + \frac{S_E}{R^2} \right) I_2 + \frac{(2\pi C)^2}{q^2 T_b} S_E I_3 \tag{7-39}$$

is a dimensionless parameter characterizing the thermal noise of the receiver. We shall call this parameter the *thermal noise characteristic* of the receiver amplifier.

Our task now is to find the minimum energy per pulse that is required to achieve a prescribed maximum bit error rate. For this we shall assume that the output voltage is approximately a gaussian variable. This is the *signal-to-noise ratio approximation*. Although the shot noise has a Poisson distribution, the inaccuracy resulting from the gaussian approximation is small.[7] The mean and variance of the gaussian output for a 1 pulse are b_{on} and σ_{on}^2, whereas for a 0 pulse they are b_{off} and σ_{off}^2. The variances σ_{on}^2 and σ_{off}^2 are defined as the worst-case values of $\langle v_N^2 \rangle$, which are obtained by substituting Eq. (7-28) or (7-29), respectively, for $\langle i_0 \rangle$ into Eq. (7-27):

$$\sigma_{on}^2 = \left(\frac{h\nu}{\eta}\right)^2 \left(\frac{\eta M^x}{h\nu} b_{on} I_2 + \frac{W}{M^2}\right) \tag{7-40}$$

$$\sigma_{off}^2 = \left(\frac{h\nu}{\eta}\right)^2 \left[\frac{\eta M^x}{h\nu} b_{on} I_2(1 - \gamma) + \frac{W}{M^2}\right] \tag{7-41}$$

If the decision threshold voltage v_{th} is set so that there is an equal error probability for 0 and 1 pulses and if we assume that there are an equal number of 0 and 1 pulses [that is, $a = b = \frac{1}{2}$ in Eq. (7-18)], then from Eqs. (7-16) and (7-17),

$$P_0(v_{th}) = P_1(v_{th}) = \tfrac{1}{2} P_e$$

Assuming that the equalizer output is a gaussian variable, the error probability P_e follows the familiar formula

$$P_e = \frac{1}{\sqrt{2\pi}\sigma_{off}} \int_{v_{th}}^{\infty} \exp\left[-\frac{(v - b_{off})^2}{2\sigma_{off}^2}\right] dv$$

$$= \frac{1}{\sqrt{2\pi}\sigma_{on}} \int_{-\infty}^{v_{th}} \exp\left[\frac{-(-v + b_{on})^2}{2\sigma_{on}^2}\right] dv \tag{7-42}$$

Defining the parameter Q as

$$Q = \frac{v_{th} - b_{off}}{\sigma_{off}} = \frac{b_{on} - v_{th}}{\sigma_{on}} \tag{7-43}$$

then Eq. (7-42) becomes

$$P_e(Q) = \frac{1}{\sqrt{\pi}} \int_{Q/\sqrt{2}}^{\infty} e^{-x^2} dx$$

$$= \frac{1}{2}\left[1 - \mathrm{erf}\left(\frac{Q}{\sqrt{2}}\right)\right] \tag{7-44}$$

where erf(x) is the *error function* which can be found in tabulated form.[21] Using such tabulated data, a plot of Q versus P_e is given in Fig. 7-7. Equation (7-44) states that relative to the noise at b_{off} the threshold voltage v_{th} must be at least Q standard deviations above b_{off}, or, equivalently, relative to the noise at b_{on} the threshold voltage must be no more than Q standard deviations below b_{on} to have the desired error rate. For example, for an error rate of $P_e = 10^{-9}$ it follows from Eq. (7-44) that Q is approximately 6.

Using the expression in Eq. (7-43) the receiver sensitivity is given by

$$b_{on} - b_{off} = Q(\sigma_{on} + \sigma_{off}) \tag{7-45}$$

If b_{off} is zero, the required energy per pulse that is needed to achieve a desired bit

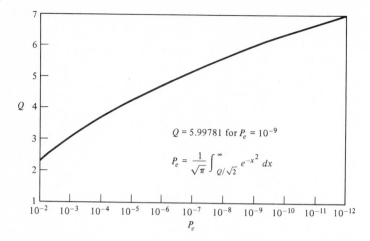

Figure 7-7 Error probability factor Q versus error probability P_e. *(Reproduced with permission from Personick,[1] copyright 1973, American Telephone and Telegraph Company.)*

error rate characterized by the parameter Q is

$$b_{\text{on}} = \frac{Q}{M}\frac{h\nu}{\eta}\left\{\left(M^{2+x}\frac{\eta}{h\nu}b_{\text{on}}I_2 + W\right)^{1/2} + \left[M^{2+x}\frac{\eta}{h\nu}b_{\text{on}}I_2(1-\gamma) + W\right]^{1/2}\right\}$$

(7-46)

We can now determine the optimum value of the avalanche gain M_{opt} by differentiating Eq. (7-46) with respect to M and putting $db_{\text{on}}/dM = 0$. After going through some lengthy but straightforward algebra, we obtain[8]

$$M_{\text{opt}}^{2+x}b_{\text{on}} = \frac{h\nu}{\eta}\frac{W}{2I_2}\left(\frac{2-\gamma}{1-\gamma}\right)K$$

(7-47)

where

$$K = -1 + \left[1 + 16\frac{1+x}{x^2}\frac{1-\gamma}{(2-\gamma)^2}\right]^{1/2}$$

(7-48)

The minimum energy per pulse necessary to achieve a bit error rate characterized by Q can then be found by substituting Eq. (7-47) into Eq. (7-46) and solving for b_{on}. Doing so yields[8]

$$b_{\text{on,min}} = Q^{(2+x)/(1+x)}\frac{h\nu}{\eta}W^{x/(2+2x)}I_2^{1/(1+x)}L$$

(7-49)

where

$$L = \left[\frac{2(1-\gamma)}{K(2-\gamma)}\right]^{1/(1+x)}\left\{\left[\frac{(2-\gamma)K}{2(1-\gamma)} + 1\right]^{1/2} + [\tfrac{1}{2}(2-\gamma)K + 1]^{1/2}\right\}^{(2+x)/(1+x)}$$

(7-50)

The parameter L is a somewhat involved expression, but it has the feature of depending only on the fraction γ of the pulse energy contained within a bit period

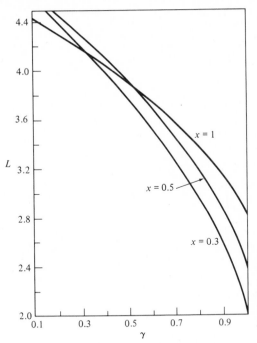

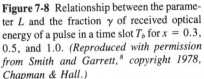

Figure 7-8 Relationship between the parameter L and the fraction γ of received optical energy of a pulse in a time slot T_b for $x = 0.3$, 0.5, and 1.0. *(Reproduced with permission from Smith and Garrett,*[8] *copyright 1978, Chapman & Hall.)*

T_b and on the avalanche photodiode factor x. Values for L are typically between 2 and 3. Recalling from Chap. 6 that x takes on values between 0 and 1.0 (for example, 0 for *pin* photodiodes, 0.5 for silicon APDs, and 1.0 for Ge APDs), we plot L as a function of γ in Fig. 7-8 for three different values of x. Note that these curves give L for any received pulse shape since L depends only on x and γ.

The optimum gain at the desired bit error rate characterized by Q can be found by substituting Eq. (7-49) into Eq. (7-47) to obtain

$$M_{\text{opt}}^{1+x} = \frac{W^{1/2}}{QI_2}\left[\frac{(2-\gamma)K}{2(1-\gamma)L}\right]^{(1+x)/(2+x)} \tag{7-51}$$

The optimum avalanche gain becomes, when $\gamma = 1$ (no intersymbol interference),

$$M_{\text{opt}}^{1+x} = \frac{2W^{1/2}}{xQI_2} \tag{7-52}$$

The proof of this is left as an exercise.

7-2-4 Performance Curves

Using Eq. (7-49) we can calculate the effect of intersymbol interference on the required energy per pulse at the optimum gain for any received and equalized pulse shape. The minimum optical power required occurs for very narrow optical input pulses.[1] Ideally this is a unit impulse or delta function. More power is necessary for other received pulse shapes. The additional or *excess power* ΔP required for pulse shapes other than impulses is normally defined as a *power penalty* measured in decibels. Thus,

$$\Delta P = 10 \log \frac{b_{on,nonimpulse}}{b_{on,impulse}} \qquad (7\text{-}53)$$

As an example we shall calculate the case for which the amplifier resistance R given by Eq. (7-10) is sufficiently large so that the term

$$\frac{(2\pi C)^2}{T_b q^2} S_E I_3$$

dominates the thermal noise in Eq. (7-39). In this case,

$$\Delta P = 10 \log \frac{I_{3,n}^{x/(2+2x)} I_{2,n}^{1/(1+x)} L_n}{I_{3,i}^{x/(2+2x)} I_{2,i}^{1/(1+x)} L_i} \qquad (7\text{-}54)$$

where the subscripts n and i refer to *nonimpulse* and *impulse*, respectively.

For the input pulse shape $h_p(t)$ to the receiver we shall choose a gaussian pulse,

$$h_p(t) = \frac{1}{\sqrt{2\pi}\alpha T_b} e^{-t^2/2\alpha^2 T_b^2} \qquad (7\text{-}55)$$

the normalized Fourier transform of which is

$$H_p'(\phi) = e^{-(2\pi\alpha\phi)^2/2} \qquad (7\text{-}56)$$

As shown in Fig. 7-9, the parameter αT_b, where T_b is the bit period, defines the variance or spread of the pulse.

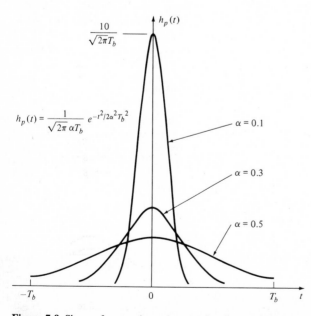

Figure 7-9 Shape of a gaussian pulse as a function of the parameter α.

For the equalizer output waveform $h_{out}(t)$ we choose the commonly used raised-cosine pulse[1, 6, 18]

$$h_{out}(t) = \frac{\sin \pi\tau}{\pi\tau} \frac{\cos \pi\beta\tau}{1 - (2\beta\tau)^2} \tag{7-57}$$

where $\tau = t/T_b$. A plot of $h_{out}(t)$ with $\beta = 0$, 0.5, and 1.0 is shown in Fig. 7-10. The normalized Fourier transform is

$$H'_{out}(\phi) = 1 \qquad \text{for } 0 < |\phi| \leq \frac{1 - \beta}{2}$$

$$H'_{out}(\phi) = \frac{1}{2}\left[1 - \sin\left(\frac{\pi\phi}{\beta} - \frac{\pi}{2\beta}\right)\right] \qquad \text{for } \frac{1 - \beta}{2} < |\phi| \leq \frac{1 + \beta}{2}$$

$$H'_{out}(\phi) = 0 \qquad \text{otherwise} \tag{7-58}$$

The parameter β varies between 0 and 1 and determines the bandwidth used by the pulse, as shown in Fig. 7-10. A β value of unity indicates the bandwidth is $2/T_b$, whereas $\beta = 0$ means that the minimum bandwidth of $1/T_b$ is used. Although less

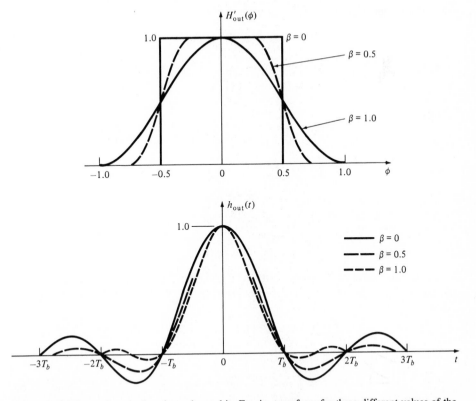

Figure 7-10 Shape of a raised cosine pulse and its Fourier transform for three different values of the parameter β.

bandwidth is used as β decreases, the tails of $h_{\text{out}}(t)$ become larger, and signal timing and equalization become more difficult. For simplicity, we shall choose $\beta = 1$ in the examples given here.

The unit impulse or Dirac delta function is characterized by

$$h_{pi}(t) = \delta(t) \tag{7-59}$$

where

$$\delta(t) = 0 \qquad \text{for } t \neq 0 \tag{7-60}$$

and

$$\int_{-\infty}^{\infty} \delta(t)\, dt = 1$$

Thus the Fourier transform of the impulse function is

$$H'_{pi}(\phi) = F[\delta(t)] = \int_{-\infty}^{\infty} \delta(t)e^{j2\pi ft}\, dt = 1 \tag{7-61}$$

Using Eqs. (7-34), (7-58), and (7-61) we have for an impulse input $H'_p(\phi)$ and a raised cosine output,

$$I_{2i} = \int_{-\infty}^{\infty} |H'_{\text{out}}(\phi)|^2\, d\phi = 1 - \frac{\beta}{4} \tag{7-62}$$

which, for $\beta = 1$, becomes

$$I_{2i} = \tfrac{3}{4} \tag{7-63}$$

From Eq. (7-35)

$$
\begin{aligned}
I_{3i} &= \int_{-\infty}^{\infty} |H'_{\text{out}}(\phi)|^2\, \phi^2\, d\phi \\
&= \frac{\beta^3}{8}\left(\frac{1}{\pi^2} - \frac{1}{6}\right) + 2\beta^2\left(\frac{1}{8} - \frac{1}{\pi^2}\right) - \frac{\beta}{16} + \frac{1}{12}
\end{aligned}
\tag{7-64}
$$

For $\beta = 1$, we have

$$I_{3i} = 0.06002$$

We now evaluate I_{2n} and I_{3n} in Eq. (7-54) for the gaussian input pulse shape given by Eq. (7-56) and for a raised-cosine output. With $\beta = 1$ in Eq. (7-58), we have

$$
\begin{aligned}
I_{2n} &= \int_{-\infty}^{\infty} \left|\frac{H'_{\text{out}}(\phi)}{H'_p(\phi)}\right|^2 d\phi \\
&= \frac{4}{\pi}\int_0^{\pi/2} e^{16\alpha^2 x^2} \cos^4 x\, dx
\end{aligned}
\tag{7-65}
$$

The results of a numerical evaluation of Eq. (7-65) as a function of α are given in

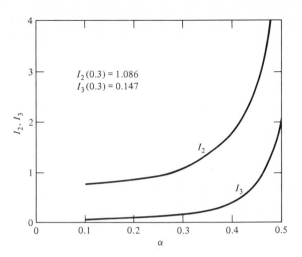

$I_2(0.3) = 1.086$
$I_3(0.3) = 0.147$

Figure 7-11 Plots of the normalized dimensionless bandwidth integrals I_2 and I_3 as a function of α for a gaussian input pulse. The output pulse is a raised cosine with $\beta = 1.0$.

Fig. 7-11. With the same assumptions,

$$I_{3n} = 2\left(\frac{2}{\pi}\right)^3 \int_0^{\pi/2} x^2 e^{16\alpha^2 x^2} \cos^4 x \, dx \tag{7-66}$$

The numerical evaluation of I_{3n} as a function of α is shown in Fig. 7-11.

Two more parameters (L_i and L_n) now remain to be evaluated in Eq. (7-54) in order to determine the receiver power penalty. Since all the pulse energy is contained within the bit period for a unit impulse input, L_i is determined by taking the limit of L in Eq. (7-50) as γ goes to unity. Thus,

$$L_i = \lim_{\gamma \to 1} L = (1 + x)\left(\frac{2}{x}\right)^{x/(1+x)} \tag{7-67}$$

The proof of this is left as an exercise.

The parameter L_n depends on the pulse energy per bit period γ and on the excess noise coefficient x of the avalanche process. For a gaussian input pulse, γ, in turn, depends on the parameter α in Eq. (7-55). The relation between γ and α is found from Eqs. (7-30) and (7-55):

$$\gamma = \int_{-T_b/2}^{T_b/2} h_p(t) \, dt = \frac{2}{\sqrt{\pi}} \int_0^{1/(2\sqrt{2}\alpha)} e^{-x^2} \, dx$$

$$= \operatorname{erf}\left(\frac{1}{2\sqrt{2}\alpha}\right) \tag{7-68}$$

where the error function erf(x) is defined in Eq. (7-44). The relationship between γ and α given by Eq. (7-68) is shown in Fig. 7-12.

We are now finally ready to evaluate Eq. (7-54). Choosing $x = 0.5$, which is characteristic of silicon avalanche photodiodes, Eq. (7-67) yields $L_i = 2.38$. What we wish to plot is the penalty in minimum received power ΔP (required for a certain

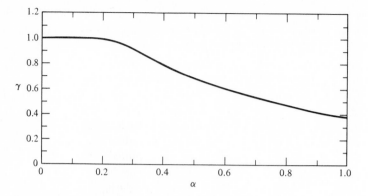

Figure 7-12 A plot of the fraction of pulse energy γ as a function of the gaussian pulse shape parameter α.

bit error rate) as a function of the fraction of pulse energy $1 - \gamma$ that has spread outside of the bit period T_b. For a given value of γ we read the corresponding value of L_n from Fig. 7-8. To find I_{2n} and I_{3n}, we first find the value of α corresponding to γ from Fig. 7-12, and then we find the values of I_{2n} and I_{3n} corresponding to this value of α from Fig. 7-11. Substituting all these values into Eq. (7-54) for various values of γ and letting $x = 0.5$, we obtain the results shown in Fig. 7-13.

The effect of intersymbol interference (or bandwidth limitation) on the receiver power penalty is readily deduced from Fig. 7-13. As the fraction of pulse energy outside the bit period increases, there is a steep rise in the power penalty curve. This curve gives a clear implication as to the effects of attempting to operate an optical fiber system at such high data rates that bandwidth limitations arise. Intersymbol interference becomes more pronounced at higher data rates, since the individual data pulses start to overlap significantly as the data rate approaches the system bandwidth limit. Since the receiver power penalty increases rapidly for larger pulse

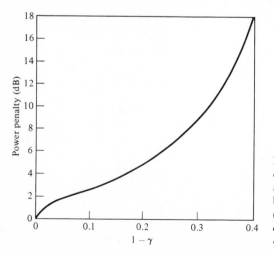

Figure 7-13 The penalty in minimum received optical power (receiver sensitivity) arising from pulse spreading outside of the bit period for gaussian received pulses. *(Reproduced with permission from Smith and Garrett,[8] copyright 1978, Chapman & Hall.)*

overlaps, operating a fiber optic system much beyond its bandwidth is generally not worthwhile, even though it may be possible to correct for intersymbol interference through the use of an equalization circuit.

7-2-5 Nonzero Extinction Ratio

In the previous section we assumed that there is no optical power incident on the photodetector during a 0 pulse so that $b_{off} = 0$. In actual systems the light source may be biased slightly on at all times in order to obtain a shorter light source turnon time (see Sec. 4-2-3). Thus some optical power is also emitted during a 0 pulse. This is of particular importance for laser diodes since it is generally desirable to bias them on to just below the lasing threshold. This means that during a zero pulse the laser acts like an LED and can launch a significant amount of optical power into a fiber.

The ratio ϵ of the optical energy emitted in the 0 pulse state to that emitted during a 1 pulse is called the *extinction ratio*

$$\epsilon = \frac{b_{off}}{b_{on}} \tag{7-69}$$

A receiver performance calculation[22] identical to that of Sec. 7-2-4 can be carried out for a nonzero extinction ratio by simply using b_{off} from Eq. (7-69) and by replacing γ by $\gamma' = \gamma(1 - \epsilon)$ in Eq. (7-29). With these replacements Eq. (7-45) yields

$$b_{on}(1 - \epsilon) = \frac{Q}{M}\frac{h\nu}{\eta}\left\{\left(M^{2+x}\frac{\eta}{h\nu}b_{on}I_2 + W\right)^{1/2} + \left[M^{2+x}\frac{\eta}{h\nu}b_{on}I_2(1 - \gamma') + W\right]^{1/2}\right\} \tag{7-70}$$

Analogous to the derivation of Eq. (7-49) we differentiate Eq. (7-70) for b_{on} with respect to M to find the minimum energy $b_{on,min}(\epsilon)$ per pulse required at the optimum gain M_{opt}, which results in[22]

$$b_{on,min}(\epsilon) = Q^{(2+x)/(1+x)}\frac{h\nu}{\eta}W^{x/(2+2x)}I_2^{1/(1+x)}L'\left(\frac{1}{1 - \epsilon}\right)^{(2+x)/(1+x)} \tag{7-71}$$

where

$$L'^{(1+x)} = \frac{2(1 - \gamma')}{K'(2 - \gamma')}\left\{\left(\frac{1}{2}\frac{2 - \gamma'}{1 - \gamma'}K' + 1\right)^{1/2} + [\tfrac{1}{2}(2 - \gamma')K' + 1]^{1/2}\right\}^{2+x} \tag{7-72}$$

with

$$K' = -1 + \left[1 + 16\frac{1 + x}{x^2}\frac{1 - \gamma'}{(2 - \gamma')^2}\right]^{1/2} \tag{7-73}$$

If a data stream has an equal probability of 1 and 0 pulses, then the minimum received power (or the receiver sensitivity) $P_{r,min}$ is given by the average energy

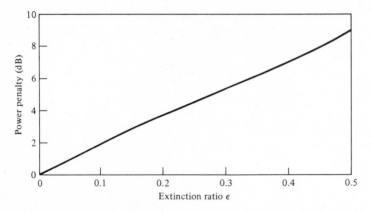

Figure 7-14 Penalty in receiver sensitivity as a function of the extinction ratio ϵ for $\gamma = 1.0$ and $x = 0.5$. *(Reproduced with permission from Hooper and White,* [22] *copyright 1978, Chapman & Hall.)*

detected per pulse times the pulse rate $1/T_b$:

$$P_{r,\min} = \frac{b_{\text{on}} + b_{\text{off}}}{2T_b} = \frac{1 + \epsilon}{2T_b} b_{\text{on}} \tag{7-74}$$

where the last equality was obtained by using Eq. (7-69). The extinction ratio penalty, that is, the penalty in receiver sensitivity as a function of the extinction ratio is

$$y(\epsilon) = \frac{P_{r,\min}(\epsilon)}{P_{r,\min}(0)} = (1 + \epsilon)\left(\frac{1}{1 - \epsilon}\right)^{(2+x)/(1+x)} \frac{L'}{L} \tag{7-75}$$

Both L and L' are given by Fig. 7-8 by using γ and γ', respectively, for the abscissa. A plot[22] of Eq. (7-75) is given in Fig. 7-14 for the special case where $\gamma = 1.0$ and $x = 0.5$.

7-3 PREAMPLIFIER DESIGN

Having examined the performance characteristics of a general class of receivers, we now turn our attention to some practical design examples. Since the sensitivity of a receiver is dominated by the noise sources at the front end (the preamplifier stage), the major emphasis in the literature has been on the design of a low-noise receiver preamplifier.

There are two basic approaches to the design of preamplifiers for fiber optic receivers. These are the *high-impedance preamplifier* shown in Fig. 7-4 and the *transimpedance preamplifier* shown in Fig. 7-15. The high-impedance preamplifier has an input bias resistance R_b with an equivalent thermal noise current $i_b(t)$. The preamplifier of the high-impedance receiver is described in terms of an arbitrarily large, noiseless, fixed gain A. The transimpedance receiver has a feedback resistance R_f with an equivalent thermal noise current $i_f(t)$ shunting the input. The preamplifier of this receiver is modeled by a noiseless open-loop gain $A(\omega)$. In

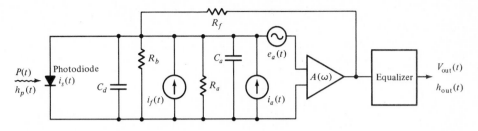

Figure 7-15 Transimpedance receiver design. *(Reproduced with permission from Personick, Rhodes, Hanson, and Chan, Proc. IEEE, **68**, 1260, Oct. 1980, copyright 1980, IEEE.)*

either case, the amplifier has an input equivalent series voltage noise source $e_a(t)$, an equivalent shunt current noise source $i_a(t)$, and an input impedance given by the parallel combination of R_a and C_a.

What we are interested in here is to evaluate the noise voltage given by Eq. (7-38) for different amplifier designs. To this end we shall examine only the parameter W of Eq. (7-39). This is a very useful figure of merit for a receiver since it measures the noisiness of the amplifier. From Eq. (7-39) it can be seen that the noise is minimized if the amplifier and bias resistances R_a and R_b are large, the total capacitance C at the amplifier input is small, and the noise current and voltage spectral heights S_I and S_E are small. In general, these parameters are not independent so that tradeoffs have to be made among them to minimize the noise. In addition, the freedom of the designer to optimize device parameters is often restricted by the limited variety of components available. We shall examine several amplifier configurations to illustrate some of the design considerations that must be taken into account.

7-3-1 High-Impedance FET Amplifiers

The design of a receiver using a high-impedance FET (field effect transistor) preamplifier has been described in detail by several authors.[6, 15, 23–26] The circuit of a simple FET amplifier is shown in Fig. 7-16. Typical FETs have very large input resistances R_a (usually greater than 10^6 Ω), so for practical purposes we can set

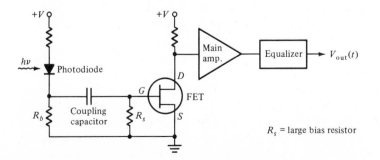

Figure 7-16 Simple FET preamplifier design.

$R_a = \infty$. The total resistance R given by Eq. (7-10) then reduces to the value of the detector bias resistor R_b.

Since the amplifier input resistance is very large, the input current noise spectral density S_I, which is the Johnson (thermal) noise of this resistance,

$$S_{I,\text{FET}} = \frac{4k_B T}{R_a} \qquad (7\text{-}76)$$

is negligible. Thus the basic noise source in an FET is the thermal noise of the conducting channel resistance, which is characterized by the transconductance g_m. The voltage noise spectral density is [27]

$$S_E = \frac{2}{3} \frac{4k_B T}{g_m} \qquad (7\text{-}77)$$

The thermal noise characteristic W [Eq. (7-39)] at the equalizer output is then

$$W = \frac{2k_B T}{q^2/T_b} \left[\left(\frac{1}{R_b} + \frac{4}{3} \frac{1}{g_m R_b^2} \right) I_2 + \left(\frac{2\pi C}{T_b} \right)^2 \frac{4}{3} \frac{I_3}{g_m} \right] \qquad (7\text{-}78)$$

For a typical FET and a good photodiode we can expect values of $C = C_a + C_d = 10$ pF and $g_m = 5000$ μS(microsiemens). To minimize the noise, the bias resistor should be very large. The effect of this is that the detector output signal is integrated by the amplifier input resistance. We can compensate for this by differentiation in the equalizing filter. This integration-differentiation approach is known as the *high-impedance amplifier design* technique. It yields low noise, but also results in a low dynamic range (the range of signal levels that can be processed with high quality). An alternative method to deal with this is described in Sec. 7-3-3.

As the signal frequency reaches a high value, the gain of an FET approaches unity. For a silicon FET, this is about 25 to 50 MHz. Much higher frequencies can be achieved with either a GaAs FET or a silicon bipolar transistor.[24,28] The feasibility of using an integrated *pin* photodiode-FET combination fabricated on InGaAs, for example, has been demonstrated.[29] This type of device is of importance for application in the longer wavelength range of 1.2 to 1.6 μm.

7-3-2 High-Impedance Bipolar Transistor Amplifiers

The circuit of a simple bipolar grounded-emitter transistor amplifier[14, 15, 24] is shown in Fig. 7-17. The input resistance of a bipolar transistor is given by[27]

$$R_{\text{in}} = \frac{k_B T}{q I_{BB}} \qquad (7\text{-}79)$$

where I_{BB} is the base bias current. For a bipolar transistor amplifier the input resistance R_a is given by the parallel combination of the bias resistors R_1 and R_2 and the transistor input resistance R_{in}. For a low-noise design R_1 and R_2 are chosen to be much greater than R_{in}, so that $R_a \simeq R_{\text{in}}$. Thus, in contrast to the FET amplifier, R_a for a transistor amplifier is adjustable by the designer.

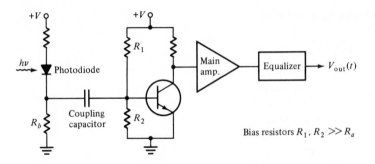

Figure 7-17 Simple bipolar transistor preamplifier design.

The spectral density of the input noise current source results from the shot noise of the base current[27]

$$S_I = 2qI_{BB} = \frac{2k_B T}{R_{in}} \qquad A^2/Hz \qquad (7\text{-}80)$$

where the last equality comes from Eq. (7-79). The spectral height of the noise voltage source is[27]

$$S_E = \frac{2k_B T}{g_m} \qquad V^2/Hz \qquad (7\text{-}81)$$

Here the transconductance g_m is related to the shot noise by virtue of the collector current I_c:

$$g_m = \frac{qI_c}{k_B T} = \frac{\beta}{R_{in}} \qquad (7\text{-}82)$$

where Eq. (7-79) has been used in the last equality to express g_m in terms of the current gain $\beta = I_c/I_{BB}$ and the input resistance R_{in}.

Substituting Eqs. (7-79) through (7-81) into Eq. (7-39), we have

$$W = \frac{T_b}{q^2} 2k_B T \left[\left(\frac{1}{R_{in}} + \frac{1}{R_b} + \frac{R_{in}}{\beta R^2} \right) I_2 + \frac{(2\pi C)^2}{T_b^2} \frac{R_{in}}{\beta} I_3 \right] \qquad (7\text{-}83)$$

The contribution C_a to C from the bipolar transistor is a few picofarads. If the photodetector bias resistor R_b is much larger than the amplifier resistance R_a, then from Eq. (7-10) $R \simeq R_a \simeq R_{in}$, so that

$$W = \frac{2k_B T}{q^2} \left[\frac{T_b}{R_{in}} \frac{\beta + 1}{\beta} I_2 + \frac{(2\pi C)^2}{\beta T_b} R_{in} I_3 \right] \qquad (7\text{-}84)$$

As is the case with a high-impedance FET preamplifier, the impedance loading the photodetector integrates the detector output signal. Again, to compensate for this, the amplified signal is differentiated in the equalizing filter.

7-3-3 Transimpedance Amplifier

Although a high-impedance design produces the lowest-noise amplifier, it has two limitations: (1) for broadband applications equalization is required and (2) it has a limited dynamic range. An alternative design is the *transimpedance amplifier*[30-33] shown in Fig. 7-15. This is basically a high-gain high-impedance amplifier with feedback provided to the amplifier input through the feedback resistor R_f. This design yields both low noise and a large dynamic range.

To compare the nonfeedback and the feedback designs, we make the restriction that both have the same transfer function $H_{out}(f)/H_p(f)$. For the transimpedance amplifier the thermal noise characteristic W_{TZ} at the equalizer output is therefore simply found by replacing R_b in Eq. (7-39) with R_b', where[30]

$$R_b' = \frac{R_b R_f}{R_b + R_f} \tag{7-85}$$

is the parallel combination of R_b and R_f. Thus, from Eq. (7-39),

$$W_{TZ} = \frac{T_b}{q^2}\left(S_I + \frac{2k_B T}{R_b'} + \frac{S_E}{(R')^2}\right)I_2 + \frac{(2\pi C)^2}{q^2 T_b} S_E I_3 \tag{7-86}$$

where, from Eq. (7-10),

$$\frac{1}{R'} = \frac{1}{R} + \frac{1}{R_f} = \frac{1}{R_a} + \frac{1}{R_b} + \frac{1}{R_f} \tag{7-87}$$

In practice, the feedback resistance R_f is much greater than the amplifier input resistance R_a. Consequently, $R' \simeq R$ in Eq. (7-87), so that

$$W_{TZ} = W_{HZ} + \frac{T_b}{q^2}\frac{2k_B T}{R_f} I_2 \tag{7-88}$$

where W_{HZ} is the high-impedance amplifier noise characteristic given by either Eq. (7-78) for FET designs or by Eq. (7-84) for the bipolar transistor case. The thermal noise of the transimpedance amplifier is thus modeled as the sum of the output noise of a nonfeedback amplifier plus the thermal noise associated with the feedback resistance. In practice, the noise considerations tend to be more involved since R_f has an effect on the frequency response of the amplifier. More details are given by Smith and Personick.[34]

We now compare the bandwidths of the two designs. From Eq. (7-9) the transfer function of the nonfeedback amplifier is

$$H(f) = \frac{AR}{1 + j2\pi RCf} \quad V/A \tag{7-89}$$

where R and C are given by Eqs. (7-10) and (7-11), respectively, and A is the frequency-independent gain of the amplifier. Using Eq. (F-10), this yields a bandwidth of $(4RC)^{-1}$. For the transimpedance amplifier the transfer function $H_{TZ}(f)$ is

$$H_{TZ} = \frac{1}{1 + j2\pi RCf/A} \tag{7-90}$$

which yields a bandwidth of

$$B_{TZ} = \frac{A}{4RC} \tag{7-91}$$

which is A times that of the high-impedance design. This makes the equalization task simpler in the feedback amplifier case.

In summary, the benefits of a transimpedance amplifier are as follows:

1. It has a wide dynamic range compared to the high-impedance amplifier.
2. Usually little or no equalization is required because the combination of R_{in} and the feedback resistor R_f is very small, which means the time constant of the detector is also small.
3. The output resistance is small so that the amplifier is less susceptible to pickup noise, cross talk, electromagnetic interference (EMI), etc.
4. The transfer characteristic of the amplifier is actually its transimpedance, which is the feedback resistor. Therefore, the transimpedance amplifier is very easily controlled and stable.
5. Although the transimpedance amplifier is less sensitive than the high-impedance amplifier (since $W_{TZ} > W_{HZ}$), this difference is usually only about 2 to 3 dB for most practical wideband designs.

7-4 ANALOG RECEIVERS

In addition to the wide usage of fiber optics for the transmission of digital signals, there are many potential applications for analog links. These range from individual 4-kHz voice channels to systems employing 5-MHz video channels. In the previous sections we discussed digital receiver performance in terms of error probability. For an analog receiver the performance fidelity is measured in terms of a *signal-to-noise ratio*. This is defined as the ratio of the mean square signal current to the mean square noise current.

The simplest analog technique is to use amplitude modulation of the source.[2] In this scheme the time-varying electric signal $s(t)$ is used to modulate directly an optical source about some bias point defined by the bias current I_B, as shown in Fig. 7-18. The transmitted optical power $P(t)$ is thus of the form

$$P(t) = P_t[1 + ms(t)] \tag{7-92}$$

where P_t is the average transmitted optical power, $s(t)$ is the analog modulation signal, and m is the modulation index defined by (see Sec. 4-3)

$$m = \frac{\Delta I}{I_B} \tag{7-93}$$

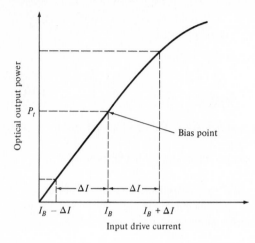

Figure 7-18 Direct analog modulation of an LED source.

Here ΔI is the variation in current about the bias point. In order not to introduce distortion into the optical signal, the modulation must be confined to the linear region of the light source output curve shown in Fig. 7-18. Also if $\Delta I > I_B$, the lower portion of the signal gets cut off and severe distortion results.

At the receiver end the photocurrent generated by the analog optical signal is

$$i_s(t) = \mathfrak{R}_0 M P_r [1 + ms(t)]$$

$$= I_p M [1 + ms(t)] \qquad (7\text{-}94)$$

where $\mathfrak{R}_0$ is the detector responsivity, P_r is the average received optical power, $I_p = \mathfrak{R}_0 P_r$ is the primary photocurrent, and M is the photodetector gain. If $s(t)$ is a sinusoidally modulated signal, then the mean square signal current at the photo-detector output is (ignoring a dc term)

$$\langle i_s^2 \rangle = \tfrac{1}{2}(\mathfrak{R}_0 M m P_r)^2 = \tfrac{1}{2}(M m I_p)^2 \qquad (7\text{-}95)$$

Recalling from Eq. (6-11) that the mean square noise current for a photodiode receiver is the sum of the mean square quantum noise current, the equivalent-resistance thermal noise current, the dark noise current, and the surface-leakage noise current, we have

$$\langle i_N^2 \rangle = 2q(I_p + I_D)M^2 F(M)B + 2qI_L B + \frac{4k_B T B}{R_{eq}} F_t \qquad (7\text{-}96)$$

where I_P = primary (unmultiplied) photocurrent = $\mathfrak{R}_0 P_r$
$\qquad I_D$ = primary bulk dark current
$\qquad I_L$ = surface leakage current
$\quad F(M)$ = excess photodiode noise factor $\simeq M^x$ $(0 < x \leq 1)$
$\qquad B$ = effective noise bandwidth
$\qquad R_{eq}$ = equivalent resistance of photodetector load and amplifier
$\qquad F_t$ = noise figure of the baseband amplifier

By a suitable choice of the photodetector the leakage current can be rendered negligible. With this assumption the signal-to-noise ratio S/N is

$$\frac{S}{N} = \frac{\langle i_s^2 \rangle}{\langle i_N^2 \rangle} = \frac{\frac{1}{2}(\Re_0 M m P_r)^2}{2q(\Re_0 P_r + I_D)M^2 F(M)B + (4k_B TB/R_{eq})F_t}$$

$$= \frac{\frac{1}{2}(I_p M m)^2}{2q(I_p + I_D)M^2 F(M)B + (4k_B TB/R_{eq})F_t} \qquad (7\text{-}97)$$

For a *pin* photodiode we have $M = 1$. When the optical power incident on the photodiode is small, the circuit noise term dominates the noise current, so that

$$\frac{S}{N} \simeq \frac{\frac{1}{2}m^2 I_p^2}{(4k_B TB/R_{eq})F_t} = \frac{\frac{1}{2}m^2 \Re_0^2 P_r^2}{(4k_B TB/R_{eq})F_t} \qquad (7\text{-}98)$$

Here the signal-to-noise ratio is directly proportional to the square of the photodiode output current and inversely proportional to the thermal noise of the circuit.

For large optical signals incident on a *pin* photodiode, the quantum noise associated with the signal detection process dominates, so that

$$\frac{S}{N} \simeq \frac{m^2 I_p}{4qB} = \frac{m^2 \Re_0 P_r}{4qB} \qquad (7\text{-}99)$$

Since the signal-to-noise ratio in this case is independent of the circuit noise, it represents the fundamental or quantum limit for analog receiver sensitivity.

When an avalanche photodiode is employed at low signal levels and with low values of gain M, the circuit noise term dominates. At a fixed low signal level, as the gain is increased from a low value, the signal-to-noise ratio increases with gain until the quantum noise term becomes comparable to the circuit noise term. As the gain is increased further beyond this point, the signal-to-noise ratio *decreases* as $F(M)^{-1}$. Thus for a given set of operating conditions, there exists an optimum value of the avalanche gain for which the signal-to-noise ratio is a maximum. Since an avalanche photodiode increases the signal-to-noise ratio for small optical signal levels, it is the preferred photodetector for this situation.

For very large optical signal levels the quantum noise term dominates the receiver noise. In this case, an avalanche photodiode serves no advantage since the detector noise increases more rapidly with increasing gain M than the signal level. This is shown in Fig. 7-19, where we compare the signal-to-noise ratio for a *pin* and an avalanche photodiode receiver as a function of the received optical power. The signal-to-noise ratio for the avalanche photodetector is at the optimum gain (see Probs. 7-20 and 7-21). The parameter values chosen for this example are $B = 5$ MHz and 25 MHz, $x = 0.5$ for the avalanche photodiode and 0 for the *pin* diode, $m = 80$ percent, $\Re_0 = 0.5$ A/W, and $R_{eq}/F_t = 10^4$ Ω. We see that for low signal levels an avalanche photodiode yields a higher signal-to-noise ratio, whereas at large received optical power levels a *pin* photodiode results in better performance.

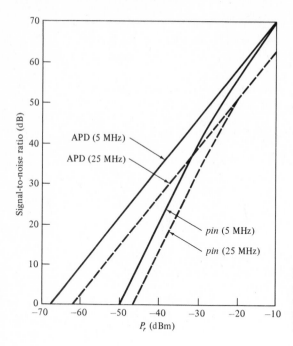

Figure 7-19 Comparison of signal-to-noise ratio for *pin* and avalanche photodiodes as a function of received optical power for bandwidths of 5 and 25 MHz.

7-5 SUMMARY

The task of an optical receiver is first to convert the optical energy emerging from the end of a fiber into an electric signal, and then to amplify this signal to a large enough level so that it can be processed by the electronics following the receiver amplifier. In these processes various noises and distortions will unavoidably be introduced, which can lead to errors in the interpretation of the received signal. The three basic stages of a receiver are a photodetector, an amplifier, and an equalizer. The design of the amplifier which follows the photodiode is of critical importance, because it is in the amplifier where the major noise sources are expected to arise. The equalizer that follows the amplifier is normally a linear frequency-shaping filter that is used to mitigate the effects of signal distortion and intersymbol interference.

In a digital receiver the amplified and filtered signal emerging from the equalizer is compared to a threshold level once per time slot to determine whether or not a pulse is present at the photodetector in that time slot. Various noises, interference from adjacent pulses, and conditions wherein the light source is not completely extinguished during a zero pulse can cause errors in the decision-making process. To calculate the error probability, we require a knowledge of the mean square noise voltage which is superimposed on the signal voltage at the decision time. Since the statistics of the output voltage at the sampling time are very complicated, approximations are used to calculate the performance of a binary optical fiber receiver. In applying these approximations, we have to make a tradeoff between computational simplicity and accuracy of the results. The simplest method is based on a gaussian approximation. In this method it is assumed that when the sequence of optical input pulses is known, the equalizer output voltage is a gaussian random variable. Thus, to calculate the error probability, we only need to know the mean and standard deviation of the output voltage.

The two basic approaches to the design of preamplifiers for fiber optic receivers are the high-impedance and the transimpedance preamplifiers. The high-impedance design produces the lowest noise, but it has two limitations: (1) for broadband applications equalization is required and (2) it has a limited dynamic range. The transimpedance amplifier is less sensitive, but it has the benefits of a wide dynamic range and little need for equalization.

For the reader who is interested in constructing some actual receivers and employing them in a data link, some detailed receiver designs and their uses and limitations are discussed in Refs. 34 through 40.

PROBLEMS

Section 7-1

7-1 (*a*) Suppose we have an ideal photodetector which produces no dark current, that is, no electron-hole pairs are generated in the absence of an optical pulse. Assume that an optical pulse of energy E falls on the photodetector in an interval τ. This can only be interpreted by the receiver as a 0 pulse if no electron-hole pairs are generated with the pulse present. Show that, if the probability of this pulse being interpreted as a zero is 10^{-9}, then

$$E = 21 \frac{h\nu}{\eta}$$

(*b*) Assume that the average power $p(t)$ detected with this photodetector is given by

$$p(t) = E \frac{B}{2}$$

the energy per pulse E times the bit rate (that is, the pulse rate) $B = 1/T_b$ divided by 2 (this assumes an equal number of 0 and 1 pulses). Show that, if $\eta = 1$, then the minimum optical power required at a 10^{-9} error probability for a 10-Mb/s (megabit per second) data rate is -76 dBm at a wavelength of 850 nm. What is the required power at 1300 nm?

Note: This is called the *quantum limit* for digital detection at 10 Mb/s at a 10^{-9} bit error rate since all system parameters are assumed ideal and the performance is only limited by the photodetection statistics.

7-2 Find the Fourier transform $h_B(t)$ of the bias circuit transfer function $H_B(f)$ given by Eq. (7-9).

7-3 Show that the following pulse shapes satisfy the normalization condition

$$\int_{-\infty}^{\infty} h_p(t) \, dt = 1$$

(*a*) Rectangular pulse ($\alpha = $ constant)

$$h_p(t) = \frac{1}{\alpha T_b} \qquad \text{for} \quad \frac{-\alpha T_b}{2} < t < \frac{\alpha T_b}{2}$$

$$h_p(t) = 0 \qquad \text{otherwise}$$

(b) Gaussian pulse

$$h_p(t) = \frac{1}{\sqrt{2\pi}} \frac{1}{\alpha T_b} e^{-t^2/2(\alpha T_b)^2}$$

(c) Exponential pulse

$$h_p(t) = \frac{1}{\alpha T_b} e^{-t/\alpha T_b} \qquad \text{for } 0 \le t < \infty$$

$$h_p(t) = 0 \qquad \text{otherwise}$$

7-4 The mathematical operation of convolving two real-valued functions of the same variable is defined as

$$p(t) * q(t) = \int_{-\infty}^{\infty} p(x)q(t-x)\, dx$$

$$= \int_{-\infty}^{\infty} q(x)p(t-x)\, dx$$

where * denotes convolution. If $P(f)$ and $Q(f)$ are the Fourier transforms of $p(t)$ and $q(t)$, respectively, show that

$$F[p(t) * q(t)] = P(f)Q(f) = F[p(t)]F[q(t)]$$

That is, the convolution of two signals in the time domain corresponds to the multiplication of their Fourier transforms in the frequency domain.

Section 7-2

7-5 Derive the expression for the mean square noise voltage at the equalizer output given by Eq. (7-21).

7-6 (a) Show that Eqs. (7-23) and (7-26) can be rewritten as Eqs. (7-36) and (7-37).
    (b) Show that Eq. (7-21) can be rewritten as Eq. (7-38).

7-7 Show that, by using Eq. (7-43), the error probability expressions given by Eq. (7-42) both reduce to Eq. (7-44).

7-8 Derive Eq. (7-47) by differentiating both sides of Eq. (7-46) with respect to M and setting $db_{on}/dM = 0$.

7-9 Verify the expression given by Eq. (7-49) for the minimum energy per pulse needed to achieve a bit error rate characterized by Q.

7-10 Show that, when there is no intersymbol interference ($\gamma = 1$), the optimum avalanche gain given by Eq. (7-51) reduces to Eq. (7-52).

7-11 Derive Eq. (7-62) for I_2 and Eq. (7-64) for I_3 (impulse input and raised-cosine output).

7-12 Verify the expressions given by Eqs. (7-65) and (7-66) for I_2 and I_3, respectively, for a gaussian input pulse and a raised-cosine output.

7-13 Derive Eq. (7-67).

7-14 Compare the values of $y(\epsilon)$ in decibels for 10 percent extinction ratio ($\epsilon = 0.10$) when (a) $\gamma = 0.90$, $x = 0.5$; (b) $\gamma = 0.90$, $x = 1.0$.

Section 7-3

7-15 (a) Plot the values of the thermal noise characteristic W for a high-impedance FET amplifier for data rates $1/T_b$ ranging from 1 to 50 Mb/s. Let $T = 300$ K, $g_m = 0.005$ S, $R_b = 10^5\ \Omega$, $C = 10$ pF, and $\gamma = 0.90$. Use Fig. 7-11 to find I_2 and I_3. Recall that I_2 and I_3 depend on α which, in turn, depends on γ.

    (b) Plot the values of W for a high-impedance bipolar transistor preamplifier for data rates ranging from 20 to 100 Mb/s. Let $T = 300$ K, $\beta = 100$, $I_{BB} = 5\ \mu$A, $C = 10$ pF, and $\gamma = 0.90$.

7-16 The receiver sensitivity P_r is given by the average energy b_{on} detected per pulse times the pulse rate $1/T_b$ (if $b_{off} = 0$):

$$P_r = \frac{b_{on}}{T_b}$$

Find the sensitivity in dBm (see App. D) of an avalanche photodiode receiver with an FET preamplifier at a 10-Mb/s data rate. Let the required bit error rate be 10^{-9} and take $T = 300$ K, $x = 0.5$, $\gamma = 0.9$, $\eta q/h\nu = 0.7$ A/W (the detector responsivity), $g_m = 0.005$ S, $R_b = 10^5$ Ω, and $C = 10$ pF.

7-17 If a transimpedance amplifier has a feedback resistance of 5000 Ω, by how much does W_{TZ} differ from W_{HZ} at a 10-Mb/s data rate? Assume $\gamma = 0.9$ and $T = 300$ K. Compared to a high-impedance amplifier, what is the decrease in receiver sensitivity (in dB) for this transimpedance amplifier at 10 Mb/s if $x = 0.5$ and $W_{HZ} = 1 \times 10^6$?

7-18 (a) To calculate the receiver sensitivity P_r as a function of gain M, we need to solve Eq. (7-46) for b_{on}. Show that, for $\gamma = 1.0$ and $b_{off} = 0$, this becomes

$$b_{on} = \frac{h\nu}{\eta}\left(M^x Q^2 I_2 + \frac{2QW^{1/2}}{M}\right)$$

(b) Consider a receiver operating at 50 Mb/s. Let the receiver have an avalanche photodiode with $x = 0.5$ and a bipolar transistor front end (preamplifier). Assume $W = 2 \times 10^6$, $Q = 6$ for a 10^{-9} bit error rate, $I_2 = 1.08$, and $\eta q/h\nu = 0.7$ A/W. Using the foregoing expression for b_{on}, plot $P_r = b_{on}/T_b$ in dBm as a function of gain M for values of M ranging from 30 to 120.

7-19 Using Eq. (F-10) for the bandwidth definition, show that the bandwidths of the transfer functions given by Eqs. (7-89) and (7-90) are $1/4RC$ and $A/4RC$.

Section 7-4

7-20 Show that the signal-to-noise ratio given by Eq. (7-97) is a maximum when the gain is optimized at

$$M_{opt}^{2+x} = \frac{4k_B T F_t / R_{eq}}{q(I_p + I_D)x}$$

7-21 (a) Show that, when the gain M is given by the expression in Prob. 7-20, the signal-to-noise ratio given by Eq. (7-97) can be written as

$$\frac{S}{N} = \frac{xm^2}{2B(2+x)}\frac{I_p^2}{[q(I_p + I_D)x]^{2/(2+x)}}\left(\frac{R_{eq}}{4k_B T F_t}\right)^{x/(2+x)}$$

(b) Show that, when I_p is much larger than I_D, the foregoing expression becomes

$$\frac{S}{N} = \frac{m^2}{2Bx(2+x)}\left[\frac{(xI_p)^{2(1+x)}}{q^2(4k_B T F_t / R_{eq})^x}\right]^{1/(2+x)}$$

7-22 Consider the signal-to-noise-ratio expression given in Prob. 7-21a. Analogous to Fig. 7-19, plot S/N in dB [that is, $10 \log (S/N)$] as a function of the received power level P_r in dBm when the dark current $I_D = 10$ nA and $x = 1.0$. Let $B = 5$ MHz, $m = 0.8$, $\mathcal{R}_0 = 0.5$ A/W, $T = 300$ K, and $R_{eq}/F_t = 10^4$ Ω. Recall that $I_p = \mathcal{R}_0 P_r$.

REFERENCES

1. S. D. Personick, "Receiver design for digital fiber optic communication systems," *Bell Sys. Tech. J.*, **52**, 843–886, July–Aug. 1973.
2. W. M. Hubbard, "Utilization of optical-frequency carriers for low and moderate bandwidth channels," *Bell Sys. Tech. J.*, **52**, 731–765, May–June 1973.

3. G. Foschini, R. D. Gitlin, and J. Salz, "Optimum direct detection for digital fiber optic communication systems," *Bell Sys. Tech. J.*, **54**, 1389–1430, Oct. 1975.

4. J. E. Mazo and J. Salz, "On optical data communication via direct detection of light pulses," *Bell Sys. Tech. J.*, **55**, 347–369, Mar. 1976.

5. S. D. Personick, P. Balaban, J. Bobsin, and P. Kumer, "A detailed comparison of four approaches to the calculation of the sensitivity of optical fiber system receivers," *IEEE Trans. Commun.*, **COM-25**, 541–548, May 1977.

6. S. D. Personick, "Receiver design for optical fiber systems," *Proc. IEEE*, **65**, 1670–1678, Dec. 1977.

7. P. Balaban, "Statistical evaluation of the error rate of the fiberguide repeater using importance sampling," *Bell Sys. Tech. J.*, **55**, 745–766, July–Aug. 1976.

8. D. R. Smith and I. Garrett, "A simplified approach to digital optical receiver design," *Opt. Quantum Electron.*, **10**, 211–221, 1978.

9. G. L. Cariolaro, "Error probability in digital fiber optic communication systems, *IEEE Trans. Inform. Theory*, **IT-24**, 213–221, Mar. 1978.

10. R. Dogliotti, A. Luvison, and G. Pirani, "Error probability in optical fiber transmission systems," *IEEE Trans. Inform. Theory*, **IT-25**, 170–178, Mar. 1979.

11. (a) J. C. Cartledge, "Receiver sensitivity of optical fiber communication systems," *IEEE Trans. Commun.*, **COM-26**, 1103–1109, July 1978.

    (b) D. G. Messerschmitt, "Minimum MSE equalization of digital fiber optic systems," *ibid.*, 1110–1118.

    (c) W. Hauk, F. Bross, and M. Ottka, "The calculation of error rates for optical fiber systems," *ibid.*, 1119–1126.

12. M. Mansuripur, J. W. Goodman, E. G. Rawson, and R. E. Norton, "Fiber optics receiver error rate prediction using the Gram-Charlier series," *IEEE Trans. Commun.*, **COM-28**, 402–407, Mar. 1980.

13. F. S. Chen and Y. S. Chen, "Sensitivity loss of digital optical receivers caused by intersymbol interference," *Bell Sys. Tech. J.*, **59**, 1877–1891, Dec. 1980.

14. S. D. Personick, "Receiver design," in *Optical Fiber Telecommunications*, S. E. Miller and A. G. Chynoweth (Eds.), Academic, New York, 1979.

15. S. D. Personick, *Optical Fiber Transmission Systems*, Plenum, New York, 1981.

16. See any basic book on communication systems, for example:

    (a) A. B. Carlson, *Communication Systems*, 2d ed., McGraw-Hill, New York, 1975.

    (b) K. S. Shanmugam, *Digital and Analog Communication Systems*, Wiley, New York, 1979.

17. (a) M. Schwartz, *Information Transmission, Modulation, and Noise*, 3d ed., McGraw-Hill, New York, 1980.

    (b) A. Van Der Ziel, *Noise: Sources, Characterization, Measurement*, Prentice-Hall, Englewood Cliffs, N.J., 1970.

18. R. W. Lucky, J. Salz, and E. J. Weldon, Jr., *Principles of Data Communications*, McGraw-Hill, New York, 1968.

19. E. A. Newcombe and S. Pasupathy, "Error rate monitoring for digital communications," *Proc. IEEE*, **70**, 805–828, Aug. 1982.

20. (a) A. Papoulis, *Probability, Random Variables, and Stochastic Processes*, McGraw-Hill, New York, 1965.

    (b) P. Z. Peebles, Jr., *Probability, Random Variables and Random Signal Principles*, McGraw-Hill, New York, 1980.

21. M. Abramowitz and I. A. Stegun, *Handbook of Mathematical Functions*, Dover, New York, 1965.

22. R. C. Hooper and R. B. White, "Digital optical receiver design for non-zero extinction ratio using a simplified approach," *Opt. Quantum Electron.*, **10**, 279–282, 1978.

23. J. E. Goell, "An optical repeater with high impedance input amplifier," *Bell Sys. Tech. J.*, **53**, 629–643, Apr. 1974.

24. J. E. Goell, "Input amplifiers for optical PCM repeaters," *Bell Sys. Tech. J.*, **53**, 1771–1793, Nov. 1974.

25. P. K. Runge, "An experimental 50 Mb/s fiber optic PCM repeater," *IEEE Trans. Comm.*, **COM-24**, 413–418, Apr. 1976.

26. T. Witkowicz, "Design of low-noise fiber optic receiver amplifiers using J-FETs," *IEEE J. Solid-State Circuits,* **SC-13,** 195–197, Feb. 1978.

27. A. van der Ziel, *Introductory Electronics,* Prentice-Hall, Englewood Cliffs, N.J., 1974.

28. T. T. Ha, *Solid State Microwave Amplifier Design,* Wiley, New York, 1981.

29. R. F. Leheny, R. E. Nahory, M. A. Pollack, A. A. Ballman, E. D. Beebe, J. C. De Winter, and R. J. Martin, "Integrated InGaAs p-i-n FET photoreceiver," *Electron. Lett.,* **16,** 353–355, Mar. 1980.

30. J. L. Hullett and T. V. Muoi, "A feedback receiver amplifier for optical transmission systems," *IEEE Trans. Commun.,* **COM-24,** 1180–1185, Oct. 1976.

31. T. L. Maione and D. D. Sell, "Experimental fiber optic transmission system for interoffice trunks," *IEEE Trans. Commun.,* **COM-25,** 517–524, May 1977.

32. T. V. Muoi and J. L. Hullett, "Receiver design for optical PPM systems," *IEEE Trans. Commun.,* **COM-26,** 295–300, Feb. 1978.

33. R. G. Smith, C. A. Brackett, and H. W. Reinbold, "Optical detector package," *Bell Sys. Tech J.,* **57,** 1809–1822, July–Aug. 1978.

34. R. G. Smith and S. D. Personick, "Receiver design for optical fiber communication systems," in *Semiconductor Devices for Optical Communications,* H. Kressel (Ed.), Springer-Verlag, New York, 1980.

35. T. L. Maione, D. D. Sell, and D. H. Wolaver, "Practical 45 Mb/s regenerator for lightwave transmission," *Bell Sys. Tech. J.,* **57,** 1837–1856, July–Aug. 1978.

36. V. L. Mirtich, "Designer's guide to fiber optic data links—parts 1, 2, 3," *EDN,* **25,** 133–140, June 20, 1980; 113–117, Aug. 5, 1980; 103–110, Aug. 20, 1980.

37. V. P. O'Neil II, "Using integrated detector/pre-amplifiers in fiber optic systems," *Electro-Opt. Sys. Design,* **13,** 35–39, Jan. 1981.

38. V. Mirtich, "Protecting optical data links from electro-magnetic interference," *Electronics,* **54,** 146–152, Jan. 13, 1981.

39. S. Moustakas, J. L. Hullett, and T. D. Stephens, "Comparison of BJT and MESFET front ends in broadband optical transimpedance amplifiers," *Opt. Quantum Electron.,* **14,** 57–60, Jan. 1982.

40. R. T. Unwin, "A high speed optical receiver," *Opt. Quantum Electron.,* **14,** 61–66, Jan. 1982.

EIGHT

TRANSMISSION LINK ANALYSES

In the preceding chapters we have presented the fundamental characteristics of the individual building blocks of an optical fiber transmission link. These include the optical fiber transmission medium, the optical source, the photodetector and its associated receiver, and the connectors used to join individual fiber cables to each other and to the source and detector. In this chapter we shall examine how these individual parts can be put together to form a complete optical fiber transmission link.

We shall first discuss in detail the simplest case of a point-to-point link. This will include examining the components that are available for a particular application and seeing how these components relate to the system performance criteria (such as dispersion and bit error rate). For a given set of components and a given set of system requirements, we then carry out a power budget analysis to determine whether the fiber optic link meets the attenuation requirements or if repeaters are needed. The final step is to perform a system rise time analysis to verify that the overall system performance requirements are met.

We next turn our attention to more complex link architectures, such as multi-terminal data bus networks and multichannel wavelength division multiplexed (WDM) systems. These types of links make more efficient use of the fiber transmission medium, but they also have limitations that are not present in simple point-to-point links. The architectures, advantages, and disadvantages of WDM systems and data buses are given in Secs. 8-2 and 8-3, respectively.

The chapter concludes with an overview of line-coding schemes that are suitable for digital data transmission over optical fibers. These coding schemes are used to introduce randomness and redundancy into the digital information stream to ensure efficient timing recovery and to facilitate error monitoring at the receiver.

8-1 POINT-TO-POINT LINKS

The simplest transmission link is a point-to-point line having a transmitter on one end and a receiver on the other, as is shown in Fig. 8-1. This type of link places the least demand on optical fiber technology and, as such, sets the basis for examining more complex system architectures.

The design of an optical link involves many interrelated variables among the fiber, source, and photodetector operating characteristics, so that the actual link design and analysis may require several iterations before they are completed satisfactorily. Since performance and cost constraints are very important factors in fiber optic communication links, the designer must carefully choose the components to ensure that the desired performance level can be maintained over the expected system lifetime without overspecifying the component characteristics.

The key system requirements needed in analyzing a link are:

1. The desired (or possible) transmission distance
2. The data rate or channel bandwidth
3. The bit error rate (BER)

To fulfill these requirements the designer has a choice of the following components and their associated characteristics:

1. Multimode or single-mode optical fiber
   (a) Core size
   (b) Core refractive-index profile
   (c) Bandwidth
   (d) Attenuation
   (e) Numerical aperture
2. LED or laser diode optical source
   (a) Emission wavelength
   (b) Spectral line width
   (c) Output power
   (d) Effective radiating area
   (e) Emission pattern
3. *pin* or avalanche photodiode
   (a) Responsivity
   (b) Operating wavelength
   (c) Speed
   (d) Sensitivity

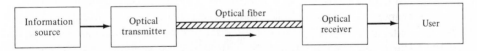

Figure 8-1 Simplex point-to-point optical fiber link.

Two analyses are usually carried out to ensure that the desired system performance can be met; these are the *link power budget* and the system *rise time budget* analyses. In the link power budget analysis one first determines the power margin between the optical transmitter output and the minimum receiver sensitivity needed to establish a specified BER. This margin can then be allocated to connector, splice, and fiber losses, plus any additional margins required for expected component degradation or temperature effects. If the choice of components did not allow the desired transmission distance to be achieved, the components might have to be changed or repeaters might have to be incorporated into the link.

Once the link power budget has been established, the designer can perform a system rise time analysis to ensure that the desired overall system performance has been met. We shall now examine these two analyses in more detail.

8-1-1 System Considerations

In carrying out a link power budget, we first decide at which wavelength to transmit and then choose components operating in this region. If the distance over which the data are to be transmitted is not too far, we may decide to operate in the 800- to 900-nm region. On the other hand, if the transmission distance is relatively long, we may want to take advantage of the lower attenuation and dispersion that occurs at wavelengths around 1.3 μm.

Having decided on a wavelength, we next interrelate the system performances of the three major optical link building blocks, that is, the receiver, transmitter, and optical fiber. Normally the designer chooses the characteristics of two of these elements and then computes those of the third to see if the system performance requirements are met. If the components have been over- or underspecified, a design iteration may be needed. The procedure we shall follow here is first to select the photodetector. We then choose an optical source and see how far data can be transmitted over a particular fiber before a repeater is needed in the line to boost up the power level of the optical signal.

In choosing a particular photodetector, we mainly need to determine the minimum optical power that must fall on the photodetector to satisfy the bit error rate (BER) requirement at the specified data rate. In making this choice, the designer also needs to take into account any design cost and complexity constraints. As we noted in Chaps. 6 and 7, a *pin* photodiode receiver is simpler, more stable with changes in temperature, and less expensive than an avalanche photodiode receiver. In addition, *pin* photodiode bias voltages are normally less than 50 V, whereas those of avalanche photodiodes are several hundred volts. However, the advantages of *pin* photodiodes may be overruled by the increased sensitivity of the avalanche photodiode if very low optical power levels are to be detected.

The system parameters involved in deciding between the use of an LED or a laser diode are signal dispersion, data rate, transmission distance, and cost. As we saw in Chap. 4, the spectral width of the laser output is much narrower than that of an LED. This is of importance in the 800- to 900-nm region, where the spectral width of an LED and the dispersion characteristics of silica fibers limit the data-rate-distance product to around 150 (Mb/s) · km. For higher values [up to 2500 (Mb/s) · km] a laser must be used at these wavelengths. At wavelengths around 1.3 μm, where signal dispersion is very low, bit-rate-distance products of at least 1500 (Mb/s) · km are achievable with LEDs. For InGaAsP lasers this figure is in excess of 25 (Gb/s) · km.

Since laser diodes typically couple from 10 to 15 dB more optical power into a fiber than an LED, greater repeaterless transmission distances are possible with a laser. This advantage and the lower dispersion capability of laser diodes may be offset by cost constraints. Not only is a laser diode itself more expensive than an LED but also the laser transmitter circuitry is much more complex, since the lasing threshold has to be dynamically controlled as a function of temperature and device aging.

For the optical fiber we have a choice between single-mode and multimode fiber, either of which could have a step- or a graded-index core. This choice depends on the type of light source used and on the amount of dispersion that can be tolerated. Multimode fibers must be used with LED sources, since very little optical power can be coupled into a single-mode fiber from an LED. As we saw in Chap. 5, the optical power that can be coupled into a fiber from an LED depends on the core-cladding index difference Δ, which, in turn, is related to the numerical aperture of the fiber (for $\Delta = 0.01$ the numerical aperture NA $\simeq 0.21$). As Δ increases, the fiber-coupled power increases correspondingly. However, since dispersion also becomes greater with increasing Δ, a tradeoff must be made between the optical power that can be launched into the fiber and the maximum tolerable dispersion.

Either a single-mode or a multimode fiber can be used with a laser diode. A single-mode fiber can provide the ultimate bit-rate-distance product, with values of 30 (Gb/s) · km being achievable. A disadvantage of single-mode fibers is that the small core size (5 to 16 μm in diameter) makes fiber splicing more difficult and critical than for multimode fibers having 50-μm core diameters.

When choosing the attenuation characteristics of a cabled fiber, the excess loss that results from the cabling process must also be considered in addition to the attenuation of the fiber itself. This must also include connector and splice losses as well as environmental-induced losses that could arise from temperature variations, radiation effects, and dust and moisture on the connectors.

8-1-2 Link Power Budget

An optical power loss model for a point-to-point link is shown in Fig. 8-2. The optical power received at the photodetector depends on the amount of light coupled into the fiber and the losses occurring in the fiber and at the connectors and splices. The link loss budget is derived from the sequential loss contributions of each element in the link. Each of these loss elements is expressed in decibels (dB) as

$$\text{Loss} = 10 \log \frac{P_{\text{out}}}{P_{\text{in}}} \tag{8-1}$$

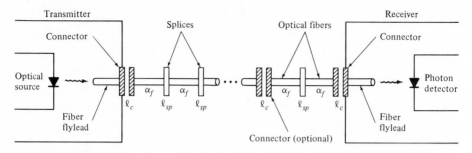

Figure 8-2 Optical power loss model for a point-to-point link. The losses occur at connectors (l_c), splices (l_{sp}), and in the fiber (α_f).

where P_{in} and P_{out} are the optical powers emanating into and out of the loss element, respectively.

In addition to the link loss contributors shown in Fig. 8-2, a link power margin is normally provided in the analysis to allow for component aging, temperature fluctuations, and losses arising from components that might be added at future dates. A link margin of 6 to 8 dB is generally used for systems that are not expected to have additional components incorporated into the link in the future.

The link loss budget simply considers the total optical power loss P_T that is allowed between the light source and the photodetector, and allocates this loss to cable attenuation, connector loss, splice loss, and system margin. Thus if P_S is the optical power emerging from the end of a fiber flylead attached to the light source, and if P_R is the receiver sensitivity, then

$$P_T = P_S - P_R$$
$$= 2l_c + \alpha_f L + \text{system margin} \qquad (8\text{-}2)$$

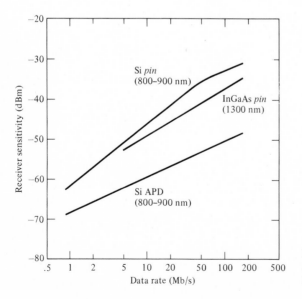

Figure 8-3 Receiver sensitivities as a function of bit rate for a 10^{-9} BER.

Here l_c is the connector loss, α_f is the fiber attenuation (dB/km), L is the transmission distance, and the system margin is nominally taken as 6 dB. Here we assume that the cable of length L has connectors only on the ends and none in between. The splice loss is incorporated into the cable loss for simplicity.

To illustrate how a link loss budget is set up, let us carry out a specific design example. We shall begin by specifying a data rate of 20 Mb/s and a bit error rate of 10^{-9} (that is, at most one error can occur for every 10^9 bits sent). For the receiver we shall choose a silicon *pin* photodiode operating at 850 nm. Figure 8-3 shows that the required receiver input signal is -42 dBm (42 dB below 1 mW). We next select a GaAlAs LED which can couple a 50-μW (-13-dBm) average optical power level into a fiber flylead with a 50-μm core diameter. We thus have a 29-dB allowable power loss. Assume further that a 1-dB loss occurs when the fiber flylead is connected to the cable and another 1-dB connector loss occurs at the cable-to-photodetector interface. Including a 6-dB system margin, the possible transmission distance for a cable with an attenuation of α_f dB/km can be found from Eq. (8-2):

$$P_T = P_S - P_R = 29 \text{ dB}$$

$$= 2 \, (1 \text{ dB}) + \alpha_f L + 6 \text{ dB}$$

If $\alpha_f = 3.5$ dB/km, then a 6.0-km transmission path is possible.

The link power budget can be represented graphically as is shown in Fig. 8-4. The vertical axis represents the optical power loss allowed between the transmitter and the receiver. The horizontal axis gives the transmission distance. Here we show a silicon *pin* receiver with a sensitivity of -42 dBm (at 20 Mb/s) and an LED with

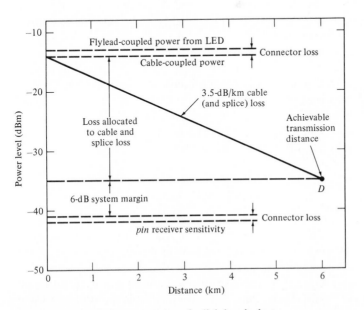

Figure 8-4 Graphical representation of a link loss budget.

an output power of −13 dBm coupled into a fiber flylead. We subtract a 1-dB connector loss at each end, which leaves a total margin of 27 dB. Subtracting a 6-dB system safety margin leaves us with a tolerable loss of 21 dB that can be allocated to cable and splice loss. The slope of the line shown in Fig. 8-4 is the 3.5-dB/km cable (and splice, in this case) loss. This line starts at the −14-dBm point (which is the optical power coupled into the cabled fiber) and ends at the −35-dB level (the receiver sensitivity minus a 1-dB connector loss and a 6-dB system margin). The intersect point D then defines the maximum possible transmission path length.

Further design examples for short- and long-wavelength systems can be found in the references [1-8] and in the problem set at the end of the chapter.

8-1-3 Rise Time Budget

A rise time budget analysis is a convenient method for determining the dispersion limitation of an optical fiber link. This is particularly useful for digital systems. In this approach the total rise time t_{sys} of the link is the root-sum-square of the rise times from each contributor t_i to the pulse rise time degradation

$$t_{\text{sys}} = \left(\sum_{i=1}^{N} t_i^2 \right)^{1/2}$$

(8-3)

The four basic elements that may significantly limit system speed are the transmitter rise time t_{tx}, the material dispersion rise time t_{mat} of the fiber, the modal dispersion rise time t_{mod} of the fiber, and the receiver rise time t_{rx}. Generally, the total transition time degradation of a digital link should not exceed 70 percent of an NRZ (non-return to zero) bit period or 35 percent of a bit period for RZ (return-to-zero) data, where one bit period is defined as the reciprocal of the data rate (NRZ and RZ data formats are discussed in more detail in Sec. 8-4).

The rise times of transmitters and receivers are generally known to the designer. The transmitter rise time is attributable primarily to the light source and its drive circuitry. The receiver rise time results from the photodetector response and the 3-dB electric bandwidth of the receiver front end. If B_{rx} is the 3-dB electric bandwidth of the receiver in megahertz, then the receiver front end rise time in nanoseconds is given by the standard empirical formula[9]

$$t_{rx} = \frac{350}{B_{rx}}$$

(8-4)

For multimode fibers the rise time depends on modal and material dispersions. Its analysis is more complicated since it is a function of the length of the fiber, the type of optical source used, and the operating wavelength. Material dispersion effects can be neglected for laser sources at both short and long wavelengths, and for LEDs at long wavelengths. Using Eq. (3-28), we see that in the 800- to 900-nm region material dispersion adds about 0.07 ns/(nm · km) to the rise time.

For cables shorter than the modal equilibrium length, the fiber bandwidth resulting from modal dispersion is inversely proportional to the cable length. As we saw in Sec. 3-5, in a long, continuous fiber which has no joints, the fiber bandwidth decreases linearly with distance for lengths less than the modal equilibrium length L_c. For lengths greater than L_c, a steady-state equilibrium condition has been established and the bandwidth then decreases as $L^{1/2}$.

In practice, an optical fiber link seldom consists of a continuous, jointless fiber. Instead, several fibers are concatenated (tandemly joined) to form a long link. The situation then becomes more complex because a modal redistribution occurs at fiber-to-fiber joints in the cable. This is a result of misaligned joints, different core index profiles in each fiber, and/or different degrees of mode mixing in individual fibers. The most important of these factors is the effect of different core index profiles in adjacent fibers. As we saw in Sec. 3-4, the value of the core index profile α influences the degree of modal-dispersion-induced pulse spreading in a fiber. The value of the index-grading parameter α that minimizes pulse dispersion depends strongly on the wavelength, so that fibers optimized for operation at different wavelengths have different values of α. Variations in α at the same wavelength thus result in overcompensated or undercompensated core index profiles (see Fig. 3-18).

The difficulty in predicting the bandwidth of a series of concatenated fibers arises from the observation[13] that the total route bandwidth can be a function of the order in which fibers are joined. For example, instead of randomly joining together arbitrary (but very similar) fibers, an improved total link bandwidth can be obtained by selecting adjoining fibers with alternating over- and undercompensated refractive-index profiles to provide some modal delay equalization. Although the ultimate concatenated fiber bandwidth can be obtained by judiciously selecting adjoining fibers for optimum modal delay equalization, in practice this is unwieldy and time consuming, particularly since the initial fiber in the link appears to control the final link characteristics.

A variety of empirical expressions for modal dispersion have thus been developed.[10-15] From practical field experience it has been found that the bandwidth B_M in a link of length L can be expressed to a reasonable approximation by the empirical relation

$$B_M(L) = \frac{B_0}{L^q} \tag{8-5}$$

where the parameter q ranges between 0.5 and 1, and B_0 is the bandwidth of a 1-km length of cable. A value of $q = 0.5$ indicates that a steady-state modal equilibrium has been reached, whereas $q = 1$ indicates little mode mixing. Based on field experience, a reasonable estimate is $q = 0.7$.

Another expression that has been proposed for B_M, based on curve fitting of experimental data, is

$$\frac{1}{B_M} = \left[\sum_{n=1}^{N} \left(\frac{1}{B_n} \right)^{1/q} \right]^q \tag{8-6}$$

where the parameter q ranges between 0.5 (quadrature addition) and 1.0 (linear addition), and B_n is the bandwidth of the nth fiber section. Alternatively, Eq. (8-6) can be written as

$$t_M(N) = \left[\sum_{n=1}^{N} (t_n)^{1/q} \right]^q \tag{8-7}$$

where $t_M(N)$ is the pulse broadening occurring over N cable sections in which the individual pulse broadenings are given by t_n.

A third empirical expression proposed by Eve[10,11] for pulse broadening in a jointed link of N fibers is

$$t_M^2(N) = \sum_{k=1}^{N} t_k^2 + \sum_{\substack{1 \\ p \neq k}}^{N} t_p t_k r_{pk} \tag{8-8}$$

where r_{pk} is a correlation coefficient between the pth and the kth fiber. Its magnitude is expected to range between 0 and 1 for strong and little mode mixing, respectively.

We now need to find the relation between the fiber rise time and the 3-dB bandwidth. For this we use a variation of the expression derived by Midwinter.[16] We assume that the optical power emerging from the fiber has a gaussian temporal response described by

$$g(t) = \frac{1}{\sqrt{2\pi}\,\sigma} e^{-t^2/2\sigma^2} \tag{8-9}$$

where σ is the rms pulse width.

The Fourier transform of this function is

$$G(\omega) = \frac{1}{\sqrt{2\pi}} e^{-\omega^2\sigma^2/2} \tag{8-10}$$

From Eq. (8-9) the time $t_{1/2}$ required for the pulse to reach its half-maximum value, that is, the time required to have

$$g(t_{1/2}) = 0.5g(0) \tag{8-11}$$

is given by

$$t_{1/2} = (2 \ln 2)^{1/2}\sigma \tag{8-12}$$

If we define the time t_{FWHM} as the full width of the pulse at its half-maximum value, then

$$t_{\text{FWHM}} = 2t_{1/2} = 2\sigma(2 \ln 2)^{1/2} \tag{8-13}$$

The 3-dB optical bandwidth $B_{3\,\text{dB}}$ is defined as the modulation frequency $f_{3\,\text{dB}}$ at which the received optical power has fallen to 0.5 of the zero frequency value. Thus from Eqs. (8-10) and (8-13) we find that the relation between the full-width half-maximum rise time t_{FWHM} and the 3-dB optical bandwidth is

$$f_{3\,\text{dB}} = B_{3\,\text{dB}} = \frac{0.44}{t_{\text{FWHM}}} \tag{8-14}$$

Using Eq. (8-5) for the 3-dB optical bandwidth of the fiber link and letting t_{FWHM} be the rise time resulting from modal dispersion, then from Eq. (8-14),

$$t_{mod} = \frac{0.44}{B_M} = \frac{0.44 \, L^q}{B_0} \qquad (8\text{-}15)$$

If t_{mod} is expressed in nanoseconds and B_M is given in megahertz, then

$$t_{mod} = \frac{440}{B_M} = \frac{440 \, L^q}{B_0} \qquad (8\text{-}16)$$

Substituting Eqs. (3-20), (8-4), and (8-16) into Eq. (8-3) gives a total system rise time of

$$t_{sys} = \left[t_{tx}^2 + D_{mat}^2 \sigma_\lambda^2 L^2 + \left(\frac{440 \, L^q}{B_0} \right)^2 + \left(\frac{350}{B_{rx}} \right)^2 \right]^{1/2} \qquad (8\text{-}17)$$

where all the times are given in nanoseconds, σ_λ is the spectral width of the optical source, and D_{mat} is the material dispersion factor of the fiber (given in nanoseconds per nanometer per kilometer). In the 800- to 900-nm region, D_{mat} is about 0.07 ns/(nm · km), but is negligible around 1300 nm (see Fig. 3-13).

As an example of a rise time budget, let us continue the analysis of the link we started to examine in Sec. 8-1-2. We shall assume that the LED together with its drive circuit has a rise time of 15 ns. Taking a typical LED spectral width of 40 nm, we have a material-dispersion-related rise time degradation of 21 ns over the 6-km link. Assuming the receiver has a 25-MHz bandwidth, then from Eq. (8-4) the contribution to the rise time degradation from the receiver is 14 ns. If the fiber we select has a 400-MHz · km bandwidth-distance product and with $q = 0.7$ in Eq. (8-5), then from Eq. (8-15) the modal-dispersion-induced fiber rise time is 3.9 ns. Substituting all these values back into Eq. (8-17) results in a link rise time of

$$\begin{aligned} t_{sys} &= (t_{tx}^2 + t_{mat}^2 + t_{mod}^2 + t_{rx}^2)^{1/2} \\ &= [(15 \text{ ns})^2 + (21 \text{ ns})^2 + (3.9 \text{ ns})^2 + (14 \text{ ns})^2]^{1/2} \\ &= 30 \text{ ns} \end{aligned}$$

This value falls below the maximum allowable 35-ns rise time degradation for our 20-Mb/s NRZ data stream. The choice of components was thus adequate to meet our system design criteria.

8-1-4 Transmission Distance versus Bit Rate

Figure 8-5 shows the attenuation and dispersion limitation on the repeaterless transmission distance as a function of data rate for the short-wavelength (800- to 900-nm) LED/*pin* combination. The BER was taken as 10^{-9} for all data rates. The fiber-coupled LED output power was assumed to be a constant −13 dBm for all data rates up to 200 Mb/s. The attenuation limit curve was then derived by using a fiber loss of 3.5 dB/km and the receiver sensitivities shown in Fig. 8-3. Since the

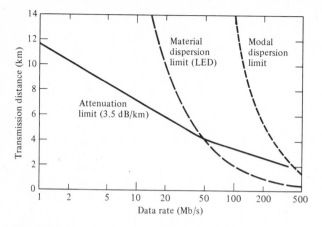

Figure 8-5 Transmission distance limits as a function of data rate for a 400-MHz · km fiber, an 800-nm LED source, and a silicon *pin* photodetector.

minimum optical power required at the receiver for a given BER becomes higher for increasing data rates, the attenuation limit curve slopes downward to the right. We have also included a 1-dB connector-coupling loss at each end and a 6-dB system operating margin.

The dispersion limit depends on material and modal dispersion, as we described in Sec. 8-1-3. Material dispersion at 800 nm is taken as 0.07 ns/(nm · km) or 3.5 ns/km for an LED with a 50-nm spectral width. The curve shown is the material dispersion limit in the absence of modal dispersion. This limit was taken to be the distance at which t_{mat} is 70 percent of a bit period. The modal dispersion was derived from Eq. (8-15) for a fiber with a 400-MHz · km bandwidth-distance product and with $q = 0.7$. The modal dispersion limit was then taken to be the distance at which t_{mod} is 70 percent of a bit period. The achievable repeaterless transmission distances are those that fall below the attenuation limit curve and to the left of the dispersion limit line. The transmission distance is attenuation-limited up to about 40 Mb/s after which it becomes material-dispersion-limited.

The derivation of similar curves for short-wavelength laser diodes and avalanche photodiodes and for long-wavelength components is left as an exercise for the reader (see Probs. 8-6 and 8-7).

8-2 WAVELENGTH DIVISION MULTIPLEXING (WDM)

Having examined point-to-point links we shall now consider more sophisticated systems which make a fuller utilization of the transmission capacity of an optical fiber. In standard point-to-point links, as shown in Fig. 8-1, a single-fiber line has one optical source at its transmitting end and one photodetector at the receiving end. Signals from different light sources require separate and uniquely assigned optical fibers. Since an optical source has a relatively narrow spectral width, this type of

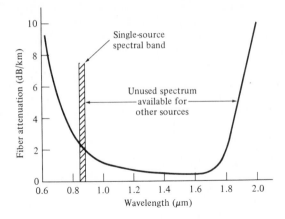

Figure 8-6 A single optical source uses a small part of the available spectral transmission band of a fiber. Wavelength division multiplexing makes simultaneous use of many spectral channels.

transmission makes use of only a very narrow portion of the transmission bandwidth capability of a fiber, as is shown in Fig. 8-6 for a typical laser diode operating at 850 nm.

From Fig. 8-6 we see that many additional spectral operating regions are possible. Ideally, a dramatic increase in the information capacity of a fiber can thus be achieved by the simultaneous transmission of optical signals over the same fiber from many different light sources having properly spaced peak emission wavelengths. By operating each source at a different peak wavelength, the integrity of the independent messages from each source is maintained for subsequent conversion to electric signals at the receiving end. This is the basis of wavelength division multiplexing[17–19] (WDM). Conceptually the WDM scheme is the same as frequency division multiplexing (FDM) used in microwave radio and satellite systems.

Two different WDM setups are shown in Figs. 8-7 and 8-8. In Fig. 8-7 a unidirectional WDM device is used to combine different signal carrier wavelengths onto a single fiber at one end and to separate them into their corresponding detectors at the other end. A bidirectional WDM scheme is shown in Fig. 8-8. This involves sending information in one direction at a wavelength λ_1 and simultaneously transmitting data in the opposite direction at a wavelength λ_2.

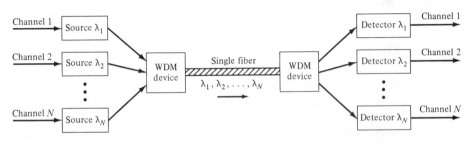

Figure 8-7 A unidirectional WDM system that combines N independent input signals for transmission over a single fiber.

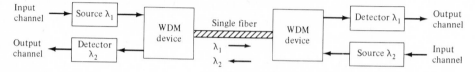

Figure 8-8 Schematic representation of a bidirectional WDM system in which two or more wavelengths are transmitted simultaneously in opposite directions over the same fiber.

Let us first examine the system requirements of the WDM technique. The three basic performance criteria are insertion loss, channel width, and cross talk. *Insertion loss* defines the amount of power loss that arises in the fiber optic line from the addition of a WDM coupling device. This includes losses occurring at the connection points of the WDM element to the fiber line and any intrinsic losses within the multiplexing element itself. In practice, designers can tolerate insertion losses of a few decibels at each end.

Channel width is the wavelength range that is allocated to a particular optical source. If laser diodes are used, channel widths of several tens of nanometers are required to ensure that no interchannel interference results from source instability (for example, drift of peak operating output wavelength with temperature changes). For LED sources, channel widths that are 10 to 20 times larger are required because of the wider spectral output width of these sources.

Cross talk refers to the amount of signal coupling from one channel to another. The tolerable interchannel cross-talk levels can vary widely depending on the application. However, in general, a -10-dB level is not sufficient whereas a -30-dB level is adequate.

In implementing a unidirectional WDM system a multiplexer is needed at the transmitting end to combine optical signals from several light sources onto a single fiber. At the receiving end a demultiplexer is required to separate the signals into appropriate detection channels. Since the optical signals that are combined generally to do not emit a significant amount of optical power outside of their designated channel spectral width, interchannel cross-talk factors are relatively unimportant at the transmitter end. The basic design problem here is that the multiplexer should provide a low-loss path from each optical source to the multiplexer output. A different requirement exists for the demultiplexer, since photodetectors are usually sensitive over a broad range of wavelengths which could include all the WDM channels. Thus, to prevent significant amounts of the wrong signal from entering each receiving channel, that is, to give good channel isolation of the different wavelengths being used, either the demultiplexer must be carefully designed or very stable optical filters with sharp wavelength cutoffs must be used.

In principle, any optical wavelength demultiplexer can also be used as a multiplexer. Thus, for simplicity, the word "multiplexer" is often used as a general term to refer to both multiplexers and demultiplexers, except when it is necessary to distinguish the two devices or functions.

The wavelength division multiplexers that are most widely used fall into two classes. These are angularly dispersive devices,[20-26] such as prisms or gratings, and

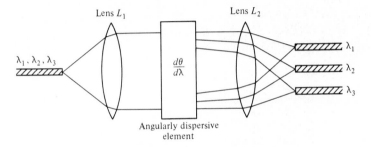

Figure 8-9 Schematic representation of an angularly dispersive WDM element shown for three wavelengths. Many wavelengths can be combined or separated with this type of device.

filter-based devices,[27-31] such as multilayer thin-film interference filters or single-mode integrated-optical devices. The basis of an angularly dispersive multiplexer is shown schematically in Fig. 8-9 for a three-wavelength system, where $d\theta/d\lambda$ is the angular dispersion of the device. When the device is used as a demultiplexer, the light emerging from the left-hand fiber is collimated by lens L_1 (the collimating lens) and passed through the dispersive element which separates the various wavelength channels into different spatially oriented beams. Lens L_2 (the focusing lens) then focuses the output beams into the appropriate receiving fibers or detectors. The linear dispersion $dx/d\lambda$ at the output fibers shown on the right-hand side of Fig. 8-9 is[17]

$$\frac{dx}{d\lambda} = f\frac{d\theta}{d\lambda}$$
(8-18)

where f is the focal length of lens L_2. In the ideal case of aberrationless optics and zero source spectral width, the intrinsic insertion loss and cross talk will be zero if the output signals are separated by more than their diameter d, that is, if

$$\frac{dx}{d\lambda}\Delta\lambda \geq d$$
(8-19)

where $\Delta\lambda$ is the spectral separation between channels. We shall assume here that all fibers (input and output) have the same diameter d and numerical aperture NA.

To collect all the light from the input fiber, the collimating lens L_1 must have a diameter b satisfying the condition[17]

$$b > 2f\frac{\text{NA}}{n'}$$
(8-20)

where n' is the refractive index of the medium between the dispersive device and the lens L_1. Combining Eqs. (8-18) through (8-20) then yields

$$b \geq \frac{2(\text{NA}/n')d}{\Delta\lambda(d\theta/d\lambda)}$$
(8-21)

In any real system the output beam spreads out because of the finite size of the light source and the angular dispersion resulting from the wavelength spread of the source spectral width. The fractional increase S in the beam diameter is approximately given by[17]

$$S = \frac{b' - b}{b} \simeq (1 + m) \frac{Wd(NA)}{b^2 n'} \tag{8-22}$$

where m is the number of wavelength channels, b' is the diameter of lens L_2, and W is the total path length from the output of lens L_2 to the input of lens L_1. To avoid overfilling the numerical aperture of the output fiber, the total beam spread must be a small fraction of the collimating lens diameter, that is, $S \ll 1$.

A large number of channels can be combined and separated with angularly dispersive multiplexing elements. Most of these devices use a grating-plus-lens combination. Sometimes a prism is used for the angularly dispersive element. Insertion losses are typically in the 1- to 3-dB range and cross-talk levels between -20 and -30 dB are routinely obtained.

The operation of a filter-type multiplexing element is shown in Fig. 8-10 for a two-wavelength operation. The filters are designed to transmit light in a specific wavelength channel and either to absorb or reflect all other wavelengths. Reflection-type filters are normally used since the losses of absorption filters tend to be high (greater than 1 dB). The reflection filter consists of a flat glass substrate upon which multiple layers of different dielectric films are deposited for wavelength selectivity. These filters can be used in series to separate additional wavelength channels. The complexity involved in stacking the filters in series and the increase in signal loss that occurs with the addition of successive multiplexers generally limit operation to two or three filters (that is, three or four channels).

In designing a WDM system, care must be taken to minimize the factors causing link margin degradation. In addition to keeping a low insertion loss of the

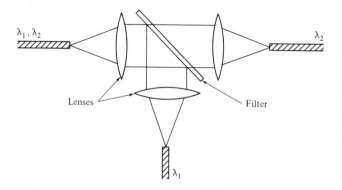

Figure 8-10 Multilayer thin-film-filter reflector used for WDM. This device is transparent at the wavelength λ_2 and reflects the wavelength λ_1.

coupling devices, the designer must minimize reflections of optical power that occur at the coupling elements or at connectors and splices, since they can give rise to signal cross talk. These effects are of particular importance in bidirectional links. For details the reader is referred to the papers by Wells[32] and by Conradi and Maciejko.[33]

8-3 DATA BUSES

An alternative to point-to-point links are data bus architectures. A *data bus* is a transmission line network over which a number of physically separated terminals can communicate with one another by multiplexed signals. This architecture can also be used to allow terminals to gain access efficiently to a variety of wideband information services on the same network, such as cable TV, graphics displays, and high-speed digital data. Conventional wideband data bus networks have been reliably and effectively implemented by using coaxial cable as the transmission medium. However, they have several disadvantages compared to fiber optic transmission lines, such as susceptibility to electromagnetic interference, larger size, and lower bandwidth.

Various types of optical fiber data buses have thus been investigated. The two general classes of optical data buses that have evolved are the *star* or *radial* configuration,[34-39] shown in Fig. 8-11, and the *in-line* or *T-coupler* bus,[40-46] shown in Fig. 8-12. Access to an optical data bus is achieved by means of a coupling element which can be either active or passive. An *active coupler* converts the optical signal on the data bus to its electric baseband counterpart before any data processing (such as injecting additional data into the signal stream or merely passing on the received data) is carried out. Its main use is for in-line configurations. A *passive coupler* employs no electronic elements. It is used passively either to tap off a portion of the optical power from the bus or to distribute optical power from an input fiber among several output fibers.

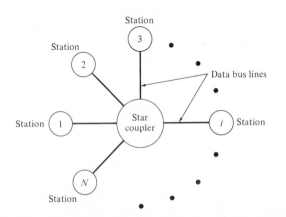

Figure 8-11 Radially- or star-configured data bus.

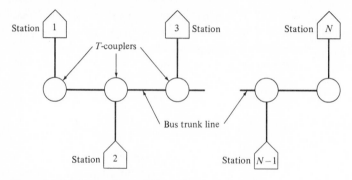

Figure 8-12 In-line or T-coupled data bus.

8-3-1 Star Configuration

A star-configured data bus uses either the transmission or the reflection star coupler shown in Fig. 8-13. These couplers are passive mixing elements; that is, the optical powers from the input ports are mixed together and then divided equally among the output ports. They may be used to combine numerous signals together, split a signal into a number of parts, or to tap optical power out of or insert optical power into a fiber optic link. Either type of star is composed of a set of input fibers, a set of output fibers, and a mixing region.

In general, the reflection star is more versatile because the relative number of input and output ports may be selected or varied after the device has been constructed (the total number of ports, in plus out, being fixed, however). By comparison, the number of transmission star input and output fibers is fixed by initial design and fabrication. The reflection star is usually less efficient since a portion of the light which has entered the coupler is injected back into the input fibers. Given the same number of input and output ports, the transmission star is twice as efficient as the reflection star. Therefore, reflection and transmission star couplers have their own particular advantages and disadvantages, and selection of a star coupler type for a particular application is largely determined by the network topology.

To see how star couplers can be applied to a given network, let us examine the various optical power losses associated with the coupler. The insertion loss L_S of

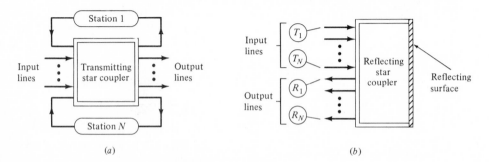

(a)   (b)

Figure 8-13 Star couplers. (a) Transmitting; (b) reflecting.

a star coupler is defined as the fraction of optical power lost in the process of coupling light from the input port to all the output ports (see Prob. 8-11 for an example). It is given in decibels by

$$L_S = 10 \log \frac{\sum_{j=1}^{N} P_j}{P_i} \tag{8-23}$$

where the P_j are the output powers from all the ports, P_i is the input power at one port, and for a transmission star coupler N is the number of outputs, whereas for a reflection star coupler N is the number of inputs plus outputs.

The optical power that enters a star coupler gets divided equally among the N output ports. That is, the optical power at any one output port is $(1/N)$th of the total optical power emerging from all the coupler outputs. This is known as the *power splitting factor* L_{sp}, which is given in decibels by

$$L_{sp} = 10 \log N$$

If P_S is the fiber-coupled output power from a source in dBm, P_R is the minimum optical power in dBm required at the receiver for a specific data rate, α_f is the fiber attenuation, and L_c is the connector loss in dB, then the balance equation for a particular link having all stations located at the same distance L from the star coupler is

$$P_S - P_R = L_S + 2\alpha_f L + 4L_c + 10 \log N + \text{ system margin} \tag{8-24}$$

where we have assumed connector losses at the transmitter, the receiver, and the input and output ports of the star coupler. For a transmission star N is the number of output ports, whereas for a reflection star N is the total number of input plus output ports.

A variety of star coupler types have been proposed in the literature[34–39] to which the reader is referred for details.

8-3-2 T-Coupler Data Buses

The bus access elements of an in-line data bus can be either active or passive couplers. Figure 8-14 gives an example of an active coupler. A photodiode receiver

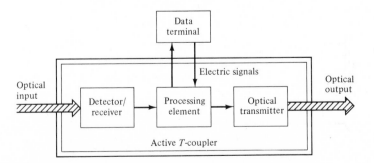

Figure 8-14 Active T-coupler.

converts the optical signal from the data bus into an electric signal. The processing element then can remove or copy part of this signal for transmission to the terminal, while the remainder is forwarded to the optical transmitter. The processor can also insert additional information from the terminal into the data stream. The transmitter, in turn, reconverts the electric signal to an optical data stream which gets sent on to the next terminal via the optical fiber bus. Couplers of this type can be easily constructed by using any of a number of commercially available photodiodes and light sources.

Passive couplers for an in-line data bus are used at each terminal to remove a portion of the optical signal from the bus trunk line or to inject additional light onto the trunk. A major problem of the passive T-coupler involves the optical power budget of the network. This difficulty arises from the fact that the optical signal is not regenerated at each terminal node, so that insertion and output losses at each tap plus the fiber losses between taps limit the network size to a small number of terminals (generally less than 10).

To evaluate the performance of an in-line data bus, let us examine the various sources of power loss along the transmission path. We shall consider this in terms of the fraction of power lost at a particular interface or within a particular component. First, over an optical fiber of length x (in kilometers), the ratio A of the received-to-transmitted power levels is given by

$$A = 10^{-(\alpha_f x/10)}$$
$$= e^{-2.3\alpha_f x/10} \tag{8-25}$$

where α_f is the fiber attenuation in decibels per kilometer.

The losses encountered in a T-coupler are shown schematically in Fig. 8-15. The coupler generally has four ports; two for connecting the device onto the fiber bus, one for receiving tapped-off data, and one for inserting data onto the line. We shall assume that a fraction F_c of optical power is lost at each port of the coupler. We take this fraction to be 20 percent, so that the connecting loss L_c is

$$L_c = -10 \log (1 - F_c) \simeq 1 \text{ dB} \tag{8-26}$$

That is, the optical power level gets reduced by 1 dB at any coupling junction.

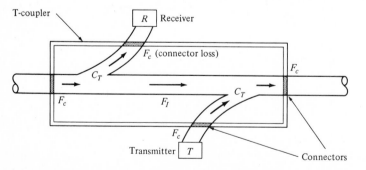

Figure 8-15 Power losses encountered in a passive T-coupler.

Let C_T represent the fraction of power removed from the bus and delivered to the detector port. Note that, in general, the fraction of power removed from the bus line in a T-coupler is actually $2C_T$, since optical power is extracted at both the receiving and the transmitting ports of the device. The power removed at the transmitting port is lost from the system. The coupling loss L_T is then given by

$$L_T = -10 \log (1 - 2C_T) \text{ dB} \tag{8-27}$$

In addition to connection and tapping losses, there is an intrinsic transmission loss L_I associated with each T-coupler. If the fraction of power lost in the coupler is F_I, then the intrinsic transmission loss L_I is

$$L_I = -10 \log (1 - F_I) \text{ dB} \tag{8-28}$$

Generally, an in-line data bus will consist of a number of stations separated by various lengths of bus line. However, here for analytical simplicity, we shall consider an in-line bus of N stations uniformly separated by a distance L. From Eq. (8-25) the fiber attenuation between adjacent stations is

$$A_0 = e^{-2.3\alpha_f L/10} \tag{8-29}$$

We shall use the notation P_{jk} to denote the optical power received at the detector of the kth station from the transmitter of the jth station. For simplicity, we shall assume that a T-coupler exists at every terminal of the data bus including the two end stations.

Because of the serial nature of the in-line data bus, the optical power available at a particular node decreases with increasing distance from the source. Thus a performance quantity of interest is the *system dynamic range*. This is the maximum optical power range to which any detector must be able to respond. The smallest difference in transmitted and received optical power occurs for adjacent stations, such as shown in Fig. 8-16, between station 1 and station 2, for example. If P_0 is the optical power launched from a source flylead and E is the coupling efficiency of optical power onto the bus line, then

$$P_{1,2} = A_0 C_T (1 - F_c)^4 E P_0 \tag{8-30}$$

The largest difference in transmitted and received optical power occurs between station 1 and station N. At the transmitting end the fractional power level

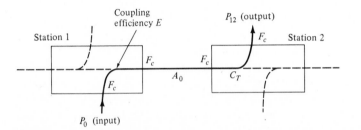

Figure 8-16 Optical path between adjacent stations for an in-line data bus.

coupled into the first length of cable is

$$F_1 = (1 - F_c)^2 E$$

At the receiving end the fraction of power from the T-coupler input port that emerges from the detector port is

$$F_N = (1 - F_c)^2 C_T$$

For each of the $N - 2$ intermediate stations the fraction of power passed through each coupler is

$$F_{\text{coup}} = (1 - F_c)^2 (1 - 2C_T)(1 - F_l)$$

Combining these factors and the transmission losses of the $N - 1$ intervening fibers, we find that the power received at station N from station 1 is

$$P_{1N} = A_0^{N-1} F_1 F_{\text{coup}}^{N-2} F_N P_0$$

$$= A_0^{N-1}(1 - F_c)^{2N}(1 - 2C_T)^{N-2} C_T (1 - F_l)^{N-2} E P_0 \qquad (8\text{-}31)$$

The worst-case dynamic range DR is then found from the ratio of Eq. (8-30) to Eq. (8-31)

$$DR = \frac{1}{[A_0(1 - F_c)^2(1 - 2C_T)(1 - F_l)]^{N-2}} \qquad (8\text{-}32)$$

A comparison of the star and in-line systems can be made by examining Eqs. (8-24) and (8-31). From Eq. (8-24) we see that the difference (in decibels) of the received and transmitted power levels in a star bus varies as log N for N terminals. If we convert Eq. (8-31) into a decibel format by taking the logarithm of both sides of the equation, we see for the in-line bus the power difference varies linearly with the number of stations. This strictly limits the number of stations on an in-line bus and imposes a much more severe requirement for the receiver dynamic range than needed for a star-configured data bus.

8-4 LINE CODING

In designing an optical fiber link, an important consideration is the format of the transmitted optical signal. This is of importance since, in any practical digital optical fiber data link, the decision circuitry in the receiver must be able to extract precise timing information from the incoming optical signal. The three main purposes of timing are to allow the signal to be sampled by the receiver at the time the signal-to-noise ratio is a maximum, to maintain the proper pulse spacing, and to indicate the start and end of each timing interval. In addition, since errors resulting from channel noise and distortion mechanisms can occur in the signal detection process, it may be desirable for the optical signal to have an inherent error-detecting capability. These features can be incorporated into the data stream by restructuring (or encoding) the signal. This is generally done by introducing extra bits into the raw data stream at the transmitter on a regular and logical basis and extracting them again at the receiver.

Signal encoding uses a set of rules for arranging the signal symbols in a particular pattern. This process is called *channel* or *line coding*. The purpose of this section is to examine the various types of line codes that are well-suited for digital transmission on an optical fiber link. The discussion here is limited to binary codes since they are the most widely used electrical codes and also because they are the most advantageous codes for optical systems.

One of the principal functions of a line code is to introduce redundancy into the data stream for the purpose of minimizing errors resulting from channel interference effects. Depending on the amount of redundancy introduced, any degree of error-free transmission of digital data can be achieved, provided that the data rate that includes this redundancy is less than the channel capacity. This is a result of the well-known Shannon channel-coding theory.[47]

Although large system bandwidths are attainable with optical fibers, the signal-to-noise considerations of the receiver discussed in Chap. 7 show that larger bandwidths result in larger noise contributions. Thus from noise considerations, minimum bandwidths are desirable. However, a larger bandwidth may be needed to have timing data available from the bit stream. In selecting a particular line code, a tradeoff must therefore be made between timing and noise bandwidth.[48] Normally these are largely determined by the expected characteristics of the raw data stream.

The three basic types of two-level binary line codes that can be used for optical fiber transmission links are the nonreturn-to-zero (NRZ) format, the return-to-zero (RZ) format, and the phase-encoded (PE) format. In NRZ codes a transmitted data bit occupies a full bit period. For RZ formats the pulse width is less than a full bit period. In the PE format both full-width and half-width data bits are present. Multilevel binary (MLB) signaling[49] is also possible, but it is used much less frequently than the popular NRZ and RZ codes. A brief description of some NRZ and RZ codes will be given here. Additional details can be found in numerous communications books.[50–53]

8-4-1 NRZ Codes

A number of different NRZ codes are widely used and their bandwidths serve as references for all other code groups. The simplest NRZ code is NRZ-level (or NRZ-L), shown in Fig. 8-17. For a serial data stream an on-off (or unipolar) signal represents a 1 by a pulse of current or light filling an entire bit period, whereas for a 0 no pulse is transmitted. These codes are simple to generate and decode, but they possess no inherent error-monitoring or correcting capabilities and they have no self-clocking (timing) features.

The minimum bandwidth is needed with NRZ coding, but the average power input to the receiver is dependent on the data pattern. For example, the high level

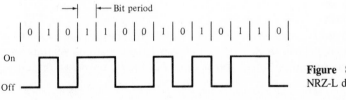

Figure 8-17 Example of an NRZ-L data pattern.

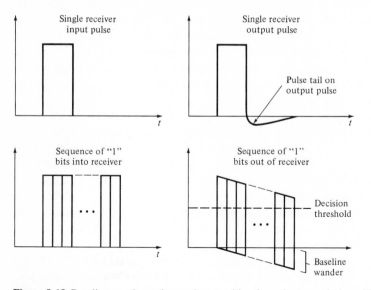

Figure 8-18 Baseline wander at the receiver resulting from the transmission of long strings of NRZ 1 bits.

of received power occurring in a long string of consecutive 1 bits can result in a *baseline wander* effect, as shown in Fig. 8-18. This effect results from the accumulation of pulse tails that arise from the low-frequency characteristics of the ac-coupling filter in the receiver.[54] If the receiver recovery to the original threshold is slow after the long string of 1 bits has ended, an error may occur if the next 1 bit has a low amplitude.

In addition a long string of NRZ ones or zeros contains no timing information since there are no level transitions. Thus, unless the timing clocks in the system are extremely stable, a long string of N identical bits could be misinterpreted as either $N - 1$ or $N + 1$ bits. However, the use of highly stable clocks increases system costs and requires a long system startup time to achieve synchronization. Two common techniques for restricting the longest time interval in which no level transitions occur are the use of block codes (which we discuss in Sec. 8-4-3) and scrambling.[55-56] *Scrambling* produces a random data pattern by the modulo-2 addition of a known bit sequence with the data stream. At the receiver the same known bit sequence is again modulo-2 added to the received data, and the original bit sequence is recovered. Although the randomness of scrambled NRZ data ensures an adequate amount of timing information, the penalty for its use is an increase in the complexity of the NRZ encoding and decoding circuitry.

8-4-2 RZ Codes

If an adequate bandwidth margin exists, each data bit can be encoded as two optical line code bits. This is the basis of RZ codes. In these codes a signal level transition occurs during either some or all of the bit periods to provide timing information. A variety of RZ code types exist, some of which are shown in Fig. 8-19. The

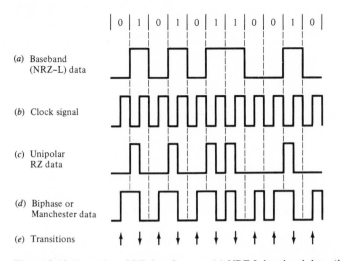

Figure 8-19 Examples of RZ data formats. (*a*) NRZ-L baseband data; (*b*) clock signal; (*c*) unipolar RZ data; (*d*) biphase or optical Manchester; (*e*) transitions occurring within a bit period for Manchester data.

baseband (NRZ-L) data are shown in Fig. 8-19*a*. In the unipolar RZ data a 1 bit is represented by a half-period optical pulse that can occur either in the first or second half of the bit period. A 0 is represented by no signal during the bit period.

A disadvantage of the unipolar RZ format is that long strings of 0 bits can cause loss of timing synchronization. A common data format not having this limitation is the *biphase* or *optical Manchester* code shown in Fig. 8-19*d*. Note that this is a unipolar code, which is in contrast to the conventional bipolar Manchester code used in wire lines. The optical Manchester signal is obtained by direct modulo-2 addition of the baseband (NRZ-L) signal and a clock signal (Fig. 8-19*b*). In this code there is a transition at the center of each bit interval. A negative-going transition indicates a 1 bit, whereas a positive-going transition means a 0 bit was sent. The Manchester code is simple to generate and decode. Since it is an RZ-type code, it requires twice the bandwidth of an NRZ code. In addition, it has no inherent error-detecting or correcting capability.

Coaxial or wire-pair cable systems commonly use the bipolar RZ or alternate mark inversion (AMI) coding scheme. These wire line codes have also been adapted to unipolar optical systems.[57,58] The two-level AMI optical pulse formats require twice the transmission bandwidth of NRZ codes, but they provide timing information in the data stream, and the redundancy of the encoded information (which is inherent in these codes) allows for direct in-service error monitoring.[59]

8-4-3 Block Codes

An efficient category of redundant binary codes is the $m\text{B}n\text{B}$ block code class.[52,60] In this class of codes, blocks of m binary bits are converted to longer blocks of $n > m$ binary bits. These new blocks are then transmitted in NRZ or RZ format. As a result of the additional redundant bits, the increase in bandwidth using this

Table 8-1 A comparison of several mBnB codes

Code	n/m	$N_{\max}$	D	W %
3B4B	1.33	4	± 3	25
6B8B	1.33	6	± 3	75
5B6B	1.20	6	± 4	28
7B8B	1.14	9	± 7	27
9B10B	1.11	11	± 8	24

scheme is given by the ratio n/m. At the expense of this increased bandwidth, the mBnB block codes provide adequate timing and error-monitoring information, and they do not have baseline wander problems since long strings of ones and zeros are eliminated.

A convenient concept used for block codes is the *accumulated* or *running disparity,* which is the cumulative difference between the number of 1 and 0 bits. A simple means of measuring this is with an up-down counter. The key factors in selecting a particular block code are low disparity and a limit in the disparity variation (the difference between the maximum and minimum values of the accumulated disparity). A low disparity allows the dc component of the signal to be canceled. A bound on the accumulated disparity avoids the low-frequency spectral content of the signal and facilitates error monitoring by detecting the disparity overflow. Generally, one chooses codes that have n even, since for odd values of n there are no coded words with zero disparity.

A comparison of several mBnB codes is given in Table 8-1. The parameters shown in this table are:

1. The ratio n/m which gives the bandwidth increase
2. The longest number $N_{\max}$ of consecutive identical symbols, where small values of $N_{\max}$ allow for easier clock recovery
3. The bounds on the accumulated disparity D
4. The percentage W of n-bit words that are not used; the detection of invalid words at the receiver permits character reframing

The most suitable codes for high data rates are the 3B4B, 5B6B, or 6B8B codes. If simplicity of the encoder and decoder circuits is the main criterion, the 3B4B is the most convenient code. The 5B6B code is the most advantageous if bandwidth reduction is the major concern.

8-5 SUMMARY

The design of an optical link involves many interrelated variables among the fiber, source, and photodetector operating characteristics. In carrying out an optical fiber link analysis, several iterations with different device characteristics may be required before it is satisfactorily completed. The key requirements needed in analyzing a link are:

1. The desired (or possible) transmission distance
2. The data rate or channel bandwidth
3. The bit error rate (BER)

Two analyses are usually carried out to ensure that the desired system performance can be met. These are the link power budget and the system rise time analysis. In the link power budget analysis one first determines the power margin between the optical transmitter output and the minimum receiver sensitivity needed to establish a specified BER. This margin can then be allocated to connector, splice, and fiber losses, plus any additional margins required for expected component degradation or temperature effects. If the choice of components did not allow the desired transmission distance to be achieved, the components might have to be changed or repeaters might have to be put into the link.

Once the link power budget has been established, the designer makes a system rise time analysis to ensure that the dispersion limit of the link has not been exceeded. The four basic elements which may significantly limit the system speed are the transmitter rise time, the material dispersion rise time of the fiber, the modal dispersion rise time of the fiber, and the receiver rise time.

The simplest optical fiber system is a point-to-point link having a transmitter on one end and a receiver on the other. Since the optical source in the transmitter has a relatively narrow spectral width, the full transmission capability of the fiber is not used in this case. A fuller utilization of the transmission capacity of a fiber can be achieved with wavelength division multiplexing (WDM). This is achieved by the simultaneous transmission of optical signals over the same fiber from a number of different light sources having properly spaced peak emission wavelengths. By operating each source at a different peak wavelength, the integrity of the independent messages from each source is maintained for subsequent conversion to electric signals at the receiving end.

An alternative to point-to-point links are data bus architectures. A data bus is a transmission line network over which a number of physically separated terminals can communicate with one another by multiplexed signals. The two general classes of optical data buses that have evolved are the star or radial configuration and the in-line or T-coupler bus. Access to an optical data bus is achieved with a coupling element that can be either active or passive. An active coupler converts the optical signal on the bus line to its electric baseband counterpart before any data processing is carried out. A passive coupler contains no electronics. It passively taps off a portion of the optical power from the bus or inserts an optical signal onto the line.

In designing an optical fiber link, line-coding schemes are often used to introduce randomness and redundancy into the digital information stream. This is done to ensure efficient timing recovery and to facilitate error monitoring at the receiver. In selecting a particular line code, a tradeoff must be made between timing and noise bandwidth. That is, to have timing data available from the bit stream, a larger bandwidth may be needed. However, larger bandwidths result in larger noise contributions. Popular line codes for optical fiber links are the nonreturn-to-zero (NRZ) format and the return-to-zero (RZ) format.

PROBLEMS

Section 8-1

8-1 Make a graphical comparison, as in Fig. 8-4, of the maximum attenuation-limited transmission distance of the following two systems operating at 20 Mb/s:

System 1 operating at 850 nm
(*a*) GaAlAs laser diode: 0-dBm (1-mW) fiber-coupled power
(*b*) Silicon avalanche photodiode: −56-dBm sensitivity
(*c*) Graded-index fiber: 3.5-dB/km attenuation at 850 nm
(*d*) Connector loss: 1 dB/connector

System 2 operating at 1300 nm
(*a*) InGaAsP LED: −13-dBm fiber-coupled power
(*b*) InGaAs *pin* photodiode: −45-dBm sensitivity
(*c*) Graded-index fiber: 1.5-dB/km attenuation at 1300 nm
(*d*) Connector loss: 1 dB/connector
Allow a 6-dB system operating margin in each case.

8-2 An engineer has the following components available:
(*a*) GaA1As laser diode operating at 850 nm and capable of coupling 1 mW (0 dBm) into a fiber
(*b*) Ten sections of cable each of which is 500 m long, has a 4-dB/km attenuation, and has connectors on both ends
(*c*) Connector loss of 2 dB/connector
(*d*) A *pin* photodiode receiver
(*e*) An avalanche photodiode receiver
Using these components, the engineer wishes to construct a 5-km link operating at 10 Mb/s. If the sensitivities of the *pin* and APD receivers are −46 and −59 dBm, respectively, which receiver should be used if a 6-dB system operating margin is required?

8-3 (*a*) Verify Eq. (8-12).
(*b*) Show that Eq. (8-14) follows from Eqs. (8-10) and (8-13).

8-4 Show that, if t_e is the full width of the gaussian pulse in Eq. (8-9) at the $1/e$ points, then the relationship between the 3-dB optical bandwidth and t_e is given by

$$f_{3\,\mathrm{dB}} = \frac{0.53}{t_e}$$

8-5 A 90-Mb/s NRZ data transmission system uses a GaAlAs laser diode having a 1-nm spectral width. The rise time of the laser transmitter output is 2 ns. The transmission distance is 7 km over a graded-index fiber having an 800-MHz · km bandwidth-distance product.
(*a*) If the receiver bandwidth is 90 MHz and the mode-mixing factor $q = 0.7$, what is the system rise time? Does this rise time meet the NRZ data requirement of being less than 70 percent of a pulse width?
(*b*) What is the system rise time if there is no mode mixing in the 7-km link, that is, $q = 1.0$?

8-6 Make a plot analogous to Fig. 8-5 of the transmission distance versus data rate of the following system. The transmitter is a GaAlAs laser diode operating at 850 nm. The laser power coupled into a fiber flylead is 0 dBm (1 mW) and the source spectral width is 1 nm. The fiber has a 3.5-dB/km attenuation at 850 nm and a bandwidth of 800 MHz · km. The receiver uses a silicon avalanche photodiode which has the sensitivity versus data rate shown in Fig. 8-3. For simplicity, the receiver sensitivity (in dBm) can be approximated from curve fitting by

$$P_R = 9 \log B - 68.5$$

where B is the data rate in Mb/s.

For the data rate range of 1 to 1000 Mb/s, plot the attenuation-limited transmission distance (including a 1-dB connector loss at each end and a 6-dB system margin), the modal dispersion limit for full mode mixing ($q = 0.5$), the modal dispersion limit for no mode mixing ($q = 1.0$), and the material dispersion limit.

8-7 Make a plot analogous to Fig. 8-5 of the transmission distance versus data rate of the following system. The transmitter is an InGaAsP LED operating at 1300 nm. The fiber-coupled power from this source is -13 dBm (50 μW) and the source spectral width is 40 nm. The fiber has a 1.5-dB/km attenuation at 1300 nm and a bandwidth of 800 MHz $\cdot$ km. The receiver uses an InGaAs *pin* photodiode which has the sensitivity versus data rate shown in Fig. 8-3. For simplicity, this receiver sensitivity P_R (in dBm) can be approximated from curve fitting by

$$P_R = 11.5 \log B - 60.5$$

where B is the data rate in Mb/s. For the data rate range of 10 to 1000 Mb/s, plot the attenuation-limited transmission distance (including a 1-dB connector loss at each end and a 6-dB system margin), the modal dispersion limit for no mode mixing ($q = 1.0$), and the modal dispersion limit for full mode mixing ($q = 0.5$). Note that the material dispersion is negligible in this case, as can be seen from Fig. 3-13.

Section 8-2

8-8 A convenient wavelength division multiplexer is the plane reflection grating which is mounted as shown in Fig. P8-8. The angular properties of this grating are given by the grating equation

$$\sin \phi + \sin \theta = \frac{k\lambda}{n'\Lambda}$$

where Λ is the grating period, k is the interference order, n' is the refractive index of the medium between the lens and grating, and ϕ and θ are the angles of the incident and reflected beams, respectively, measured normal to the grating.

(a) Using the grating equation, show that the angular dispersion is given by

$$\frac{d\theta}{d\lambda} = \frac{k}{n'\Lambda \cos \theta} = \frac{2 \tan \theta}{\lambda}$$

where the last equality was derived from the condition $\phi \simeq \theta$ (this condition minimizes distortion).

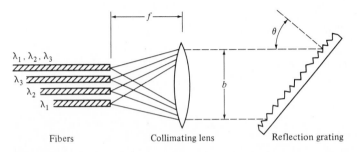

Figure P8-8

(b) Show that the minimum collimating lens diameter is

$$b \geq \frac{\lambda(NA/n')d}{\Delta\lambda \tan \theta}$$

(c) Letting W in Eq. (8-22) be the maximum path length of the collimated beam, show that the fractional beam spread is given by

$$S = 2(1 + m) \tan^2 \theta \frac{\Delta\lambda}{\lambda}$$

(d) Given that $\Delta\lambda = 26$ nm, $\lambda = 850$ nm, and $m = 3$, what is the upper limit on θ for beam spreading of less than 1 percent?

8-9 A four-channel WDM system using laser diodes has the optical level parameters shown in Table P8-9. The system operates over 10 km at a 20-Mb/s rate with a 10^{-9} BER. What is the system operating margin for each channel? Assume connector losses at the multiplexer, source, and detector are 1.0 dB.

Table P8-9

Source wavelength (nm)	810	835	860	890
Source output (dBm)	0.0	−1.0	−0.5	−1.8
Multiplexer insertion loss (dB)	2.9	3.5	4.1	3.1
Fiber attenuation (dB/km)	3.9	3.6	3.4	3.5
Demultiplexer insertion loss (dB)	3.6	3.1	3.6	3.8
Receiver sensitivity (dBm)	−56.0	−55.5	−55.9	−55.0

Section 8-3

8-10 Consider a star data bus network with 16 inputs and 16 outputs operating at 10 Mb/s. Assume this system has the following parameters: connector loss $L_c = 1.0$ dB, star coupler insertion loss $L_S = 5$ dB, and a fiber loss of 2 dB/km. Let the sources be InGaAsP LEDs having an output of -16 dBm from a fiber flylead, and assume InGaAs *pin* photodiodes with a -49-dBm sensitivity are used. Assume that a 6-dB system margin is required.

(a) What is the maximum transmission distance if a transmission star coupler is used?

(b) What is the maximum distance if a reflection star coupler is used?

8-11 Consider a transmission star coupler having seven inputs and seven outputs. Suppose the coupler is constructed by arranging the seven fibers in a circular pattern (a ring of six with one in the center) and butting them against the end of a glass rod which serves as the mixing element.

(a) If the fibers have 50-μm core diameters and 125-μm outer cladding diameters, what is the coupling loss resulting from light escaping between the fiber cores? Let the rod diameter be 300 μm. Assume the fiber cladding is not removed.

(b) What is the coupling loss if the fiber ends are arranged in a row and a 50- by 800-μm glass plate is used as the star coupler?

8-12 Repeat Prob. 8-11 for seven fibers having 200-μm core diameters and 400-μm outer cladding diameters. What should the sizes of the glass rod and the glass plate be in this case?

8-13 An engineer wishes to construct an in-line data bus operating at 10 Mb/s. The stations are to be separated by 100 m, for which optical fibers with a 3-dB/km attenuation are used. The optical sources are laser diodes having an output of 1 mW from a fiber flylead and the detectors are avalanche photodiodes with a -58-dBm (1.6-nW) sensitivity. The couplers have a power-coupling efficiency $E = 10$ percent, a power tapoff factor of $C_T = 5$ percent, and a 10 percent fractional intrinsic loss F_I. The power loss at the connectors is 20 percent (1 dB).

(a) Make a plot of P_{IN} in dBm as a function of the number of stations N from 0 to -58 dBm.

(*b*) What is the system operating margin for eight stations?

(*c*) What is the worst-case dynamic range for the maximum allowable number of stations if a 6-dB power margin is required?

REFERENCES

1. E. E. Basch, H. A. Carnes, and R. F. Kearns, "Calculate performance into fiber-optic links," *Electron. Design,* **28,** 161–166, Aug. 16, 1980.
2. J. Bliss and D. W. Stevenson, "Fiber optics: A designer's guide to system budgeting," *Electro-Opt. Sys. Design,* **13,** 23–32, Aug. 1981.
3. C. Kleekamp and B. Metcalf, "Designer's guide to fiber optics," *EDN,* **23,** 44–51, Jan. 5, 1978 (pt. 1); 46–54, Jan. 20, 1978 (pt. 2); 61–68, Feb. 20, 1978 (pt. 3); 51–62, Mar. 5, 1978 (pt. 4).
4. S. E. Miller and A. G. Chynoweth, *Optical Fiber Communications,* Academic, New York, 1979, chap. 20.
5. Tech. Staff of CSELT, *Optical Fibre Communication,* McGraw-Hill, New York, 1980, pt. 4, chap. 2.
6. A. Bender and S. Storozum, "Charts simplify fiber optic system design," *Electron.,* **51,** 135–142, Nov. 23, 1978.
7. J. Conradi, F. P. Kapron, and J. C. Dyment, "Fiber optical transmission between 0.8 and 1.4 μm," *IEEE Trans. Electron Devices,* **ED-25,** 180–193, Feb. 1978.
8. P. B. Ozmar, "A look at optical fiber CATV," *IEEE Trans. Cable TV,* **CATV-4,** 43–46, Jan. 1979.
9. M. S. Ghausi, *Principles and Design of Linear Active Circuits,* McGraw-Hill, New York, 1965, chap. 16.
10. M. Eve, "Statistical model for the prediction of the bandwidth of an optical route," *Electron. Lett.,* **13,** 315–316, May 1977.
11. M. Eve, "Multipath time dispersion theory of an optical network," *Opt. Quantum Electron.,* **10,** 45–51, Jan. 1978.
12. F. P. Kapron, F. M. E. Sladen, P. M. Garel-Jones, and D. G. Kneller, "Attenuation and pulse broadening along concatenated fiber links," in *Tech. Digest—Symp. on Opt. Fiber Measurements,* NBS Special Publ. 597, 63–66, Oct. 1980.
13. M. J. Buckler, "Optimization of concatenated fiber bandwidth via differential mode delay," in *Tech. Digest—Symp. on Opt. Fiber Measurements,* NBS Special Publ. 597, 59–62, Oct. 1980.
14. K. I. Kitayama, S. Seikai, and N. Uchida, "Impulse response prediction based on experimental mode coupling coefficient in a 10-km long graded index fiber," *IEEE J. Quantum Electron.,* **QE-16,** 356–362, Mar. 1980.
15. M. Eriksrud, A. Hordvik, N. Ryen, and G. Nakken, "Comparison between measured and predicted transmission characteristics of 12 km spliced graded index fibers," *Opt. Quantum Electron.,* **11,** 517–523, 1979.
16. J. Midwinter, *Optical Fibers for Transmission,* Wiley, New York, 1979, app. 5.
17. W. J. Tomlinson, "Wavelength multiplexing in multimode optical fibers," *Appl. Opt.,* **16,** 2180–2194, Aug. 1977.
18. K. Ito, Y. Umeda, Y. Sugiyama, K. Nakajima, K. Oshima, and M. Nunoshita, "Bidirectional fibre optic loop-structured network," *Electron. Lett.,* **17,** 84–86, Jan. 1981.
19. J. Conradi, R. Maciejko, J. Straus, I. Few, G. Duck, W. Sinclair, A. J. Springthorpe, and J. C. Dyment, "Laser based WDM multichannel video transmission system," *Electron. Lett.,* **17,** 91–92, Jan. 1981.
20. S. Sugimoto, K. Minemura, K. Kobayashi, M. Seki, M. Shikada, A. Ueki, T. Yanase, and T. Miki, "High-speed digital-signal transmission experiments by optical WDM," *Electron. Lett.,* **13,** 680–682, Oct. 1977.
21. W. J. Tomlinson and G. D. Aumiller, "Optical multiplexer for multimode fiber transmission systems," *Appl. Phys. Lett.,* **31,** 169–171, Aug. 1977.

22. W. J. Tomlinson and C. L. Lin, "Optical wavelength-division multiplexer for the 1-1.4 μm spectral region," *Electron. Lett.*, **14**, 345–347, May 1978.
23. K. Aoyama and J. Minowa, "Low-loss optical demultiplexer for WDM systems in the 0.8 μm wavelength region," *Appl. Opt.*, **18**, 2834–2836, Aug. 1979.
24. R. Watanabe, K. Nosu, T. Harada, and T. Kita, "Optical demultiplexer using concave grating in 0.7-0.9 μm wavelength region," *Electron. Lett.*, **16**, 106–108, Jan. 1980.
25. R. Watanabe, K. Nosu, and Y. Fujii, "Optical grating multiplexer in the 1.1–1.5 μm wavelength region," *Electron. Lett.*, **16**, 108–109, Jan. 1980.
26. K. Kobayashi and M. Seki, "Microoptic grating multiplexers and optical isolaters for fiber optic communications," *IEEE J. Quantum Electron.*, **QE-16**, 11–22, Jan. 1980.
27. S. Sugimoto, K. Minemura, K. Kobayashi, M. Shikada, H. Nomura, K. Kaeda, A. Ueki, and S. Matsushita, "Wavelength division two-way fibre-optic transmission experiments using micro-optic duplexers," *Electron. Lett.*, **14**, 15–17, Jan. 1978.
28. K. Nosu, H. Ishio, and K. Hashimoto, "Multireflection optical multi-demultiplexer using inter-ference filters," *Electron. Lett.*, **15**, 414–415, July 1979.
29. W. J. Tomlinson, "Applications of GRIN-rod lenses in optical fiber communication systems," *Appl. Opt.*, **19**, 1127–1138, Apr. 1980.
30. G. Winzer, H. F. Mahlein, and A. Reichelt, "Single-mode and multimode all-fiber directional couplers for WDM," *Appl. Opt.*, **20**, 3128–3135, Sept. 1981.
31. S. K. Sheem and R. P. Moeller, "Single-mode fiber wavelength multiplexer," *J. Appl. Phys.*, **51**, 4050–4052, Aug. 1980.
32. W. H. Wells, "Crosstalk in a bidirectional optical fiber," *Fiber Int. Opt.*, **1**, 243–287, 1978.
33. J. Conradi and R. Maciejko, "Digital optical receiver sensitivity degradation caused by crosstalk in bidirectional fiber optic systems," *IEEE Trans. Commun.*, **COM-29**, 1012–1016, July 1981.
34. M. C. Hudson and F. L. Thiel, "The star coupler: a unique interconnection component for multimode optical waveguide communication systems," *Appl. Opt.*, **13**, 2450–2545, Nov. 1974.
35. B. S. Kawasaki and K. O. Hill, "Low-loss access coupler for multimode optical fiber distribution networks," *Appl. Opt.*, **16**, 1794–1795, July 1977.
36. D. H. McMahon and R. L. Gravel, "Star repeaters for fiber optic links," *Appl. Opt.*, **16**, 501–503, Feb. 1977.
37. E. G. Rawson and R. M. Metcalfe, "Fibernet: Multimode optical fibers for local computer networks," *IEEE Trans. Commun.*, **COM-26**, 983–990, July 1978.
38. M. Stockmann and H. H. Witte, "Planar star coupler for multimode fibers," *Appl. Opt.*, **19**, 2584–2588, Aug. 1980.
39. H. H. Witte and V. Kulich, "Branching elements for optical data buses," *Appl. Opt.*, **20**, 715–718, Feb. 1981.
40. R. A. Andrews, A. F. Milton, and T. G. Giallorenzi, "Military applications of fiber optics and integrated optics," *IEEE Trans. Microwave Theory Tech.*, **MTT-21**, 763–769, Dec. 1973.
41. H. F. Taylor, W. M. Caton, and A. L. Lewis, "Data busing with fiber optics," *Naval Res. Rev.*, **28**, 12–25, Feb. 1975.
42. A. F. Milton and A. B. Lee, "Optical access couplers and a comparison of multiterminal fiber communication systems," *Appl. Opt.*, **15**, 244–252, Jan. 1976.
43. M. D. Drake, "Multimode fiber-optics coupler with low insertion loss," *Appl. Opt.*, **17**, 3248–3252, Oct. 1978.
44. E. Weidel and J. Guttmann, "Asymmetric T-couplers for fibre optic data buses," *Electron. Lett.*, **16**, 673–674, Aug. 1980.
45. M. Nunoshita and Y. Nomura, "Optical bypass switch for fiber optic data bus system," *Appl. Opt.*, **19**, 2574–2577, Aug. 1980.
46. D. E. Altman and H. F. Taylor, "An eight terminal fiber optics data bus using Tee couplers," *Fiber Integ. Opt.*, **1**, no. 2, 135–152, 1977.
47. C. E. Shannon, "A mathematical theory of communication," *Bell Sys. Tech. J.*, **27**, 379–423, July 1948 and **27**, 623–656, Oct. 1948; "Communication in the presence of noise," *Proc. IRE*, **37**, 10–21, Jan. 1949.
48. R. C. Houts and T. A. Green, "Comparing bandwidth requirements for binary baseband signals," *IEEE Trans. Commun.*, **COM-21**, 776–781, June 1973.

49. T. V. Muoi and J. L. Hullett, "Receiver design for multilevel digital optical fiber systems," *IEEE Trans. Commun.*, **COM-23,** 987–994, Sept. 1975.

50. W. C. Lindsey and M. K. Simon, *Telecommunication Systems Engineering,* Prentice-Hall, Englewood Cliffs, N.J., 1973.

51. D. Tugal and O. Tugal, *Data Transmission,* McGraw Hill, New York, 1982.

52. CSELT Technical Staff, *Optical Fibre Communication,* McGraw-Hill, New York, 1981.

53. S. Haykin, *Communication Systems,* Wiley, New York, 1978.

54. S. D. Personick, *Optical Fiber Transmission Systems,* Plenum, New York, 1981, chap. 2.

55. J. E. Savage, "Some simple self-synchronizing digital data assemblers," *Bell Sys. Tech. J.,* **46,** 449–487, Feb. 1967.

56. R. D. Gitlin and J. F. Hayes, "Timing recovery and scramblers in data transmission," *Bell Sys. Tech. J.,* **54,** 569–593, Mar. 1975.

57. Y. Takasaki, M. Tanaka, N. Maeda, K. Yamashita, and K. Nagano, "Optical pulse formats for fiber optic digital communications," *IEEE Trans. Commun.,* **COM-24,** 404–413, Apr. 1976.

58. R. Petrovic, "New transmission code for digital optical communications," *Electron. Lett.,* **14,** 541–542, Aug. 1978.

59. E. A. Newcombe and S. Pasupathy, "Error rate monitoring for digital communications," *Proc. IEEE,* **70,** 805–828, Aug. 1982.

60. M. Rousseau, "Block codes for optical fibre communication," *Electron. Lett.,* **12,** 478–479, Sept. 1976.

NINE

MEASUREMENTS

The design and installation of any optical fiber communication system require measurement techniques for verifying the operational characteristics of the system components. Of particular importance are accurate and precise measurements of the optical fiber since this component cannot readily be replaced once a transmission system has been installed. Two basic groups of people are interested in fiber characterization. These are the manufacturers, who are concerned with the material composition and fabrication effects on fiber properties, and the system engineer, who must have sufficient data on the fiber to perform meaningful design calculations and to evaluate systems during installation and operation.

During the design, installation, and operation of a fiber optical transmission system the fiber parameters of interest are the core and cladding diameters, the refractive-index profile, the fiber attenuation, and the dispersion or bandwidth. Generally, the values of all these parameters can be supplied by the fiber manufacturer. The fiber geometry, refractive-index profile, and numerical aperture are not expected to change during cable manufacture, installation, or operation. Thus, once these parameters are known, there is no need to remeasure them.

However, the attenuation and bandwidth of a fiber can change during fiber cabling and cable installation. For example, microbending can cause additional loss in the fiber, and modal redistribution at fiber joints can significantly affect the bandwidth when several sections of fiber cables are connected in series. Measurement procedures for these two parameters are thus of interest to the user. In addition, the user is concerned with methods for locating breaks and faults in optical fiber cables.

The main emphasis in this chapter will be from the user point of view. This will include descriptions of measurement methods for fiber attenuation, break location, and bandwidth. For completeness we briefly mention several techniques used for determining the refractive-index profile. A comprehensive treatment of optical fiber measurement principles can be found in the work by Marcuse,[1] a general overview of measurement techniques has been given by Barnoski and Personick,[2] and Krahn et al.,[3] discuss measuring and test equipment for field usage. Detailed reviews of measurement setups for fiber characterizations of interest to both manufacturers and users can be found in Refs. 4 through 6.

In addition to measuring optical fiber performance parameters, systems engineers are also interested in examining the operational characteristics of optical sources. These include factors such as the rise and fall times of the optical output in response to an electric pulse, intermodulation distortion, and harmonic distortion. Some relatively simple test setups for measuring these parameters are presented in Sec. 9-5.

Once an optical fiber link has been installed, a measurement of its overall performance is usually necessary. A simple but powerful method for assessing the data transmission characteristics of the link is the eye-pattern technique. This method has been extensively used for wire systems, and its application to optical fiber links is described in Sec. 9-6.

9-1 ATTENUATION MEASUREMENTS

As we saw in Chap. 3, attenuation of optical power in a fiber waveguide is a result of absorption processes, scattering mechanisms, and waveguide effects. The manufacturer is generally interested in the magnitudes of the individual contributions to attenuation, whereas the system engineer who uses the fiber is more concerned with the total transmission loss of a fiber. Here we shall treat only measurement techniques for total transmission loss. Details of separate scattering and absorption loss measurements in both bulk glasses and optical fibers are given in Refs. 4 through 6.

One of two basic methods is normally used for determining attenuation in multimode fibers. The earliest and most common approach involves measuring the optical power transmitted through a long and a short length of the same fiber by using identical input couplings.[7,8] This method is known as the cutback technique. The other technique involves the use of an optical time domain reflectometer (OTDR).[9,10]

9-1-1 Cutback Attenuation Measurement Method

The *cutback technique,* which is a destructive method requiring access to both ends of the fiber, is illustrated in Fig. 9-1. To find the transmission loss, the optical power is first measured at the output (or far end) of the fiber. Then without disturbing the input condition, the fiber is cut off a few meters from the source, and the output power at this near end is measured. If P_F and P_N represent the output powers of the far and near ends of the fiber, respectively, the average loss α in

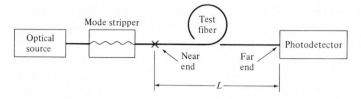

Figure 9-1 Schematic experimental setup for determining fiber attenuation by the cutback technique. The optical power is first measured at the far end, the fiber is then cut at the near end, and the power output there is measured.

decibels per kilometer is given by

$$\alpha = \frac{10}{L} \log \frac{P_N}{P_F} \qquad (9\text{-}1)$$

where L (in kilometers) is the separation of the two measurement points. The reason for following these steps is that it is extremely difficult to calculate the exact amount of optical power launched into a fiber. By using the cutback method the optical power emerging from the short fiber length is basically the input power to the fiber of length L, provided certain precautions are taken. These include index-matching the fiber output end to the detector surface, avoiding any instabilities in the optical source, and illuminating the same spot on the detector surface.

In carrying out this measuring technique, special attention must be paid to how optical power is launched into the fiber. This is because different launch conditions can yield different loss values.[11] The effects on modal distributions in the fiber resulting from changes in the numerical aperture and spot size on the launch end

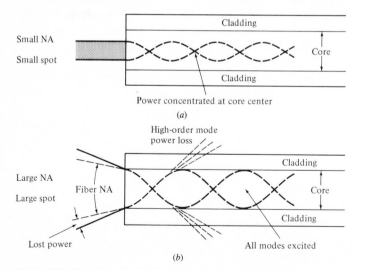

Figure 9-2 The effects of launch numerical aperture and spot size on the modal distribution. Underfilling the fiber excites only lower-order modes; an overfilled fiber has excess attenuation from higher-order mode loss.

of the fiber are shown in Fig. 9-2. If the spot size is small and its NA is less than that of the fiber core, the optical power in the fiber is concentrated in the center of the core. In this case, the attenuation contribution arising from higher-order-mode power loss is negligible. In Fig. 9-2b the spot size is larger than the fiber core and the spot NA is larger than that of the fiber. For this overfilled condition, those parts of the incident light beam that fall outside the fiber core and outside the fiber NA are lost. In addition, there is a large contribution to the attenuation arising from higher-mode power loss.

The importance of matching the spot size and NA of the incident beam and the fiber is shown in Fig. 9-3, which gives the attenuation of an 0.23-NA Ge-B-doped graded-index fiber as a function of wavelength for several different launch NAs.[12] Since differences on the order of 1 dB/km can easily be obtained, knowing the exact launch conditions when interpreting fiber attenuation data is quite important.

In general, accurate loss data which can be extrapolated to arbitrary fiber lengths are only achievable under a steady-state equilibrium-mode distribution. An equilibrium-mode distribution occurs when the far-end and near-end radiation patterns are identical. An example of this is shown in Fig. 9-4 for the same fiber described in Fig. 9-3. The radiation patterns widen along the fiber for under-excitation (small launch NA) and narrow for overexcitation (large launch NA). For the steady-state excitation condition, the far- and near-end radiation patterns are essentially constant, which corresponds to a launch NA of 0.23 in this case.

Steady-state equilibrium-mode distributions can be achieved either by matched-beam or mandrel-wrap excitation methods.[13] In the *matched-beam* method both the source NA and the spot size are well controlled so that a definite modal pattern is launched. In the *mandrel-wrap* excitation technique, excess higher-order modes launched by initially overexciting the fiber are filtered out by wrapping several turns of fiber around a mandrel which is about 1.0 to 1.5 cm in diameter.

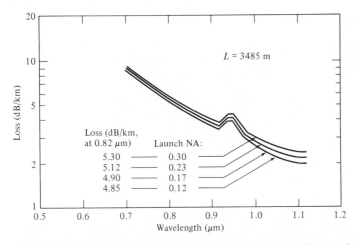

Figure 9-3 Attenuation of an 0.23-NA Ge-B-doped graded-index fiber as a function of wavelength for different launch NAs using matched-spot-size excitation. *(Reproduced with permission from Cohen, Kaiser, and Lin,[12] copyright 1980, IEEE.)*

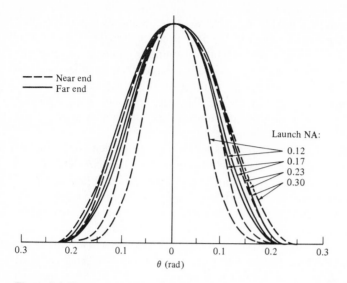

Figure 9-4 Near- and far-end radiation patterns as a function of launch NA using matched-spot-size excitation. *(Reproduced with permission from Cohen, Kaiser, and Lin,* [12] *copyright 1980, IEEE.)*

9-1-2 Optical Time Domain Reflectometer

Unlike the foregoing attenuation measurement technique, the *optical time domain reflectometer* (OTDR) method is not destructive and requires access to only one end of the fiber. In addition the OTDR method gives information about the variation of attenuation with length, which is not provided by the other measurement techniques.

The basis of the OTDR technique is shown in Fig. 9-5. An OTDR is fundamentally an optical radar. It operates by periodically launching narrow laser pulses into one end of a fiber under test by using either a directional coupler or a beam splitter. The properties of the optical fiber are then determined by analyzing the amplitude and temporal characteristics of the waveform of the back-scattered light.

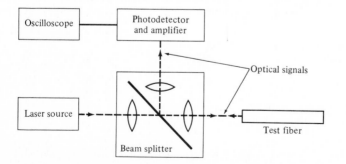

Figure 9-5 Operating principle of an optical time domain reflectometer (OTDR).

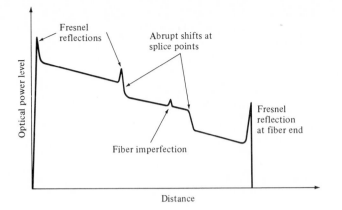

Figure 9-6 A typical OTDR return waveform.

A typical return waveform is shown in Fig. 9-6. This back-scattered waveform has four distinct features:

1. A large initial pulse resulting from Fresnel reflection at the input end of the fiber
2. A long decaying tail resulting from Rayleigh scattering in the reverse direction as the input pulse travels along the fiber
3. Abrupt shifts in the curve caused by optical loss at joints in the fiber line
4. Pulses arising from Fresnel reflections at the far end of the fiber, at fiber joints, and at fiber imperfections

The back-scattered light is principally produced by Fresnel reflection and Rayleigh scattering. Fresnel reflection occurs when light enters a medium having a different index of refraction. For a glass-air interface the reflected power P_{ref} for light (of power P_0) incident perpendicular to the interface is

$$P_{\text{ref}} = P_0\left(\frac{n_{\text{fiber}} - n_{\text{air}}}{n_{\text{fiber}} + n_{\text{air}}}\right)^2 \tag{9-2}$$

where n_{fiber} and n_{air} are the refractive indices of the fiber core and of air. A perfect fiber end reflects about 4 percent of the power incident on it. However, since fiber ends are generally not perfectly cleaved, the reflected power tends to be much lower than the maximum possible value.

Throughout the length of the fiber light is reflected in all directions by Rayleigh scattering. This factor is the dominant loss mechanism in most high-quality fibers. The optical power that is Rayleigh-scattered in the reverse direction inside the fiber can be used to determine the attenuation of the fiber.

The optical power in the fiber at a distance x from the input coupler can be written as

$$P(x) = P(0) \exp\left[-\int_0^x \beta(y)\, dy\right] \tag{9-3}$$

Here $P(0)$ is the fiber input power and $\beta(y)$ is the fiber loss coefficient, which may be position-dependent; that is, the loss may not be uniform along the fiber length. The parameter β is measured in natural units called *nepers* which are related to the loss $\alpha(y)$ in decibels per kilometer through the relationship

$$\beta(y) = \frac{\alpha(y)}{10 \log e}$$

Under the assumption that the scattering is the same at all points along the optical waveguide and is independent of the modal distribution, the power $P_R(x)$ scattered in the reverse direction at the point x is[14]

$$P_R(x) = SP(x) \qquad (9\text{-}4)$$

Here S is the fraction of the total power which is scattered in the backward direction and trapped in the fiber. Thus the back-scattered power from point x which is seen by the photodetector is

$$P_D(x) = P_R(x) \exp\left[-\int_0^x \beta_R(y)\, dy \right] \qquad (9\text{-}5)$$

where $\beta_R(x)$ is the loss coefficient for the reverse-scattered light. Since the modes in the fiber excited by the back-scattered light can be different from those launched in the forward direction, the parameter $\beta_R(x)$ may be different from $\beta(x)$.

Substituting Eqs. (9-3) and (9-4) into Eq. (9-5) yields

$$P_D(x) = SP(0) \exp\left[\frac{-2\bar{\alpha}(x)x}{10 \log e} \right] \qquad (9\text{-}6)$$

where the average attenuation coefficient $\bar{\alpha}(x)$ is defined as

$$\bar{\alpha}(x) = \frac{\int_0^x [\alpha(y) + \alpha_R(y)]\, dy}{2x} \qquad (9\text{-}7)$$

Using this equation, the average attenuation coefficient can be found from experimental semilog data plots such as the one shown in Fig. 9-6. For example, the average attenuation between two points x_1 and x_2, where $x_2 > x_1$ is

$$\bar{\alpha} = \frac{-10[\log P_D(x_2) - \log P_D(x_1)]}{2(x_2 - x_1)} \qquad \text{dB}/\text{km} \qquad (9\text{-}8)$$

By using analog signal recovery techniques the dynamic range of laboratory-based OTDRs is limited to about 25 dB of one-way fiber loss. For commercially available OTDRs the dynamic range is approximately 15 dB.

9-2 FIBER FAULT LOCATION

In addition to measuring attenuation, an OTDR can also be used to locate breaks and imperfections in an optical fiber. The fiber length L (and, hence, the position of the break or fault) can be calculated from the time difference between the reflected pulses from the front and the far ends of the fiber. If this time difference is t, then the length L is given by

$$L = \frac{ct}{2n_1} \tag{9-9}$$

where n_1 is the core refractive index of the fiber. For 5-ns-wide laser input pulses, the length resolution of this method is approximately 1 m.

A difficulty that may arise when using this method is that the quality of the far-end face of the fiber may be such that it reflects only a small portion of the 4 percent theoretical maximum of the incident optical power. In such cases a break (or a fiber end) can only be located with complete certainty by observing when the back-scatter signal ends. Since this becomes increasingly difficult as the fiber becomes longer (because of the feebleness of the back-scattered light), a photon-counting technique has been devised.[15] The details of this method are beyond the scope of this book, but it essentially consists of sampling photons from the return waveform of an input laser pulse and comparing them with a background wave-form. When the photon count rate drops sharply, the end of the fiber has been reached. With this method the OTDR dynamic range can be extended to more than 40 dB as compared to about 15 dB with the conventional OTDR method.

9-3 DISPERSION MEASUREMENTS

As we saw in Chap. 3, light pulses become wider as they propagate along a fiber because of material, modal, and waveguide dispersion effects. This widening will cause two closely spaced pulses to overlap eventually and become indistinguishable after they have traveled a certain distance along the fiber. The degree of pulse spreading determines the information-carrying capacity, or maximum bit rate, of a fiber. The limit of this capacity is usually expressed with reference to kilometer lengths in units of MHz · km. Step-index fibers are generally limited to bandwidths of about 20 MHz · km. At wavelengths for which the refractive index of the core is optimum, graded-index fibers can have bandwidths of several GHz · km, whereas single-mode fibers can have capacities well in excess of this.

As we saw in Chap. 2, the detailed effects of dispersion in fibers are complex and not completely understood. However, for practical purposes, the fiber can be considered as a filter characterized either by an impulse response $h(t)$ or by a power transfer function $H(f)$ which is the Fourier transform of the impulse response.[16] Either of these can be measured to determine the pulse dispersion. The impulse response measurements are made in the time domain, whereas the power transfer function is measured in the frequency domain.

Both time domain and frequency domain dispersion measurements assume that the fiber behaves quasilinearly in power, that is, the individual overlapping output pulses from an optical waveguide can be treated as adding linearly.[16] The behavior of such a system in the time domain is described simply as

$$p_{out}(t) = h(t) * p_{in}(t) = \int_{-T/2}^{T/2} p_{in}(t - \tau)h(\tau)\, d\tau \qquad (9\text{-}10)$$

That is, the output pulse response $p_{out}(t)$ of the fiber can be calculated through the convolution (denoted by $*$) of the input pulse $p_{in}(t)$ and the power impulse response function $h(t)$ of the fiber. The period T between the input pulses should be taken to be wider than the expected time spread of the output pulses.

In the frequency domain, Eq. (9-10) can be expressed as the product (see App. F)

$$P_{out}(f) = H(f)P_{in}(f) \qquad (9\text{-}11)$$

Here $H(f)$, the power transfer function of the fiber at the baseband frequency f, is the Fourier transform of $h(t)$

$$H(f) = \int_{-\infty}^{\infty} h(t)e^{-j2\pi ft}\, dt \qquad (9\text{-}12)$$

and $P_{out}(f)$ and $P_{in}(f)$ are the Fourier transforms of the output and input pulse responses $p_{out}(t)$ and $p_{in}(t)$

$$P(f) = \int_{-\infty}^{\infty} p(t)e^{-j2\pi ft}\, dt \qquad (9\text{-}13)$$

The transfer function of a fiber optic cable is of importance because it contains the bandwidth information of the system. In order for pulse dispersion to be negligible in digital systems, one of the following approximately equivalent conditions should be satisfied:[17] (1) The fiber transfer function must not roll off to less than 0.707 of its low-frequency value for frequencies up to half the desired bit rate, or (2) the rms width of the fiber impulse response must be less than one-fourth the pulse spacing.

9-3-1 Time Domain Dispersion Measurements

The simplest approach for making pulse dispersion measurements in the time domain is to inject a narrow pulse of optical energy into one end of an optical fiber and to detect the broadened output pulse at the other end.[18] From the output pulse shape an rms pulse width σ, as defined in Fig. 9-7, can be calculated by

$$\sigma^2 = \frac{\int_{-\infty}^{\infty} (t - \bar{t})^2 p_{out}(t)\, dt}{\int_{-\infty}^{\infty} p_{out}(t)\, dt} \qquad (9\text{-}14)$$

where the center time $\bar{t}$ of the pulse is determined from

$$\bar{t} = \frac{\int_{-\infty}^{\infty} t p_{out}(t)\, dt}{\int_{-\infty}^{\infty} p_{out}(t)\, dt} \qquad (9\text{-}15)$$

The evaluation of Eq. (9-14) requires a numerical integration. An easier

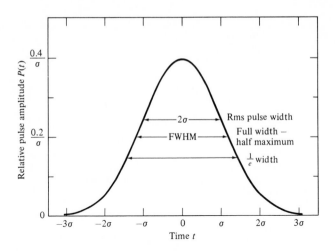

Figure 9-7 Definitions of pulse shape parameters.

method frequently used is to assume that the output response of a fiber can be approximated by a gaussian described by

$$p_{\text{out}}(t) = \frac{1}{\sigma\sqrt{2\pi}} e^{-t^2/2\sigma^2} \qquad (9\text{-}16)$$

where the parameter σ determines the pulse width, as shown in Fig. 9-7. This figure also shows the $1/e$ time width τ_e which is equal to $2\sigma\sqrt{2}$. The optical bandwidth of the fiber can be defined through a Fourier transform. This is normally done in terms of the 3-dB bandwidth, which is the modulation frequency at which the optical power has fallen to one-half the value of the zero frequency modulation (dc value). It can readily be shown that this is

$$f_{-3\text{ dB optical}} = \frac{0.53}{\tau_e} = \frac{0.188}{\sigma} \text{ Hz} \qquad (9\text{-}17)$$

where "-3dB optical" means a 50 percent optical power reduction. Electric bandwidths are related to optical bandwidths by $1/\sqrt{2}$, so that

$$f_{-3\text{ dB electric}} = \frac{0.375}{\tau_e} = \frac{0.133}{\sigma} \text{ Hz} \qquad (9\text{-}18)$$

One major difficulty with the direct-detection technique is that the time resolution of most photodetectors is on the order of tenths of a nanosecond. Thus, to measure accurately pulse spreading in low-dispersion fibers, very long fiber lengths are required. In addition, since the rate of pulse spreading depends on mode mixing and differential mode attenuation, both of which vary with fiber length, dispersion measurement over a fixed fiber length cannot generally be directly extrapolated to other fiber lengths.

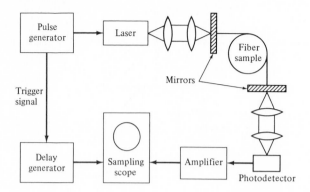

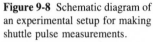

Figure 9-8 Schematic diagram of an experimental setup for making shuttle pulse measurements.

To overcome these limitations the shuttle pulse arrangement[19] shown in Fig. 9-8 was developed. In this setup a given length of fiber is terminated on both ends by partially reflecting mirrors. Light pulses launched into the fiber through one of the reflectors then shuttle back and forth between the mirrors. These pulses can be sampled at the far end after each of the $2N - 1$ transits through the fiber, where $N = 1, 2, 3, \ldots$. The dispersion can be measured by comparing the widths of pulses returning from successive roundtrips through the fiber. This method is extremely useful since it can yield data on mode coupling by using relatively short samples of high-quality fibers.

The number of shuttle pulses N that can be measured with this setup depends on the dynamic range D of the system. If D (in decibels) is defined as ten times the logarithm of the ratio between the power launched from the source that can be coupled into the fiber and the minimum detectable power at the receiver, then the maximum number of detectable pulses N can be determined from the equation

$$D + 20 \log T - \alpha L_0 = (-20 \log R + 2\alpha L_0)(N - 1) \qquad (9\text{-}19)$$

Here T and R are the optical power transmission and reflection coefficients of the mirrors, α is the fiber attenuation in decibels per kilometer, and L_0 is the actual fiber length in kilometers between the source and the detector. This equation takes into account that every pulse sampled by the detector is attenuated by T^2, because each pulse passes through two mirrors, and that each successive pulse is attenuated by the square of the mirror reflectivity R^2, because two reflections occur per roundtrip.

For N pulses the extrapolated fiber length L is given by

$$L = (2N - 1)L_0 \qquad (9\text{-}20)$$

For example, if $L_0 = 100$ m and 10 pulses can be detected, then the pulse spreading can be extrapolated to 1900 m.

9-3-2 Frequency Domain Dispersion Measurements

Frequency domain dispersion measurements yield information on amplitude-versus-frequency response and phase-versus-frequency response. These data are often more useful for system designers than time domain pulse dispersion mea-

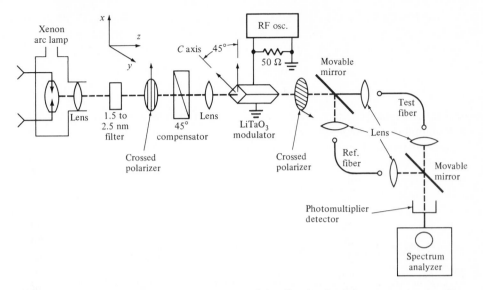

Figure 9-9 An experimental arrangement for determining fiber baseband frequency response. *(Reproduced with permission from Cohen, Astle, and Kaminow,[21] copyright 1976, The American Telephone & Telegraph Co.)*

surements, especially if equalization techniques are to be performed on the detected signal at the receiver. The dispersion measurements can be made by sinusoidally modulating a continuous-wave (CW) light signal about a fixed level.[20-22] The baseband frequency response is then found from the ratio of the sine wave amplitudes at the beginning and end of the fiber.

There are a number of advantages to the frequency domain measurement technique:

1. The fiber transfer function is found directly without Fourier-transforming the time domain data. To find the impulse response in the time domain requires a cumbersome deconvolution process of the output pulse with the input pulse.
2. Photodetector linearity over a wide range is not required as in the time domain because of the small signal modulation about a constant light level.
3. It is generally easier to modulate sinusoidally an optical source at high frequencies than it is to produce a series of narrow output pulses.

An experimental arrangement[21] for finding fiber baseband frequency response is shown in Fig. 9-9. After passing through a set of narrowband interference filters, light from a xenon arc lamp is focused into a $LiTaO_3$ electro-optic crystal the intensity of which modulates the incoherent CW xenon light carrier on a frequency-tunable sinusoidal envelope. The modulated beam is focused into either the fiber under test or into a short-length (approximately 2 m) reference fiber. A photomultiplier detects the fiber output and the baseband modulation components are shown on a spectrum analyzer.

The component at the modulating frequency f from the reference fiber is taken as $P_{in}(f)$, and the same component from the test fiber is $P_{out}(f)$. The reduction of the sine wave amplitude P_{out}/P_{in} resulting from transmission in the fiber yields

$$\frac{P_{out}(f)}{P_{in}(f)} = |H(f)| \tag{9-21}$$

which is the magnitude of the power transfer function defined in Eq. (9-12). Although this technique does not measure the input-to-output phase change of the sinusoidal modulation, unless there is extreme phase distortion the information capacity of the fiber will be given by the magnitude of the transfer function. The 3-dB bandwidth of the fiber is characterized by the frequency for which the power transfer function in Eq. (9-21) is decreased by 3 dB relative to its dc value.

9-4 REFRACTIVE-INDEX-PROFILE MEASUREMENTS

In multimode fibers the signal bandwidth capacity depends critically on the refractive-index profile of the fiber core, as is shown in Fig. 9-10. The fiber bandwidth can be maximized at a specific operating wavelength by optimizing the shape of the refractive-index profile of the core. Since very slight departures from the optimum profile value can dramatically decrease the fiber bandwidth (see Chap. 3), very precise methods are required for measuring the index profile.

One interesting and very advantageous property of the refractive-index profile is that it can be measured either on the fiber itself or on the preform before the fiber is pulled. A number of different methods based on different physical principles have been used for this.[1,23] The major techniques applied to fibers are:

1. End-reflection technique
2. Transmitted near-field scanning method

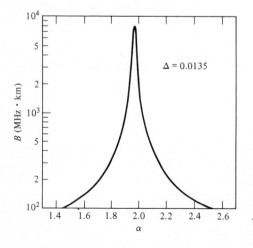

Figure 9-10 Variations in bandwidth resulting from slight deviations in the refractive-index profile for a graded-index fiber with $\Delta = 0.0135$. *(Reproduced with permission from D. Marcuse, Appl. Opt., **19**, 188, Jan. 1980.)*

3. Refracted near-field technique
4. Interferometric slab method
5. Transverse interferometer method
6. Focusing method

9-4-1 End-Reflection Technique

The end-reflection technique[24, 25] is a simple concept in which a focused laser beam running parallel to the fiber axis is incident on the fiber end face. The Fresnel reflection of this light from the glass-to-air interface is measured. For normal incident optical power the reflectivity in the core area is given by

$$\frac{P_r}{P_i} = \left[\frac{n(r) - 1}{n(r) + 1}\right]^2 \tag{9-22}$$

where P_r and P_i are the reflected and incident light powers, respectively. In the cladding region the ratio of power P_c reflected from the cladding to incident power is

$$\frac{P_c}{P_i} = \left(\frac{n_2 - 1}{n_2 + 1}\right)^2 \tag{9-23}$$

If $F = P_r/P_c$ is the ratio of these two equations, then the difference between the core refractive-index distribution $n(r)$ and the cladding index n_2 can be found from the relationship

$$n(r) - n_2 = \frac{(n_2^2 - 1)(\sqrt{F} - 1)}{(n_2 + 1) - (n_2 - 1)\sqrt{F}} \tag{9-24}$$

Although this technique directly gives the refractive-index profile, it requires very accurate alignment and measurement since the optical powers reflected from the core and the cladding are nearly identical. In addition, the physical quality and cleanliness of the fiber end face can produce erroneous results.

9-4-2 Transmitted Near-Field Scanning Method

The transmitted near-field scanning method[26] is based on the observation[27] that, when all bound modes in a fiber are equally excited, there is a close resemblance between the near-field intensity distribution (the field distribution at the immediate output of the fiber) and the refractive-index profile. Experimentally a lambertian incoherent source is butted against one end of a fiber. The lambertian distribution assures that all modes in the fiber are equally excited. The light intensity from the source must be uniform over the core area and must emit light uniformly over the range of acceptance angles of the fiber. Lambertian sources satisfy these conditions.

At the fiber output a sharply focused microscope is used to scan the near-field power distribution across the fiber end face. The refractive index $n(r)$ at radius r

can be related to the near-field optical power $P(r)$ by

$$\frac{n(r) - n_2}{n(0) - n_2} \simeq \frac{n^2(r) - n_2^2}{n^2(0) - n_2^2} = \frac{P(r)}{P(0)} \tag{9-25}$$

Here n_2 is the cladding index, and $n(0)$ and $P(0)$ are the refractive index and optical power, respectively, at the core center.

9-4-3 Refracted Near-Field Technique

The refracted near-field technique[23,28–32] is an attractive method since it can be applied to any length of fiber, requires no special end preparation or cleanliness, and no correction for leaky modes is needed. In contrast to the transmitted near-field technique, this method determines the index profile by measuring the power that is refracted from the core through the cladding.

The principle of this method can be seen from the ray trajectories in Fig. 9-11. The fiber is immersed in a liquid (index-matching oil) having an index equal to that of the cladding. If a ray lying outside the fiber acceptance angle enters the fiber core from the matching oil at an angle θ', it will be refracted out of the core and into the cladding. The angle at which it leaves the core is θ'' and it enters the cladding at an angle θ. Applying Snell's law twice to find the relationships between θ' and θ and between θ and θ'' yields

$$n_2 \sin \theta' = n(r) \sin \theta \tag{9-26a}$$

$$n(r) \cos \theta = n_2 \cos \theta'' \tag{9-26b}$$

Eliminating θ from these two equations then gives the relation between the input and output angles in terms of the cladding index n_2 and the core refractive index at the entrance position r:

$$n_2 \sin \theta' = \sqrt{n^2(r) - n_2^2 + n_2^2 \sin^2 \theta''} \tag{9-27}$$

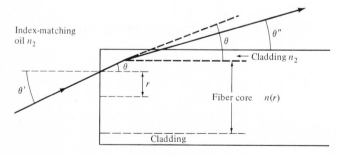

Figure 9-11 Basis of the refracted near-field technique for refractive-index-profile measurement. *(Reproduced with permission from Marcuse and Presby,[23] copyright 1980, IEEE.)*

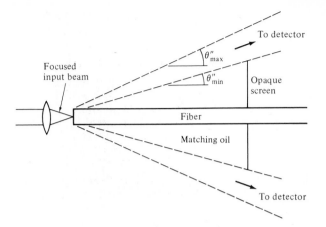

Figure 9-12 Experimental arrangement for the refracted near-field technique. *(Reproduced with permission from Marcuse and Presby,*[23] *copyright 1980, IEEE.)*

A schematic of an experimental setup is shown in Fig. 9-12. A laser beam having a numerical aperture significantly larger than the acceptance angle of the fiber is focused on a very small spot on the fiber end. The light escaping from the sides of the core consists of that portion of the incident light falling outside of the fiber acceptance angle and leaky mode loss from the captured optical power. Since only purely refracting modes are desired at the detector, an opaque circular screen is used to block off all light below a minimum angle θ''_{min}.

To determine the refractive index the following quantities must first be measured: the refractive index of the cladding, the minimum angle θ''_{min} defined by the size of the opaque screen, and the maximum angle θ'_{max} of the incident light, which is defined either by a focusing method or by an aperture placed in the input beam. With the use of the optical power $P(r)$ reaching the detector, the refractive index of the core can then be determined from[23] (see Prob. 9-12):

$$n(r) - n_2 = n_2 \cos \theta''_{min} (\cos \theta''_{min} - \cos \theta'_{max}) \frac{P(a) - P(r)}{P(a)} \qquad (9\text{-}28)$$

where $P(a)$ is the output power measured when the input light is focused into the cladding, that is, at the point where $n(r) = n_2$.

9-4-4 Interferometric Methods

Interferometry is a very accurate method for determining the refractive-index profiles of both fibers and preforms. The two types of interferometric procedures used for this are the slab method[23, 33] and the transverse method.[34, 35] The slab method requires cutting a thin circular slice out of the fiber or preform to be tested. The faces of this slab must be polished, flat, and very accurately parallel. Light

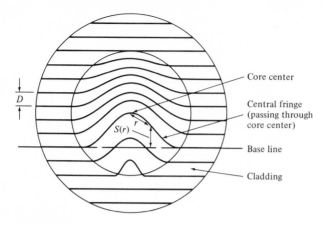

Figure 9-13 Fringe pattern observed in the field of view of an interference microscope when viewing a graded-index slab sample.

passing through this slab undergoes a phase shift which is dependent on the optical path length [the product of the slab thickness and the refractive index $n(r)$]. Upon emerging from the slab, this phase-shifted light is compared with the incident light, whereby interference fringes such as shown in Fig. 9-13 are formed.

The refractive-index difference between the core and cladding can be calculated from the fringe shift $S(r)$ and the parallel fringe spacing D, according to the relationship

$$n(r) - n_2 = \frac{\lambda S(r)}{Dd} \tag{9-29}$$

where λ is the wavelength of the measuring light and d is the slab thickness. The parameter $S(r)$ is the central fringe deviation at a distance r as measured from the baseline connecting the same cladding fringe at both sides of the core, as shown in Fig. 9-13.

The transverse interferometric method does not require cutting the fiber and preparing a polished slab. The fiber is immersed in index-matching oil and placed transversely under an interference microscope. Light illuminates the fiber at right angles with each ray passing through regions of varying refractive indices. The total optical path length must, therefore, be expressed as an integral, and an integral equation must be solved to obtain the index distribution from the measured fringe shift. With the use of an automatic computer-controlled system[34] to solve this integral equation, the index distribution can be obtained within minutes of fiber fabrication. In contrast, the more accurate but time-consuming slab method requires several days for the same information.

9-4-5 Focusing Method

Another scheme which uses transverse illumination but is not associated with interferometry is the focusing method,[1, 36, 37] shown in Fig. 9-14. This method can

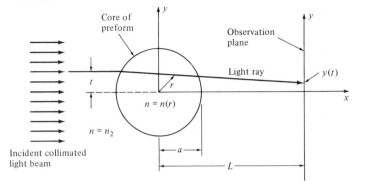

Figure 9-14 Concept schematic of focusing method for refractive-index profiling of optical fibers and preforms. *(Reproduced with permission from Marcuse and Presby,[23a] copyright 1980, IEEE.)*

be applied to both fibers and preforms. Here the core or preform is illuminated at right angles to its cylindrical axis with a collimated beam of incoherent, filtered light of uniform intensity. The fiber core or preform is immersed in an index-matching fluid to prevent light refraction at the outer cladding boundary. The core acts as a cylindrical lens having a low f number to focus the light onto an observation plane. By using the power density distribution measured in the observation plane, the following integral equation can be solved numerically to yield the refractive index

$$n(r) - n_2 = \frac{n_2}{\pi L} \int_r^a \frac{t - y(t)}{(t^2 - r^2)^{1/2}} \, dt$$

This equation takes into account the relationship between the entrance distance t of a ray and its position $y(t)$ in the observation plane. Details on this method can be found in the literature.[1, 36, 37]

9-5 MEASUREMENTS OF OPTICAL SOURCE CHARACTERISTICS

Optical fiber links can be employed for either digital or analog data transmission. In digital applications the performance parameter of interest is the rise time of the optical output of a light source in response to an applied electric pulse. This is important since the rise and fall times of the optical source output constrain the bandwidth in an optical fiber system. Two important parameters to consider for analog applications are harmonic distortion and intermodulation distortion, as discussed in Sec. 4-3. In general, intermodulation distortion is a more severe problem than harmonic distortion because the harmonics are usually outside of the signal bandpass and can thus be eliminated by electrical filtering. However, if the optical fiber link is used in a frequency-division-multiplexed system, such as the combination of several video channels, for example, cross talk could result if the harmonics that are generated in any particular frequency band overlap with an adjacent channel band.

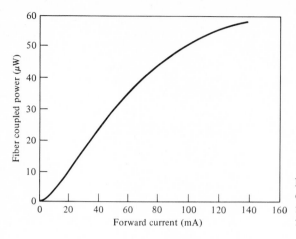

Figure 9-15 Typical optical power emitted into a 50-μm core-diameter fiber from an LED as a function of forward current.

In this section we shall look at some simple methods for measuring LED response time, harmonic distortions, and intermodulation distortion. Similar measurement techniques can also be applied to laser diodes. As a background for these measurements, let us first examine the optical power output versus current and the current versus voltage relationships for LEDs. As we noted in Chap. 4 and as is shown in Fig. 9-15, an LED is a nonlinear device particularly at low and high current levels. There are two reasons for these nonlinearities. At low current injection levels nonradiative band-to-band tunneling currents lead to low radiative efficiency. This nonradiative component saturates at higher current levels and is negligible compared to the radiative components. However, at high drive current levels the effects of thermal heating in the active region of the device also lead to a decrease in radiative efficiency. This results in the leveling off of the optical output with increasing drive current, as shown in Fig. 9-15.

A second parameter of interest is the current-versus-voltage characteristic of an LED. This has a typical diode shape, as shown in Fig. 9-16. The knee of the curve

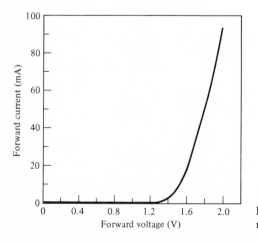

Figure 9-16 Typical current-versus-voltage relationship for an LED.

is typically 1.3 to 1.4 V. The slope of the curve gives the dynamic resistance of the device, with typical values being a few ohms. An LED thus acts as a nonlinear, low-valued resistor over most of its operating range.

9-5-1 LED Response Time

A test setup for measuring rise and fall times of an LED is shown in Fig. 9-17. In this setup a variable dc bias can be applied to the LED while it is being driven by pulses from a square-wave generator. The sum of the resistances R_2 and R_3 plus the dynamic resistance of the LED should match the signal generator output impedance. For example, for a 2-Ω LED resistance we could have $R_2 = 39\ \Omega$ and $R_3 = 10\ \Omega$ to present a 51-Ω load to the signal generator output. A current-limiting resistor R_1 (say, 1 kΩ) is placed in series with the voltage source to prevent unexpected current surges from damaging the LED.

Alternatively, the LED could be biased from a current source, but one must use extreme caution in this case. The primary rule is never to connect an optical source to a current source when the power is on. Most constant-current supplies develop high voltages at their output when they are in an unloaded condition. This voltage charges up capacitances associated with the power supply, the test circuit, and/or the cable. Connecting a low-resistance LED to such a supply then discharges the capacitors. The resultant current surge could readily exceed the rated current-handling capability of the LED and may damage or destroy the device. If an LED is biased with a current source, its output voltage should be limited to a few volts. However, the preferred approach is to use a voltage source with a current-limiting resistor.

When a signal from the square-wave generator is applied to the LED, the optical output pulses are measured with a high-speed analog receiver. A fast

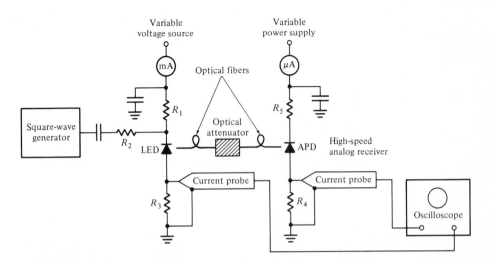

Figure 9-17 Test setup for measuring LED rise and fall times.

responding avalanche photodiode (APD) is generally used in the receiver for these applications. If short fiber lengths are used in the coupling link, the optical attenuator should be adjusted so that the optical power level at the receiver input causes no distortion of the signal as seen on the oscilloscope.

Two measurements are needed to determine the rise time of the LED optical output. The first is the 10 to 90 percent rise time τ_{Li} of the LED current. This waveform can be measured either with a current probe or as a voltage drop across resistor R_3. The other measurement is the 10 to 90 percent rise time τ_{meas} of the receiver output waveform which is measured across resistor R_4. By using the manufacturer-supplied avalanche photodiode rise time τ_{APD}, the LED optical output rise time τ_{LED} can be computed from the relation

$$\tau_{LED} = (\tau_{meas}^2 - \tau_{Li}^2 - \tau_{APD}^2)^{1/2} \tag{9-30}$$

9-5-2 Harmonic Distortion

Harmonic distortion appearing in the optical output of a sinusoidally modulated light source can be measured with the test circuitry shown in Fig. 9-18. The setup and precautions for biasing the LED are the same as described in Sec. 9-5-1. The bias level of the source must be such that the modulation depth of the signal is less than one. This is explained in Sec. 4-3 and shown in Fig. 4-17 for LEDs and laser diodes.

When a sinusoidally modulated signal is applied to the source, the optical output is coupled to an analog receiver through a properly attenuated optical link. The attenuator is adjusted so that the optical power level at the receiver input causes no distortion of the signal as observed on the monitoring oscilloscope. The signal from the receiver is then measured at the desired frequency with a wave analyzer.

The harmonic distortion can be approximated by making a Taylor series expansion of the optical power $P(t)$ in terms of the modulating drive current $i(t)$ in the vicinity of the source bias point. For example, for small excursions about the bias

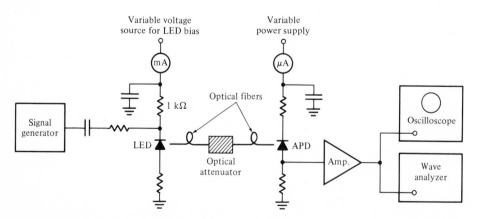

Figure 9-18 Test circuitry for measuring harmonic distortion.

point, the optical power may be represented as

$$P(t) \simeq a_1 i(t) + a_2 i^2(t) + a_3 i^3(t) + a_4 i^4(t) \tag{9-31}$$

where we have truncated the series after the fourth term. If we modulate the light source about the bias level with a cosine wave of amplitude I_0 and frequency ω,

$$i(t) = I_0 \cos \omega t \tag{9-32}$$

then the optical output power is given by

$$P(t) = a_1 I_0 \cos \omega t + a_2 I_0^2 \cos^2 \omega t + a_3 I_0^3 \cos^3 \omega t + a_4 I_0^4 \cos^4 \omega t \tag{9-33}$$

Making use of trigonometric identities for the $\cos^n \omega t$ terms and collecting coefficients of the various harmonic terms result in

$$P(t) = (\tfrac{1}{2} a_2 I_0^2 + \tfrac{3}{8} a_4 I_0^4) + (a_1 I_0 + \tfrac{3}{4} a_3 I_0^3) \cos \omega t$$
$$+ (\tfrac{1}{2} a_2 I_0^2 + \tfrac{1}{2} a_4 I_0^4) \cos 2\omega t + \tfrac{1}{4} a_3 I_0^3 \cos 3\omega t$$
$$+ \tfrac{1}{8} a_4 I_0^4 \cos 4\omega t \tag{9-34}$$

The first term in parenthesis in Eq. (9-34) represents a dc level and is not of interest here. The coefficients of the $\cos n\omega t$ terms are read off of the wave analyzer tuned to the frequencies $nf = n\omega/2\pi$. The second-order harmonic distortion is given by the ratio of the coefficient of the $\cos 2\omega t$ term to the coefficient of the fundamental $\cos \omega t$ term. Thus the second-order harmonic distortion is

$$20 \log \frac{\tfrac{1}{2} a_2 I_0^2 + \tfrac{1}{2} a_4 I_0^4}{a_1 I_0 + \tfrac{3}{4} a_3 I_0^3} \quad \text{dB} \tag{9-35}$$

Similarly, the third-order harmonic distortion is found from the ratio of the $\cos 3\omega t$ and the $\cos \omega t$ coefficients. The third-order harmonic distortion is

$$20 \log \frac{\tfrac{1}{4} a_3 I_0^3}{a_1 I_0 + \tfrac{3}{4} a_3 I_0^3} \quad \text{dB} \tag{9-36}$$

9-5-3 Intermodulation Distortion

Intermodulation distortion is a type of cross talk that is of concern in systems where a number of independent signals operating at different carrier frequencies are electrically multiplexed and then transmitted over an optical fiber link. A method for measuring this type of distortion is shown in Fig. 9-19 for two independent carrier frequencies (signal sources). Equal-amplitude sinusoidally varying signals at frequencies f_1 and f_2 are first combined with a power combiner. The output of the power combiner is used to drive a properly biased optical source.

The optical output of the light source is coupled to an avalanche-photodiode-based receiver through an optical fiber. As we noted in Sec. 9-5-1, the setting (or characteristic) of the optical attenuator is such that the signal level at the receiver does not produce a distorted waveform as viewed on the oscilloscope. The received signal detected by the photodiode is amplified and measured with a spectrum

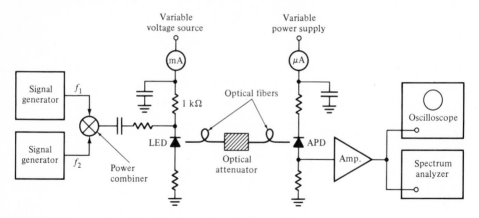

Figure 9-19 Test circuitry for measuring intermodulation distortion.

analyzer. A spectrum analyzer is a swept receiver that gives a visual display of the signal amplitude in dBm as a function of frequency over a selected frequency range. The differences between the fundamental-frequency signal level and the distortion-component signal levels at the desired frequencies can thus be read off of the instrument directly in decibels.

To determine the intermodulation distortions the received signal amplitude is measured on the spectrum analyzer at $|f_1 \pm f_2|$ for the second-order components, and at $2f_1 - f_2$ and $2f_2 - f_1$ for the third-order component. These levels are then compared to the fundamental-frequency signal level at f_1 or f_2 (since both fundamentals have the same amplitude). As we noted in Sec. 4-3, the second-order components usually fall outside of the system passband and can be eliminated with appropriate filters in the receiver. The third-order intermodulation distortion products are the most troublesome since they can fall within the bandpass channels of the receiver.

If three or more signal sources are used in the setup shown in Fig. 9-19, second-order distortion components appear at frequencies $|f_i \pm f_j|$ and third-order components arise at $2f_i \pm f_j$ and $|f_i \pm f_j \pm f_k|$, where f_i, f_j, and f_k are the frequencies of any of the signal sources.

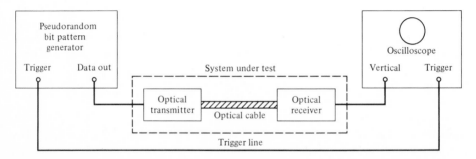

Figure 9-20 Basic equipment used for eye-pattern generation.

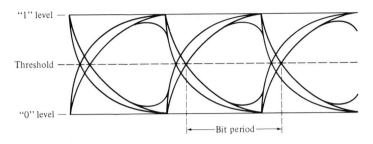

Figure 9-21 Eye-pattern diagram.

9-6 EYE PATTERN

The eye-pattern technique[38–41] is a simple but powerful experimental method for assessing the data-handling ability of a digital transmission system. This method has been used extensively for evaluating the performance of wire systems and can also be applied to optical fiber data links. The eye-pattern measurements are made in the time domain and allow the effects of waveform distortion to be shown immediately on an oscilloscope.

An eye-pattern measurement can be made with the basic equipment shown in Fig. 9-20. The output from a pseudorandom data pattern generator is applied to the vertical input of an oscilloscope and the data rate is used to trigger the horizontal sweep. This results in the type of pattern shown in Fig. 9-21, which is called the *eye pattern* because the display shape resembles a human eye. To see how the display pattern is formed, consider the eight possible 3-bit-long NRZ combinations shown in Fig. 9-22. When these eight combinations are superimposed simultaneously, an eye pattern as shown in Fig. 9-21 is formed.

To measure system performance with the eye-pattern method, a variety of word patterns should be provided. A convenient approach is to generate a *random data*

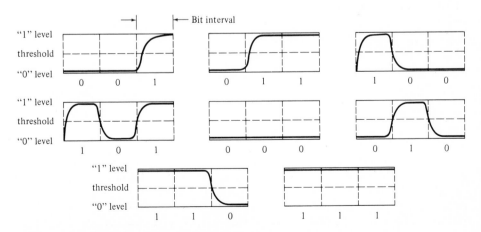

Figure 9-22 Eight possible 3-bit-long NRZ combinations.

signal, because this is the characteristic of data streams found in practice. This type of signal generates ones and zeros at a uniform rate but in a random manner. A variety of pseudorandom pattern generators are available for this purpose. The word *pseudorandom* means that the generated combination or sequence of ones and zeros will eventually repeat but that it is sufficiently random for test purposes. A pseudorandom bit sequence comprises four different 2-bit-long combinations, eight different 3-bit-long combinations, sixteen different 4-bit-long combinations, and so on (that is, sequences of 2^N different N-bit-long combinations) up to a limit set by the instrument. After this limit has been generated, the data sequence will repeat.

A great deal of system performance information can be deduced from the eye-pattern display. To interpret the eye pattern, consider the simplified drawing shown in Fig. 9-23. The following information regarding the signal amplitude distortion, timing jitter, and system rise time can be derived:

1. The width of the eye opening defines the time interval over which the received signal can be sampled without error from intersymbol interference. The best time to sample the received waveform is when the height of the eye opening is largest.
2. The height of the eye opening is reduced as a result of amplitude distortion in the data signal. The maximum distortion is given by the vertical distance between the top of the eye opening and the maximum signal level. The greater the eye closure becomes the more difficult it is to detect the signal.
3. The height of the eye opening at the specified sampling time shows the noise margin or immunity to noise. *Noise margin* is the percentage ratio of the peak signal voltage V_1 for an alternating bit sequence (defined by the height of the eye opening) to the maximum signal voltage V_2 as measured from the threshold level, as shown in Fig. 9-23. that is,

$$\text{Noise margin (percent)} = \frac{V_1}{V_2} \times 100 \text{ percent} \qquad (9\text{-}37)$$

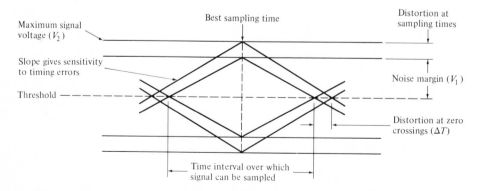

Figure 9-23 Simplified eye-pattern diagram.

4. The rate at which the eye closes as the sampling time is varied (that is, the slope of the eye pattern sides) determines the sensitivity of the system to timing errors. The possibility of timing errors increases as the slope becomes more horizontal.

5. Timing jitter (also referred to as *edge jitter* or *phase distortion*) in an optical fiber system arises from noise in the receiver and pulse distortion in the optical fiber. If the signal is sampled in the middle of the time interval (that is, midway between the times when the signal crosses the threshold level), then the amount of distortion ΔT at the threshold level indicates the amount of jitter. Timing jitter is thus given by

$$\text{Timing jitter (percent)} = \frac{\Delta T}{T_b} \times 100 \text{ percent} \qquad (9\text{-}38)$$

where T_b is one bit interval.

6. The 10 to 90 percent rise and fall times of the signal can be easily measured by using the 0 and 100 percent reference levels produced by long strings of zeros and ones, respectively.

7. Any nonlinearities of the channel transfer characteristics will create an asymmetry in the eye pattern. If a purely random data stream is passed through a truly linear system, all the eye openings will be identical and symmetrical.

9-7 SUMMARY

In this chapter we have presented some measurement methods for characterizing optical fibers, light sources, and data link performance. Measurements of optical fiber parameters are of particular importance since the optical fiber is the most difficult component to replace once a transmission system has been installed. Techniques for measuring optical fiber attenuation and bandwidth are important since these two parameters have a major effect on the overall system performance, and because they could vary during the fiber-cabling process or as a result of cable installation.

Attenuation in a fiber can be measured either by using a cutback technique or by means of an optical time domain reflectometer. The cutback technique is a destructive method that requires access to both ends of the fiber. To find the transmission loss, the optical power is first measured at the output end of the fiber. Then without disturbing the input-power-launching condition, the fiber is cut off a few meters from the source and the output power is again measured. The average attenuation of the fiber can thus be calculated from the power change over the measured fiber length.

In contrast to the cutback technique, the optical time domain reflectometer (OTDR) method is not destructive and requires access to only one end of the fiber. An OTDR is fundamentally an optical radar. It operates by periodically launching narrow laser pulses into one end of a fiber. The properties of the optical fiber under test are then determined by analyzing the amplitude and temporal characteristics of the waveform of the back-scattered light. Some useful features of the OTDR method include giving information about the variation of attenuation with length, measuring the optical loss arising at fiber joints, and locating fiber breaks in a cable.

Light pulses become wider as they propagate along a fiber because of material, modal, and waveguide dispersion effects. This widening may cause two closely spaced pulses to overlap eventually and become indistinguishable. Since the degree of pulse spreading determines the information-carrying capacity of a fiber, dispersion measurements are important. To determine pulse distortion, a fiber can be considered as a filter characterized either by an impulse response $h(t)$ or by a power transfer function $H(f)$, which is the Fourier transform of the impulse response. The impulse response measurements are made in the time domain, whereas the power transfer function is measured in the frequency domain. A simple approach for making pulse dispersion measurements in the time domain is to inject a narrow optical pulse into one end of the fiber and to detect the broadened output pulse at the other end. Since this method generally requires very long fibers for accurate results, a shuttle pulse arrangement has been found to be more practical. Frequency domain measurements yield information on the amplitude versus frequency response of a fiber, which is often the main parameter of interest to system designers. These measurements can be made by sinusoidally modulating a light source about a fixed level. The baseband frequency response is then found from the ratio of the sine wave amplitudes at the beginning and end of the fiber.

The shape of the refractive-index profile of a fiber core is of great importance since very slight departures from an optimum value at a given wavelength can dramatically decrease the fiber bandwidth. Very precise methods are thus required for measuring the core index profile. Among these are:

1. The end-reflection technique
2. The transmitted near-field scanning method
3. The refracted near-field technique
4. The interferometric slab method
5. The transverse interferometer scheme
6. The focusing method

Optical source measurements of interest include the rise time and the linearity of the optical output power. Rise and fall times are important in digital applications since they constrain the bandwidth in an optical fiber system. For digital applications, nonlinearities in the optical output of a luminescent source are not a major concern. However, nonlinearities are of importance for analog applications because harmonic and intermodulation distortions can arise and degrade the signal-to-noise ratio. Simple test setups for these measurements are described in Sec. 9-5.

Once an optical fiber link has been installed and assembled, a measurement of its overall performance is usually desired. The eye-pattern technique is a simple method that provides a great deal of information for assessing the data-handling capability of a digital transmission system. We described the principles, an experimental setup, and the measurement data interpretation of this technique in Sec. 9-6.

PROBLEMS

Section 9-1

9-1 The optical power in a fiber at a distance x from the input end is given by Eq. (9-3). By assuming that the loss coefficient is uniform along the fiber, use this equation to derive Eq. (9-1).

9-2 The attenuation of a 1.8-km fiber having an 0.23 NA is measured by using a matched-beam excitation technique. For a launch NA of 0.23 the far-end optical power is measured as 5 μW and the near-end power emerging from the fiber is 60 μW. What is the average attenuation of this fiber in decibels per kilometer? How would the attenuation measurement be affected for a larger launch NA? For a smaller launch NA?

9-3 Three 500-m-long fibers have been spliced together in series. An OTDR is used to measure the attenuation of the resultant fiber. The reduced data of the OTDR oscilloscope display is shown in Fig. P9-3. What are the attenuations in decibels per kilometer of the three individual fibers? What are the splice losses in decibels? What are some possible reasons for the large splice loss occurring between the second and third fiber?

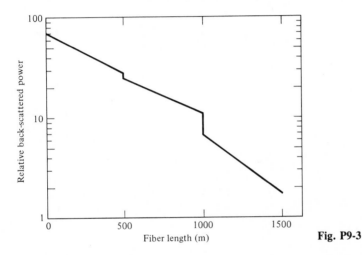

Fig. **P9-3**

9-4 Assuming that Rayleigh scattering is approximately isotropic (uniform in all directions), show that the fraction of scattered light trapped in the fiber in the backward direction is given by

$$S \simeq \frac{\pi(NA)^2}{4\pi n^2} = \frac{1}{4}\left(\frac{NA}{n}\right)^2$$

where NA is the fiber numerical aperture, n is the core refractive index, and NA/n represents the half-angle of the cone of captured rays. If NA = 0.20 and n = 1.50, what fraction of the scattered light is recaptured by the fiber in the reverse direction?

Section 9-2

9-5 Show that, when using an OTDR, an optical pulse width of 5 ns or less is required to locate a fiber fault to within 1 m of its true position.

Section 9-3

9-6 Derive Eq. (9-17).

9-7 Derive Eq. (9-19).

9-8 A shuttle pulse setup has a 50-dB dynamic range. The mirrors have a transmission coefficient $T = 0.1$ and a reflection coefficient $R = 0.9$. If $\alpha = 5$ dB/km, how many pulses can be detected by using a 200-m-long fiber? How many pulses can be detected if $T = 0.3$ and $R = 0.7$? What are the extrapolated fiber lengths corresponding to these two cases?

9-9 A gaussian approximation of $|H(f)|$ in the form

$$|H(f)| \simeq e^{-(2\pi f\sigma)^2/2}$$

has been found to be accurate to at least the 0.75-amplitude point in the frequency domain. Using this relationship, plot $P(f)/P(0)$ as a function of frequency from 0 to 1000 MHz for fibers having impulse responses of full rms pulse widths 2σ equal to 2.0, 1.0, and 0.5 ns. What are the 3-dB bandwidths of these fibers?

9-10 Derive Eq. (9-24).

9-11 Figure 9-11 shows the trajectory of a light ray passing through two different dielectric interfaces, these being the index fluid-to-core boundary and the core-to-cladding boundary. Using Snell's law at each of these interfaces, derive Eqs. (9-26a) and (9-26b). Show that Eq. (9-27) can be obtained from these two equations.

9-12 Derive Eq. (9-28). *Hint:* Assuming a uniform angular input intensity I, the light collected by the detector is

$$P(r) = I \int_0^{2\pi} d\phi \int_{\theta'_{min}}^{\theta'_{max}} \sin\theta \, d\theta$$

Express θ'_{min} in terms of the minimum angle θ''_{min} defined by the screen. Use the fact that $n(r) - n_2 \ll 1$ to eliminate the square-root expression, and find the unknown source intensity I from the expression for $P(r)$ when $r = a$.

Section 9-5

9-13 Show that Eq. (9-34) results from the substitution of Eq. (9-32) into Eq. (9-31).

9-14 A particular frequency division multiplexed system is designed to transmit five independent channels in a passband ranging from 40 to 85 MHz. The five carrier frequencies are at 46, 53, 62, 70, and 80 MHz. The information bandwidth of each carrier is 1.5 MHz. At what frequencies do the second- and third-order intermodulation and harmonic distortion components occur? Which of these components are of concern?

REFERENCES

1. D. Marcuse, *Principles of Optical Fiber Measurements*, Academic, New York, 1981.
2. M. K. Barnoski and S. D. Personick, "Measurements in fiber optics," *Proc. IEEE*, **66**, 429–441, Apr. 1978.
3. F. Krahn, W. Meininghaus, and D. Rittich, "Measuring and test equipment for optical cable," *Philips Telecom. Rev.*, **37**, 241–249, Sept. 1979.
4. L. G. Cohen, P. Kaiser, P. L. Lazay, and H. M. Presby, "Fiber characterization," in *Optical Fiber Telecommunications*, S. E. Miller and A. G. Chynoweth (Eds.), Academic, New York, 1979.
5. J. E. Midwinter, *Optical Fibers for Transmission*, Wiley, New York, 1979.
6. G. W. Day and D. L. Franzen (Eds.), *Technical Digest—Symposium on Optical Fiber Measurements, 1980*, NBS Special Publ. 597, Oct. 1980.
7. D. B. Keck and A. R. Tynes, "Spectral response of low loss optical waveguides," *Appl. Opt.*, **11**, 1502–1506, July 1972.
8. P. Kaiser and H. W. Astle, "Measurements of spectral total and scattering losses in unclad optical fibers," *J. Opt. Soc. Amer.*, **64**, 469–474, Apr. 1974.

9. M. K. Barnoski and S. M. Jensen, "Fiber waveguides: a novel technique for investigating attenuation characteristics," *Appl. Opt.*, **15**, 2112–2115, Sept. 1976.

10. M. K. Barnoski, M. D. Rourke, S. M. Jensen, and R. T. Melville, "An optical time-domain reflectometer," *Appl. Opt.*, **16**, 2375–2379, Sept. 1977.

11. R. Olshansky and S. M. Oaks, "Differential mode attenuation measurements in graded index fibers," *Appl. Opt.*, **17**, 1830–1835, June 1978.

12. L. G. Cohen, P. Kaiser, and C. Lin, "Experimental techniques for evaluation of fiber transmission loss and dispersion," *Proc. IEEE*, **68**, 1203–1209, Oct. 1980.

13. P. Kaiser, "Loss measurements of graded index fibers: accuracy versus convenience," in *Technical Digest-Symposium on Optical Fiber Measurements, 1980*, NBS Special Publ. 597, Oct. 1980, pp. 11–14.

14. S. D. Personick, "Photon probe—An optical fiber time-domain reflectometer," *Bell Sys. Tech. J.*, **56**, 355–366, Mar. 1977.

15. P. Healey and P. Hensel, "Optical time domain reflectometry by photon counting," *Electron. Lett.*, **16**, 631–633, July 1980.

16. S. D. Personick, "Baseband linearity and equalization in digital fiber optical communication systems," *Bell Sys. Tech. J.*, **52**, 1175–1194, Sept. 1973.

17. S. D. Personick, "Receiver design for digital fiber optic communication systems, I and II," *Bell Sys. Tech. J.*, **52**, 843–874, July–Aug. 1973.

18. D. Gloge, E. L. Chinnock, and T. P. Lee, "Self pulsing GaAs laser for fiber dispersion measurements," *IEEE J. Quantum Electron.*, **QE-8**, 844–846, Nov. 1972.

19. L. G. Cohen and H. M. Presby, "Shuttle pulse measurements of pulse spreading in a low-loss graded index fiber," *Appl. Opt.*, **14**, 1361–1363, June 1975.

20. S. D. Personick, W. M. Hubbard, and W. S. Holden, "Measurements of the baseband frequency response of a 1-km fiber," *Appl. Opt.*, **13**, 266–268, Feb. 1974.

21. L. G. Cohen, H. W. Astle, and I. P. Kaminow, "Frequency domain measurements of dispersion in multimode optical fibers," *Bell Sys. Tech. J.*, **55**, 1509–1523, Dec. 1976.

22. I. Kobayashi, "Bandwidth measurement in multimode optical fibers," in *Technical Digest-Symposium on Optical Fiber Measurements, 1980*, NBS Special Publ. 597, Oct. 1980, pp. 49–54.

23. D. Marcuse and H. M. Presby, "Index profile measurements of fibers and their evaluation," *Proc. IEEE*, **68**, 666–688, June 1980; "Index-profile characterization of fiber preforms and drawn fibers," *ibid.*, 1198–1203, Oct. 1980.

24. W. Eickhoff and E. Weidel, "Measuring method for the refractive index profile of optical glass fibers," *Opt. Quantum Electron.*, **7**, 103–113, 1975.

25. M. Ikeda and H. Yoshikiyo, "Refractive index profile of a graded index fiber: measurement by a reflection method," *Appl. Opt.*, **14**, 814–815, Apr. 1975.

26. F. M. E. Sladen, D. N. Payne, and M. J. Adams, "Determination of optical fiber refractive index profiles by a near-field scanning technique," *Appl. Phys. Lett.*, **28**, 255–258, Mar. 1976.

27. D. Gloge and E. A. J. Marcatili, "Multimode theory of graded core fibers," *Bell Sys. Tech. J.*, **52**, 1563–1578, Nov. 1973.

28. W. J. Stewart, "A new technique for measuring the refractive index profiles of graded optical fibers," in *Tech. Dig. 1977 Int. Conf. Integrated Optics and Optical Fiber Communications*, Tokyo, Japan, July 18–20, 1977.

29. K. I. White, "Practical application of the refracted near-field technique for the measurement of optical fiber refractive index profiles," *Opt. Quantum Electron.*, **11**, 185–196, 1979.

30. M. Young, "Calibration technique for refracted near-field scanning of optical fibers," *Appl. Opt.*, **19**, 2479–2480, Aug. 1980.

31. M. J. Saunders, "Optical fiber profiles using the refracted near-field technique: a comparison with other methods," *Appl. Opt.*, **20**, 1645–1651, May 1981.

32. M. Young, "Optical fiber index profiles by the refracted-ray method (refracted near-field scanning)," *Appl. Opt.*, **20**, 3415–3422, Oct. 1981.

33. H. M. Presby, W. Mammel, and R. M. Derosier, "Refractive index profiling of graded index optical fibers," *Rev. Sci. Instrum.*, **47**, 348–352, Mar. 1976.

34. L. M. Boggs, H. M. Presby, and D. Marcuse, "Rapid automatic index profiling of whole fiber samples: part 1," *Bell Sys. Tech. J.*, **58**, 867–882, Apr. 1979.

35. H. M. Presby, D. Marcuse, H. W. Astle, and L. M. Boggs, "Rapid automatic index profiling of whole fiber samples: part II," *Bell Sys. Tech. J.,* **58,** 883–902, Apr. 1979.
36. D. Marcuse, "Refractive index determination by the focusing method," *Appl. Opt.,* **18,** 9–13, Jan. 1979.
37. D. Marcuse and H. M Presby, "Focusing method for nondestructive measurement of optical fiber index profiles," *Appl. Opt.,* **18,** 14–22, Jan. 1979.
38. R. W. Lucky, J. Salz, and E. J. Weldon, Jr., *Principles of Data Communication,* McGraw-Hill, New York, 1968.
39. D. Tugal and O. Tugal, *Data Transmission,* McGraw-Hill, New York, 1982.
40. K. S. Shanmugam, *Digital and Analog Communication Systems,* Wiley, New York, 1979.
41. R. L. Ohlhaber, "Eye pattern testing of fiber optic systems," *FOC-78 Proceedings* (Information Gatekeeper, Inc.) 195–199, Chicago, Sept. 1978.

CHAPTER TEN

OPTICAL FIBER FABRICATION AND CABLING

In Chap. 2 we discussed the different structures of optical fibers and we examined the concepts of how optical energy propagates along a dielectric optical fiber waveguide. The structure and material composition of a fiber also dictate how and to what degree optical signals get attenuated and distorted as they propagate along a fiber. This was examined in detail in Chap. 3. To conclude the discussion of optical fibers, we shall show here how fibers are manufactured and what materials are used. We shall also analyze their mechanical strengths and illustrate how they can be incorporated into cable structures which protect the glass waveguides from the external environment.

10-1 FIBER MATERIALS

In selecting materials for optical fibers, a number of requirements must be satisfied. For example:

1. It must be possible to make long, thin, flexible fibers from the material.
2. The material must be transparent at a particular optical wavelength in order for the fiber to guide light efficiently.
3. Physically compatible materials having slightly different refractive indices for the core and cladding must be available.

Materials satisfying these requirements are glasses and plastics.

The majority of fibers are made of glass consisting either of silica (SiO_2) or a silicate. The variety of available glass fibers ranges from high-loss glass fibers with large cores used for short-transmission distances to very transparent (low-loss) fibers employed in long-haul applications. Plastic fibers are less widely used because of their substantially higher attenuation compared to glass fibers. The main use of plastic fibers is in short-distance applications and in abusive environments, where the greater mechanical strength of plastic fibers offers an advantage over the use of glass fibers.

10-1-1 Glass Fibers

Glass is made by fusing mixtures of metal oxides, sulfides, or selenides.[1-4] The resulting material is a randomly connected molecular network rather than a well-defined ordered structure as found in crystalline materials. A consequence of this random order is that glasses do not have well-defined melting points. When glass is heated up from room temperature, it remains a hard solid up to several hundred degrees centigrade. As the temperature is increased further, the glass gradually begins to soften until at very high temperatures it becomes a viscous liquid. The expression "melting temperature" is commonly used in glass manufacture. This term refers only to an extended temperature range in which the glass becomes fluid enough to free itself fairly quickly of gas bubbles.

The largest category of optically transparent glasses from which optical fibers are made consists of the oxide glasses. Of these the most common is silica (SiO_2) which has a refractive index of 1.458 at 850 nm. To produce two similar materials having slightly different indices of refraction for the core and cladding, either fluorine or various oxides (referred to as *dopants*) such as B_2O_3, GeO_2, or P_2O_5 are added to the silica. As shown in Fig. 10-1 the addition of GeO_2 or P_2O_5 increases the refractive index whereas doping the silica with fluorine or B_2O_3 decreases it. Since the cladding must have a lower index than the core, examples of fiber compositions are:

1. GeO_2-SiO_2 core; SiO_2 cladding
2. P_2O_5-SiO_2 core; SiO_2 cladding

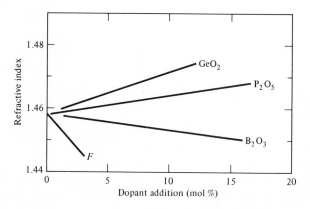

Figure 10-1 Variation in refractive index as a function of doping concentration in silica glass. *(Fluorine data from K. Rau et al.,* Top. Meet. Opt. Fiber Trans., *Williamsburg, Va., 1977, Paper TuC4; GeO_2, P_2O_5 and B_2O_3 data reprinted with permission from C. R. Hammond, and S. R. Norman,* Opt. Quantum Electron., **9**, *399-409, copyright 1977, Chapman & Hall.)*

3. SiO_2 core; B_2O_3-SiO_2 cladding
4. GeO_2-B_2O_3-SiO_2 core; B_2O_3-SiO_2 cladding

Here the notation GeO_2-SiO_2, for example, denotes a GeO_2-doped silica glass.

The principal raw material for silica is sand. Glass composed of pure silica is referred to either as *silica glass, fused silica,* or *vitreous silica.* Some of its desirable properties are a resistance to deformation at temperatures as high as 1000°C, a high resistance to breakage from thermal shock because of its low thermal expansion, good chemical durability, and high transparency in both the visible and infrared regions of interest to fiber optic communication systems. Its high melting temperature is a disadvantage if the glass is prepared from a molten state. However, this problem is partially avoided when using vapor deposition techniques, which we shall discuss in Sec. 10-2.

An alternative to the silica glasses are the low-melting silicates.[2] Typical glasses used for optical fibers are the soda-lime silicates, the germanosilicates, and various borosilicates. The soda-lime silicates, for example, are combinations of silica, an alkaline oxide such as Na_2O or Li_2O, and a second oxide such as CaO (lime), MgO, ZnO, or BaO. These glasses are relatively easy to melt and fabricate. The raw materials of soda-lime silicates are ultrapure powdered forms of oxides (SiO_2, GeO_2, and B_2O_3) and carbonates (Na_2CO_3, K_2CO_3, $CaCO_3$, and $BaCO_3$).

In 1978 investigations were reported[5,6] on materials having extremely low light transmission loss in the 2- to 5-μm wavelength region. The principal materials in this mid-infrared range are various halide crystals[5-7] (for example, TlBrI, TlBr, KCl, CsI, and AgBr) and fluoride glasses[7] composed of mixtures of GdF_3, BaF_2, ZrF_4, and AlF_3. One limitation of these materials was that only short lengths of fibers could be fabricated because conventional glass-drawing techniques cannot be used for crystalline materials. However, investigations are now actively continuing in achieving longer lengths by extrusion methods. Another problem is that, since these materials are inherently weaker than oxide glasses, special cabling must be done to strengthen the fiber and to protect it from moisture.

10-1-2 Plastic-Clad Glass Fibers

Optical fibers constructed with glass cores and glass claddings are very important for long-distance applications where the very low losses achievable in these fibers are needed. For short-distance applications (up to several hundred meters), where higher losses are tolerable, the less expensive plastic-clad silica fibers can be used. These fibers are composed of silica cores with the lower refractive-index cladding being a polymer (plastic) material. These fibers are often referred to as PCS (plastic-clad silica) fibers.

A common material source for the silica core is selected high-purity natural quartz. A common cladding material is a silicone resin having a refractive index of 1.405 at 850 nm. Silicone resin is also frequently used as a protective coating for other types of fibers. Another popular plastic cladding material[8] is perfluoronated ethylene propylene (Teflon FEP). The low refractive index of 1.338 of this material results in fibers with potentially large numerical apertures.

Plastic claddings are only used for step-index fibers. The core diameters are larger (150 to 600 μm) than the standard 50-μm-diameter core of all-glass graded-index fibers, and the larger difference in the core and cladding indices results in a high numerical aperture. This allows low-cost large-area light sources to be used for coupling optical power into these fibers, thereby yielding comparatively inexpensive but lower-quality systems which are quite satisfactory for many applications.

10-1-3 Plastic Fibers

All-plastic multimode step-index fibers are good candidates for fairly short (up to about 100 m) and low-cost links. Although they exhibit considerably greater optical signal attenuations than glass fibers, the toughness and durability of plastic allow these fibers to be handled without special care. The high refractive-index differences that can be achieved between the core and cladding materials yield numerical apertures as high as 0.6 and large acceptance angles of up to 70°. In addition, the mechanical flexibility of plastic allows these fibers to have large cores, with typical diameters ranging from 110 to 1400 μm. These factors permit the use of inexpensive large-area light-emitting diodes which, in conjunction with the less expensive plastic fibers, make an economically attractive system.

Examples of plastic fiber constructions are:

1. A polysterene core ($n_1 = 1.60$) and a methyl methacrylate cladding ($n_2 = 1.49$) to give an NA of 0.60
2. A polymethyl methacrylate core ($n_1 = 1.49$) and a cladding made of its co-polymer ($n_2 = 1.40$) to give an NA of 0.50

10-2 FIBER FABRICATION

Two basic techniques[4, 9, 10] are used in the fabrication of all-glass optical waveguides. These are the vapor phase oxidation processes and the direct-melt methods. The direct-melt method follows traditional glass-making procedures in that optical fibers are made directly from the molten state of purified components of silicate glasses. In the vapor phase oxidation process, highly pure vapors of metal halides (e.g., $SiCl_4$ and $GeCl_4$) react with oxygen to form a white powder of SiO_2 particles. The particles are then collected on the surface of a bulk glass by one of three different commonly used processes and are sintered (transformed to a homogeneous glass mass by heating without melting) by one of a variety of techniques to form a clear glass rod or tube (depending on the process). This rod or tube is called a *preform*. It is typically around 10 mm in diameter and 60 to 90 cm long. Fibers are made from the preform[11, 12] by using the equipment shown in Fig. 10-2. The

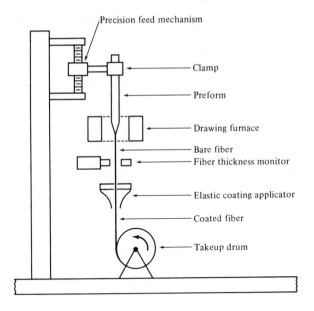

Precision feed mechanism

— Clamp

— Preform

— Drawing furnace

— Bare fiber
— Fiber thickness monitor

— Elastic coating applicator

— Coated fiber

— Takeup drum

Figure 10-2 Schematic of fiber-drawing apparatus.

preform is precision-fed into a circular heater called the *drawing furnace*. Here the preform end is softened to the point where it can be drawn into a very thin filament which becomes the optical fiber. The turning speed of the takeup drum at the bottom of the draw tower determines how fast the fiber is drawn. This, in turn, will determine the thickness of the fiber, so that a precise rotation rate must be maintained. An optical fiber thickness monitor is used in a feedback loop for this speed regulation. To protect the bare glass fiber from external contaminants such as dust and water vapor, an elastic coating is applied to the fiber immediately after it is drawn.

We shall now briefly examine some details of the direct-melt method and three vapor phase oxidation processes. For comprehensive reviews of these and other fiber fabrication processes, the reader is referred to the literature.[4, 9, 10]

10-2-1 Outside Vapor Phase Oxidation

The first fiber to have a loss of less than 20 dB/km was made at the Corning Glass Works[13, 14] by the *outside vapor phase oxidation* (OVPO) process. This method is illustrated in Fig. 10-3. First, a layer of SiO_2 particles called a *soot* is deposited from a burner onto a rotating graphite or ceramic mandrel. The glass soot adheres to this bait rod and, layer by layer, a cylindrical, porous glass preform is built up. By properly controlling the constituents of the metal halide vapor stream during the deposition process, the glass compositions and dimensions desired for the core and cladding can be incorporated into the preform. Either step- or graded-index preforms can thus be made.

When the deposition process is completed, the mandrel is removed and the porous tube is then vitrified in a dry atmosphere at a high temperature (above

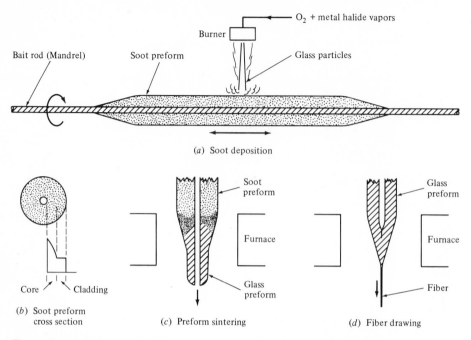

Figure 10-3 Basic steps in preparing a preform by the OVPO process. (*a*) Bait rod rotates and moves back and forth under the burner to produce a uniform deposition of glass soot particles along the rod; (*b*) profiles can be step or graded index; (*c*) following deposition, the soot preform is sintered into a clear glass preform; (*d*) fiber being drawn from the glass preform. *(Reproduced with permission from Schultz.[4])*

1400°C) to a clear glass preform. This clear preform is subsequently mounted in a fiber-drawing tower and made into a fiber, as shown in Fig. 10-2. The central hole in the tube preform collapses during this drawing process.

10-2-2 Vapor Phase Axial Deposition

The OVPO process described in Sec. 10-2-1 is a lateral deposition method. Another OVPO-type process is the *vapor phase axial deposition* method[15] (VAD), illustrated in Fig. 10-4. In this method the SiO_2 particles are formed in the same way as described in the OVPO process. As these particles emerge from the torches, they are deposited onto the end surface of a silica glass rod which acts as a seed. A porous preform is grown in the axial direction by moving the rod upward. The rod is also continuously rotated to maintain cylindrical symmetry of the particle deposition. As the porous preform moves upward, it is transformed into a solid, transparent rod preform by zone melting (heating in a narrow localized zone) with the carbon ring heater shown in Fig. 10-4. The resultant preform can then be drawn into a fiber by heating it in another furnace, as shown in Fig. 10-2.

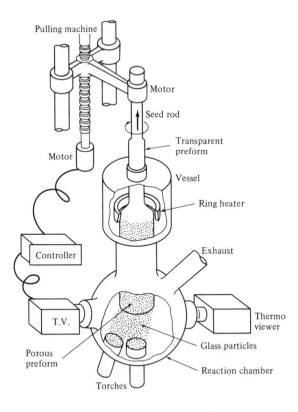

Figure 10-4 Apparatus used for the VAD (vapor phase axial deposition) process. *(Reproduced with permission from Izawa and Inagaki,[15] copyright 1980, IEEE.)*

Both step- and graded-index fibers in either multimode or single-mode varieties can be made by the VAD method. The advantages of the VAD method are: (1) the preform has no central hole as occurs in the OVPO process; (2) the preform can be fabricated in continuous lengths which can affect process costs and product yields; and (3) the fact that the deposition chamber and the zone-melting ring heater are tightly connected to each other in the same enclosure allows the achievement of a clean environment.

10-2-3 Modified Chemical Vapor Deposition

The *modified chemical vapor deposition* (MCVD) process shown in Fig. 10-5 was pioneered at Bell Laboratories[16, 17] and widely adopted elsewhere to produce very-low-loss graded-index fibers. The glass vapor particles arising from the reaction of the constituent metal halide gases and oxygen flow through the inside of a revolving silica tube. As the SiO_2 particles are deposited, they are sintered to a clear glass layer by an oxyhydrogen torch which travels back and forth along the tube. When the desired thickness of glass has been deposited, the vapor flow is shut off and the tube is heated strongly to cause it to collapse into a solid rod preform. The fiber that

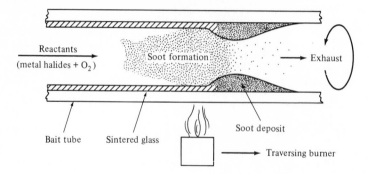

Figure 10-5 Schematic view of MCVD (modified chemical vapor deposition) process. *(Reproduced with permission from Schultz.[4])*

is subsequently drawn from this preform rod will have a core that consists of the vapor-deposited material and a cladding consisting of the original silica tube.

10-2-4 Double-Crucible Method

Multicomponent glasses are generally drawn into fibers by using a *direct-melt* technique.[18, 19] In this method, glass rods for the core and cladding materials are first made separately by melting mixtures of purified powders to make the appropriate glass composition. These rods are then used as feed stock for each of two concentric crucibles, as shown in Fig. 10-6. The inner crucible contains molten core glass and the outer one contains the cladding glass. The fibers are drawn from the molten state through orifices in the bottom of the two concentric crucibles in a continuous production process.

Although this method has the advantage of being a continuous process, careful attention must be paid to avoid contaminants during the melting process. The main

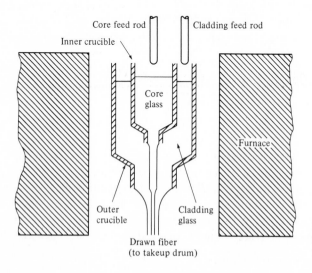

Figure 10-6 Double-crucible arrangement for drawing fibers from the molten state of glass.

sources of contamination arise from the furnace environment and from the crucible. Silica crucibles are usually used in preparing the glass feed rods, whereas the concentric double crucibles used in the drawing furnace are made from platinum. A detailed description of the crucible design and an analysis of the fiber-drawing process can be found in the book by Midwinter.[19]

10-3 MECHANICAL PROPERTIES OF FIBERS

In addition to the transmission properties of optical waveguides, their mechanical characteristics play a very important role when they are used as the transmission medium in optical communication systems.[20-22] Fibers must be able to withstand the stresses and strains that occur during the cabling process and the loads induced during the installation and service of the cable. During cable manufacture and installation the loads applied to the fiber can be either impulsive or gradually varying. Once the cable is in place, the service loads are usually slowly varying ones, which can arise from temperature variations or a general settling of the cable following installation.

Strength and *static fatigue* are the two basic mechanical characteristics of glass optical fibers. Since the sight and sound of shattering glass are quite familiar, one intuitively suspects that glass is not a very strong material. However, the longitudinal breaking stress of pristine glass fibers is comparable to that of metal wires. The cohesive bond strength of the constituent atoms of a glass fiber governs its theoretical intrinsic strength. Maximum tensile strengths of 14 GPa (2×10^6 lb/in^2) have been observed in short-gauge-length glass fibers. This is close to the 20-GPa tensile strength of steel wire. The difference between glass and metal is that, under an applied stress, glass will extend elastically up to its breaking strength, whereas metals can be stretched plastically well beyond their true elastic range. Copper wires, for example, can be elongated plastically by more than 20 percent before they fracture. For glass fibers elongations of only about 1 percent are possible before fracture occurs.

In practice the existence of stress concentrations at surface flaws or microcracks limits the median strength of long glass fibers to the 700- to 3500-MPa (1 to 5×10^5 lb/in^2) range. The fracture strength of a given length of glass fiber is determined by the size and geometry of the severest flaw (the one that produces the largest stress concentration) in the fiber. A hypothetical, physical flaw model is shown in Fig. 10-7. This elliptically shaped crack is generally referred to as a

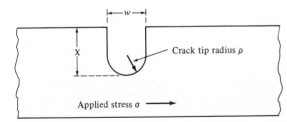

Figure 10-7 A hypothetical model of a microcrack in an optical fiber.

Griffith microcrack.[23] It has a width w, a depth χ, and a tip radius ρ. The strength of the crack for silica fibers follows the relation[23, 24]

$$K = Y\chi^{1/2}\sigma \tag{10-1}$$

where the stress intensity factor K is given in terms of the stress σ in megapascals applied to the fiber, the crack depth χ is given in millimeters, and Y is a dimensionless constant that depends on flaw geometry. For surface flaws, which are the most critical in glass fibers, $Y = \sqrt{\pi}$. From this equation the maximum crack size allowable for a given applied stress level can be calculated. The maximum values of K depend upon the glass composition but tend to be in the range of 0.6 to 0.9 $MN/m^{3/2}$.

Since an optical fiber generally contains many flaws having a random distribution of size, the fracture strength of a fiber must be viewed statistically. If $F(\sigma, L)$ is defined as the cumulative probability that a fiber of length L will fail below a stress level σ, then, under the assumption that the flaws are independent and randomly distributed in the fiber and that the fracture will occur at the most severe flaw, we have

$$F(\sigma, L) = 1 - e^{-LN(\sigma)} \tag{10-2}$$

where $N(\sigma)$ is the cumulative number of flaws per unit length with a strength less than σ. A widely used form for $N(\sigma)$ is the empirical expression proposed by Weibull[25]

$$N(\sigma) = \frac{1}{L_0}\left(\frac{\sigma}{\sigma_0}\right)^m \tag{10-3}$$

where m, σ_0, and L_0 are constants related to the initial inert strength distribution. This leads to the so-called *Weibull expression*

$$F(\sigma, L) = 1 - \exp\left[-\left(\frac{\sigma}{\sigma_0}\right)^m \frac{L}{L_0}\right] \tag{10-4}$$

A plot of a Weibull expression is shown in Fig. 10-8 for measurements performed on long-fiber samples.[26] These data were obtained by testing to destruction a large number of fiber samples. The fact that a single curve can be drawn through the data indicates that the failures arise from a single type of flaw. Earlier works[27] showed a double-curve Weibull distribution with different slopes for short and long fibers. This is indicative of flaws arising from two sources, one from the fiber manufacturing process and the other from fundamental flaws occurring in the glass preform and the fiber. By careful environmental control of the fiber-drawing furnace, numerous 1-km lengths of silica fiber having a single failure distribution and a maximum strength of 3500 MPa have been fabricated.[26–28]

In contrast to strength which deals with instantaneous failure under an applied load, *static fatigue* relates to the slow growth of preexisting flaws in the glass fiber under humid conditions and tensile stress.[29–32] This gradual flaw growth causes the fiber to fail at a lower stress level than that which could be reached under a strength test. A flaw such as the one shown in Fig. 10-7 propagates through the fiber because of chemical erosion of the fiber material at the flaw tip. The primary cause of this

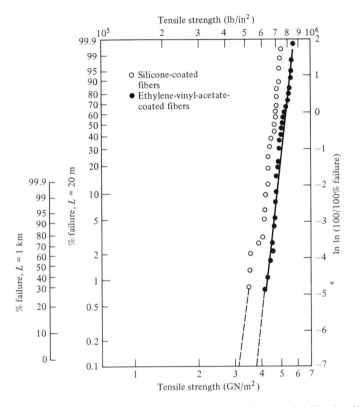

Figure 10-8 A Weibull-type plot showing the cumulative probability that fibers of 20-m and 1-km lengths will fracture at the indicated applied stresses. *(Reproduced with permission from Miller et al.[26])*

erosion is the presence of water in the environment which reduces the strength of the SiO_2 bonds in the glass. The speed of the growth reaction is increased when the fiber is put under stress. However, based on experimental investigations, it is generally believed (but not yet fully substantiated) that static fatigue does not occur if the stress level is less than approximately 0.20 of the inert strength (in a dry environment, such as a vacuum). Certain fiber materials are more resistant to static fatigue than others, with fused silica being the most resistant of the glasses in water. In general, coatings which are applied to the fiber immediately during the manufacturing process afford a good degree of protection against environmental corrosion.[33]

Another important factor to consider is dynamic fatigue. When an optical cable is being installed in a duct, it experiences repeated stress owing to surging effects. The surging is caused by varying degrees of friction between the optical cable and the duct or guiding tool in a manhole on a curved route. Varying stresses also arise in aerial cables that are set into transverse vibration by the wind. Theoretical and experimental investigations[34] have shown that the time to failure under these conditions is related to the maximum allowable stress by the same lifetime parameters that are found from the cases of static stress and stress that increases at a constant rate.

A high assurance of fiber reliability can be provided by proof testing.[21, 35] In this method, an optical fiber is subjected to a tensile load greater than that expected at any time during the cable manufacturing, installation, and service. Any fibers which do not meet the proof test are rejected. Empirical studies[30] of slow crack growth show that the growth rate $d\chi/dt$ is approximately proportional to a power of the stress intensity factor, that is

$$\frac{d\chi}{dt} = AK^b \tag{10-5}$$

Here A and b are material constants and the stress intensity factor is given by Eq. (10-1). For most glasses b ranges between 15 and 50.

If a proof test stress σ_p is applied for a time t_p, then from Eq. (10-5) we have

$$B(\sigma_i^{b-2} - \sigma_p^{b-2}) = \sigma_p^b t_p \tag{10-6}$$

where σ_i is the initial inert strength and

$$B = \frac{2}{b-2} \left(\frac{K}{Y}\right)^{2-b} \frac{1}{AY^b} \tag{10-7}$$

When this fiber is subjected to a static stress σ_s after proof testing, the time to failure t_s is found from Eq. (10-5) to be

$$B(\sigma_p^{b-2} - \sigma_s^{b-2}) = \sigma_s^b t_s \tag{10-8}$$

Combining Eqs. (10-6) and (10-8) yields

$$B(\sigma_i^{b-2} - \sigma_s^{b-2}) = \sigma_p^b t_p + \sigma_s^b t_s \tag{10-9}$$

To find the failure probability F_s of a fiber after a time t_s after proof testing, we first define $N(t, \sigma)$ to be the number of flaws per unit length which will fail in a time t under an applied stress σ. Assuming that $N(\sigma_i) \gg N(\sigma_s)$, then,

$$N(t_s, \sigma_s) \simeq N(\sigma_i) \tag{10-10}$$

Solving Eq. (10-9) for σ_i and substituting into Eq. (10-3), we have, from Eq. (10-10),

$$N(t_s, \sigma_s) = \frac{1}{L_0} \left\{ \frac{[(\sigma_p^b t_p + \sigma_s^b t_s)/B + \sigma_s^{b-2}]^{1/(b-2)}}{\sigma_0} \right\}^m \tag{10-11}$$

The failure number $N(t_p, \sigma_p)$ per unit length during proof testing is found from Eq. (10-11) by setting $\sigma_s = \sigma_p$ and letting $t_s = 0$, so that

$$N(t_p, \sigma_p) = \frac{1}{L_0} \left[\frac{(\sigma_p^b t_p/B + \sigma_p^{b-2})^{1/(b-2)}}{\sigma_0} \right]^m \tag{10-12}$$

Letting $N(t_x, \sigma_x) = N_x$, the failure probability F_s for a fiber after it has been proof-tested is given by

$$F_s = 1 - e^{-L(N_s - N_p)} \tag{10-13}$$

Substituting Eqs. (10-11) and (10-12) into Eq. (10-13), we have

$$F_s = 1 - \exp\left(-N_p L\left\{\left[\left(1 + \frac{\sigma_s^b t_s}{\sigma_p^b t_p}\right)\frac{1}{1 + C}\right]^{m/(b-2)} - 1\right\}\right) \qquad (10\text{-}14)$$

where $C = B/(\sigma_p^2 t_p)$, and where we have ignored the term

$$\left(\frac{\sigma_s}{\sigma_p}\right)^b \frac{B}{\sigma_s^2 t_p} \ll 1 \qquad (10\text{-}15)$$

This holds since typical values of the parameters in this term are[35] $\sigma_s/\sigma_p \simeq 0.3$ to 0.4, $t_p \simeq 10$ s, $b > 15$, $\sigma_p = 350$ MN/m^2, and $B \simeq 0.05$ to 0.5 (MN/m^2)^{2}s.

The expression for F_s given by Eq. (10-14) is valid only when the proof stress is unloaded immediately, which is not the case in actual proof testing of optical fibers. When the proof stress is released within a finite duration, the C value should be rewritten as[35, 36]

$$C = \gamma \frac{B}{\sigma_p^2 t_p} \qquad (10\text{-}16)$$

where γ is a coefficient of slow-crack-growth effect arising during the unloading period.

10-4 FIBER OPTIC CABLES

In any practical application of optical waveguide technology, the fibers need to be incorporated in some type of cable structure.[37-42] The cable structure will vary greatly, depending on whether the cable is to be pulled into underground or intrabuilding ducts, buried directly in the ground, installed on outdoor poles, or submerged under water. Different cable designs are required for each type of application, but certain fundamental cable design principles will apply in every case. The objectives of cable manufacturers have been that the optical fiber cables should be installable with the same equipment, installation techniques, and precautions as those used in conventional wire cables. This requires special cable designs because of the mechanical properties of glass fibers.

One important mechanical property is the maximum allowable axial load on the cable since this factor determines the length of cable that can be reliably installed. In copper cables the wires themselves are generally the principal load-bearing members of the cable, and elongations of more than 20 percent are possible without fracture. On the other hand, extremely strong optical fibers tend to break at 4 percent elongation, whereas typical good-quality fibers exhibit long-length breaking elongations of about 0.5 to 1.0 percent. Since static fatigue occurs very quickly at stress levels above 40 percent of the permissible elongation and very slowly below 20 percent of the breaking limit, fiber elongations during cable manufacture and installation should be limited to 0.1 to 0.2 percent.

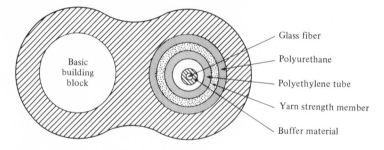

Figure 10-9 A hypothetical two-fiber cable design. The basic building block on the left is identical to that shown for the right-hand fiber.

Steel wire which has a Young's modulus of 2×10^4 MPa has been extensively used for reinforcing conventional electric cables and can also be employed for optical fiber cables. For some applications it is desirable to use nonmetallic constructions, either to avoid the effects of electromagnetic induction or to reduce cable weight. In this case plastic strength members and high-tensile-strength organic yarns such as Kevlar® (a product of the DuPont Chemical Corporation) are used. With good fabrication practices the optical fibers are isolated from other cable components, they are kept close to the neutral axis of the cable, and room is provided for the fibers to move when the cable is flexed or stretched.

Another factor to consider is fiber brittleness. Since glass fibers do not deform plastically, they have a low tolerance for absorbing energy from impact loads. Hence, the outer sheath of an optical cable must be designed to protect the glass fibers inside from impact forces. In addition, the outer sheath should not crush when subjected to side forces, and it should provide protection from corrosive environmental elements. In underground installations, a heavy-gauge-metal outer sleeve may also be required to protect against potential damage from burrowing rodents, such as gophers.

In designing optical fiber cables, several types of fiber arrangements are possible and a large variety of components could be included in the construction. The simplest designs are one- or two-fiber cables intended for indoor use. In a hypothetical two-fiber design shown in Fig. 10-9, a fiber is first coated with a buffer material and placed loosely in a tough, oriented polymer tube, such as polyethylene. For strength purposes this tube is surrounded by strands of aramid yarn which, in turn, is encapsulated in a polyurethane jacket. A final outer jacket of polyurethane, polyethylene, or nylon binds the two encapsulated fiber units together.

Larger cables can be created by stranding several basic fiber building blocks (as shown in Fig. 10-9) around a central strength member. This is illustrated in Fig. 10-10 for a six-fiber cable. The fiber units are bound onto the strength member with paper or plastic binding tape, and then surrounded by an outer jacket. If repeaters are required along the route where the cable is to be installed, it may be advantageous to include wires within the cable structure for powering these repeaters. The wires can also be used for fault isolation or as an engineering order wire for voice communications during cable installation.

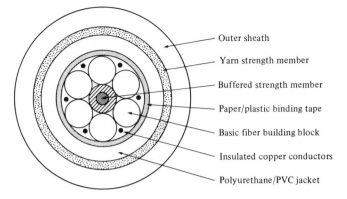

Outer sheath

Yarn strength member

Buffered strength member

Paper/plastic binding tape

Basic fiber building block

Insulated copper conductors

Polyurethane/PVC jacket

Figure 10-10 A typical six-fiber cable created by stranding six basic fiber-building blocks around a central strength member.

10-5 SUMMARY

Glasses and plastics are the principal materials of which fibers are made. Of these, silica-based glass is the most widely used material for the following reasons:

1. It is possible to make long, thin, flexible fibers from this material (and also from plastic).
2. Pure silica glass is highly transparent in the 800- to 1600-nm wavelength range. This is necessary in order for the fiber to guide light efficiently. Plastic fibers are less widely used because of their higher attenuation compared to highly pure glass fibers.
3. By adding trace amounts of certain elements (known as dopants) to the silica, physically compatible materials having slightly different refractive indices for the core and cladding are made available.

All-glass optical fibers are generally made either by a vapor phase oxidation process or by a direct-melt method. The direct-melt method follows traditional glass-making procedures in that optical fibers are made directly from the molten state of purified glasses. Although this method has the advantage of being a continuous process, great care must be taken during the melting process to avoid contaminants (which can give rise to high optical attenuation). In the vapor phase oxidation process, highly pure vapors of metal halides react with oxygen to form a white powder of SiO_2 particles. These particles are collected on the surface of a bulk glass by one of three processes, and are sintered to form a clear glass rod or tube (depending on the process). This rod or tube is called a preform. Once the preform is made, one end is softened to the point where it can be drawn into a very thin filament which becomes the optical fiber.

The mechanical characteristics of optical fibers must also be carefully considered when fibers are used as the transmission medium in optical communication systems. The optical fibers must be able to withstand the stresses and strains that occur during the cabling process and during the installation and service life of the cable. Strength and static fatigue are the two basic mechanical characteristics of glass optical fibers. Strength deals with the instantaneous failure of a fiber under an applied load, and static fatigue relates to the slow growth of preexisting flaws in the glass fiber under humid conditions and tensile stress.

Since an optical fiber generally contains many flaws having a random distribution of sizes, the fracture strength must be viewed statistically. If $F(\sigma, L)$ is the cumulative probability that a fiber of length L will fail below a stress level σ, then a useful and widely used expression for $F(\sigma, L)$ is the empirical Weibull formula

$$F(\sigma, L) = 1 - \exp\left[-\left(\frac{\sigma}{\sigma_0}\right)^m \frac{L}{L_0}\right]$$

where m, σ_0, and L_0 are constants related to the initial, inert strength distribution.

A high assurance of fiber reliability can be provided by proof testing. In this procedure an optical fiber is subjected for a short time to a tensile load greater than that expected at any time during the cable manufacturing, installation, and service. Any fibers that do not meet the proof test are rejected. Given a knowledge of the initial flaw strength and population, fracture mechanics theory can be used to predict the optical fiber cable failure rate (or equivalently, the cable lifetime) in field conditions where a static stress is expected to be imposed on the cable. If the initial failure distribution has a Weibull form, then the stress corrosion failure distribution in a field environment also has a Weibull form.

Special cable designs are required because of the mechanical properties of glass. These designs can vary greatly depending on where and how the cable is to be used. In general, however, the objectives of cable manufacturers have been to make the optical fiber cables in such a way that they should be installable with the same equipment, installation techniques, and precautions as those used in conventional wire cables.

PROBLEMS

Section 10-1

10-1 Calculate the numerical apertures of: (a) a plastic step-index fiber having a core refractive index of $n_1 = 1.60$ and a cladding index $n_2 = 1.49$; (b) a step-index fiber having a silica core ($n_1 = 1.458$) and a silicone resin cladding ($n_2 = 1.405$).

Section 10-2

10-2 When a preform is drawn into a fiber, the principle of conservation of mass must be satisfied under steady-state drawing conditions. Show that for a solid rod preform this is represented by the expression

$$s = S\left(\frac{D}{d}\right)^2$$

where D and d are the preform and fiber diameters, and S and s are the preform feed and fiber-draw speeds, respectively. A typical drawing speed is 1.2 m/s for a 125-μm outer-diameter fiber. What is the preform feed rate in cm/min for a 9-mm-diameter preform?

10-3 A silica tube with inside and outside radii of 3 and 4 mm, respectively, is to have a certain thickness of glass deposited on the inner surface. What should the thickness of this glass deposition be if a fiber having a core diameter of 50 μm and an outer cladding diameter of 125 μm is to be drawn from this preform?

10-4 (*a*) The density of fused silica is 2.6 g/cm^3. How many grams are needed for a 1-km-long 50-μm-diameter fiber core?

(*b*) If the core material is to be deposited inside of a glass tube at a 0.5-g/min deposition rate, how long does it take to make the preform for this fiber?

Section 10-3

10-5 During fabrication of optical fibers, dust particles incorporated into the fiber surface are prime examples of surface flaws which can lead to reduced fiber strength. What size dust particles are tolerable if a glass fiber having a 20-N/mm$^{3/2}$ stress intensity factor is to withstand a 700-MN/m^2 stress?

10-6 Static fatigue in a glass fiber refers to the condition where a fiber is stressed to a level σ_a, which is much less than the fracture stress associated with the weakest flaw. Initially the fiber will not fail but, with time, cracks in the fiber will grow as a result of chemical erosion at the crack tip. One model for the growth rate of a crack of depth χ assumes a relation of the form given in Eq. (10-5).

(*a*) Using this equation, show that the time required for a crack of initial depth χ_i to grow to its failure size χ_f is given by

$$t = \frac{2}{(b-2)A(Y\sigma)^b}(\chi_i^{(2-b)/2} - \chi_f^{(2-b)/2})$$

(*b*) For long, static fatigue times (on the order of 20 yr) $K_i^{2-b} \ll K_f^{2-b}$ for large values of b. Show that under this condition the failure time is

$$t = \frac{2K_i^{2-b}}{(b-2)A\sigma^2Y^2}$$

10-7 Derive Eq. (10-9) by starting with Eq. (10-5).

10-8 Derive Eq. (10-14) by using the expressions given in Eqs. (10-11) and (10-12) for the number of flaws per unit length failing in a time t. Verify the relationship given in Eq. (10-15).

10-9 Consider two similar fiber samples of lengths L_1 and L_2 subjected to stress levels of σ_1 and σ_2, respectively. If σ_{1c} and σ_{2c} are the corresponding fast-fracture stress levels for equal failure probability, show that

$$\frac{\sigma_{1c}}{\sigma_{2c}} = \left(\frac{L_2}{L_1}\right)^{1/m}$$

From Fig. 10-8 estimate the value of m for a 10 percent failure probability of these particular ethylene-vinyl-acetate-coated fibers.

REFERENCES

1. G. W. Morey, *The Properties of Glass,* Reinhold, New York, 1954.
2. C. J. Phillips, *Glass: Its Industrial Applications,* Reinhold, New York, 1960.
3. B. C. Bagley, C. R. Kurkjian, J. W. Mitchell, G. E. Peterson, and A. R. Tynes, "Materials, properties, and choices," in *Optical Fiber Telecommunications,* S. E. Miller and A. G. Chynoweth (Eds.), Academic, New York, 1979.
4. P. C. Schultz, "Progress in optical waveguide process and materials," *Appl. Opt.,* **18**, 3684–3693, Nov. 1979.
5. D. A. Pinnow, A. L. Gentile, A. G. Standler, A. T. Timper, and L. M. Holbrook, "Polycrystalline fiber optical waveguides for infrared transmission," *Appl. Phys. Lett.,* **33**, 28–29, July 1978.

6. L. G. Van Uitert and S. H. Wemple, "ZnCl₂ glass: potential ultralow-loss optical fiber material," *Appl. Phys. Lett.,* **33,** 57–59, July 1978.

7. S. Mitachi, S. Shibata, and T. Manabe, "Teflon FEP-clad fluoride glass fiber," *Electron. Lett.,* **17,** 128–129, Feb. 1981.

8. P. Kaiser, A. C. Hart, Jr., and L. L. Blyler, Jr., "Low-loss FEP-clad silica fibers," *Appl. Opt.,* **14,** 156–162, Jan. 1975.

9. P. C. Schultz, "Vapor phase materials and processes for glass optical waveguides," in *Fiber Optics,* B. Bendow and S. S. Mitra (Eds.), Plenum, New York, 1979.

10. W. G. French, R. E. Jaeger, J. B. MacChesney, S. R. Nagel, K. Nassau, and A. D. Pearson, "Fiber preform preparation," in *Optical Fiber Telecommunications,* S. E. Miller and A. G. Chynoweth (Eds.), Academic, New York, 1979.

11. D. L. Myers and F. P. Partus, "Preform fabrication and fiber drawing by Western Electric Product Engineering Control Center," *Bell Sys. Tech. J.,* **57,** 1735–1744, July–Aug. 1978.

12. R. E. Jaeger, A. D. Pearson, J. C. Williams, and H. M. Presby, "Fiber drawing and control," in *Optical Fiber Telecommunications,* S. E. Miller and A. G. Chynoweth (Eds.), Academic, New York, 1979.

13. F. P. Kapron, D. B. Keck, and R. D. Maurer, "Radiation losses in glass optical waveguides," *Appl. Phys. Lett.,* **17,** 423–425, Nov. 1970.

14. P. C. Schultz, "Fabrication of optical waveguides by the outside vapor deposition process," *Proc. IEEE,* **68,** 1187–1190, Oct. 1980.

15. T. Izawa and N. Inagaki, "Materials and processes for fiber preform fabrication—vapor-phase axial deposition," *Proc. IEEE,* **68,** 1184–1187, Oct. 1980.

16. J. B. MacChesney, P. B. O'Conner, and H. M. Presby, "A new technique for preparation of low-loss and graded-index optical fibers," *Proc. IEEE,* **62,** 1280–1281, Sept. 1974.

17. J. B. MacChesney, "Materials and processes for preform fabrication—modified chemical vapor deposition and plasma chemical vapor deposition," *Proc. IEEE,* **68,** 1181–1184, Oct. 1980.

18. K. J. Beales, C. R. Day, A. G. Dunn, and S. Partington, "Multicomponent glass fibers for optical communications," *Proc. IEEE,* **68,** 1191–1194, Oct. 1980.

19. J. E. Midwinter, *Optical Fibers for Transmission,* Wiley, New York, 1979.

20. B. K. Tariyal and D. Kalish, "Mechanical behavior of optical fibers," in *Fracture Mechanics of Ceramics,* R. C. Bradt, D. P. H. Hasselman, and F. F. Lange (Eds.), vol. 3, Plenum, New York, 1978.

21. D. Kalish, D. L. Key, C. R. Kurkjian, B. K. Tariyal, and T. T. Wang, "Fiber characterization—mechanical," in *Optical Fiber Telecommunications,* S. E. Miller and A. G. Chynoweth (Eds.), Academic, New York, 1979.

22. J. E. Midwinter, *Optical Fibers for Transmission,* Wiley, New York, 1979, chap. 12.

23. A. A. Griffith, "The phenomena of rupture and flow in solids," *Philos. Trans. Roy. Soc. (London),* **221A,** 163–198, Oct. 1920.

24. R. Olshansky and R. D. Maurer, "Tensile strength and fatigue of optical fibers," *J. Appl. Phys.,* **47,** 4497–4499, Oct. 1976.

25. W. Weibull, "A statistical theory of the strength of materials," *Ing. Vetenskaps Akad. Handl. (Proc. Roy. Swed. Inst. Eng. Res.),* no. 151, 1939; "The phenomenon of rupture in solids," *ibid.,* no. 153, 1939.

26. T. J. Miller, A. C. Hart, W. I. Vroom, and M. J. Bowden, "Silicone and ethylene-vinyl-acetate-coated laser-drawn silica fibers with tensile strengths > 3.5 GN/m² (500 kpsi) in > 3 km lengths," *Electron. Lett.,* **14,** 603–605, Aug. 1978.

27. H. Schonhorn, C. R. Kurkjian, R. E. Jaeger, H. N. Vazirani, R. V. Albarino, and F. V. DiMarcello, "Epoxy-acrylate coated fused silica fibers with tensile strengths greater than 500 kpsi (3.5 GN/m²) in 1 km gauge lengths," *Appl. Phys. Lett.,* **29,** 712–714, Dec. 1976.

28. F. V. DiMarcello and A. C. Hart, "Furnace-drawn silica fibers with tensile strengths > 3.5 GN/m² (500 kpsi) in 1 km lengths," *Electron. Lett.,* **14,** 578–579, Aug. 1978.

29. T. T. Wang, H. N. Vazirani, H. Schonhorn, and H. M. Zupko. "Effects of water and moisture on strengths of optical glass (silica) fibers coated with a uv-cured epoxy acrylate," *J. Appl. Polym. Sci.,* **23,** 886–892, Feb. 1979.

30. A. G. Evans and S. M. Wiederhorn, "Proof testing of ceramic materials—an analytical basis for failure predictions," *Int. J. Fracture,* **10,** 379–392, Sept. 1974.

31. J. E. Ritter, Jr., "Probability of fatigue failure in glass fibers," *Fiber Integ. Opt.*, **1**, 387–399, 1978.

32. J. E. Ritter, Jr., J. M. Sullivan, Jr., and K. Jakus, "Application of fracture mechanics theory to fatigue failure of optical glass fibers," *J. Appl. Phys.*, **49**, 4779–4782, Sept. 1978.

33. D. A. Pinnow, G. D. Robertson, Jr., and J. A. Wysocki, "Reductions in static fatigue of silica fibers by hermetic jacketing," *Appl. Phys. Lett.*, **34**, 17–19, Jan. 1979.

34. Y. Katsuyama, Y. Mitsunaga, H. Kobayashi, and Y. Ishida, "Dynamic fatigue of optical fiber under repeated stress," *J. Appl. Phys.*, **53**, 318–321, Jan. 1982.

35. Y. Mitsunaga, Y. Katsuyama, and Y. Ishida, "Reliability assurance for long-length optical fibre based on proof testing," *Electron. Lett.*, **17**, 567–568, Aug. 1981.

36. A. G. Evans, "Proof testing—the effect of slow crack growth," *Mater. Sci. Eng.*, **19**, 69–77, May 1975.

37. J. A. Olszewski, G. H. Foot, and Y. Y. Huang, "Development and installation of an optical fiber cable for communications," *IEEE Trans. Comm.*, **COM-26**, 991–998, July 1978.

38. F. Frahn, G. Olejak, and W. Weidhaas, "The manufacture of optical cables," *Philips Telecommun. Rev.*, **37**, 231–240, Sept. 1979.

39. Y. Katsuyama, S. Mochizuki, K. Ishihara, and T. Miyashita, "Single mode optical fiber cable," *Appl. Opt.*, **18**, 2232–2236, July 1979.

40. Details on cable designs are mostly presented in conference proceedings such as:
    (*a*) *Proc. Int. Wire & Cable Symp.* (USA), annual conf.
    (*b*) *Proc. European Conf. Opt. Commun.*, 1st (1974) to present
    (*c*) *OSA/IEEE Topical Meetings on Opt. Fiber Commun.*, Williamsburg, Va., Jan. 1975 and Feb. 1977; Washington, D.C., March 1979; Phoenix, Ariz., Apr. 1982.

41. G. Galliano and F. Tosco, "Optical fibre cables," in *Optical Fibre Communication*, Technical Staff of CSELT, McGraw-Hill, New York, 1980.

42. Y. Katsuyama, Y. Ishida, C. Tanaka, K. Harada, S. Tanaka, O. Ichikawa, and S. Suzuki, "Study on mechanical and transmission characteristics of optical fiber cable during installation," *J. Opt. Commun.*, **3**, 2–7, Mar. 1982.

A

INTERNATIONAL SYSTEM OF UNITS

Quantity	Unit	Symbol	Dimensions
Length	meter	m	
Mass	kilogram	kg	
Time	second	s	
Temperature	kelvin	K	
Current	ampere	A	
Frequency	hertz	Hz	$1/s$
Force	newton	N	$kg \cdot m/s^2$
Pressure	pascal	Pa	N/m^2
Energy	joule	J	$N \cdot m$
Power	watt	W	J/s
Electric charge	coulomb	C	$A \cdot s$
Potential	volt	V	J/C
Conductance	siemens	S	A/V
Resistance	ohm	Ω	V/A
Capacitance	farad	F	C/V
Magnetic flux	weber	Wb	$V \cdot s$
Magnetic induction	tesla	T	Wb/m^2
Inductance	henry	H	Wb/A

USEFUL MATHEMATICAL RELATIONS

Some of the mathematical relations encountered in this text are listed for convenient reference. More comprehensive listings are available in various handbooks.[1-4]

B-1 TRIGONOMETRIC IDENTITIES

$$e^{\pm j\theta} = \cos\theta \pm j\sin\theta$$
$$\sin^2\theta + \cos^2\theta = 1$$
$$\cos^2\theta - \sin^2\theta = \cos 2\theta$$
$$4\sin^3\theta = 3\sin\theta - \sin 3\theta$$
$$4\cos^3\theta = 3\cos\theta + \cos 3\theta$$
$$8\sin^4\theta = 3 - 4\cos 2\theta + \cos 4\theta$$
$$8\cos^4\theta = 3 + 4\cos 2\theta + \cos 4\theta$$
$$\sin(\alpha \pm \beta) = \sin\alpha\cos\beta \pm \cos\alpha\sin\beta$$
$$\cos(\alpha \pm \beta) = \cos\alpha\cos\beta \mp \sin\alpha\sin\beta$$
$$\tan(\alpha \pm \beta) = \frac{\tan\alpha \pm \tan\beta}{1 \mp \tan\alpha\tan\beta}$$

B-2 VECTOR ANALYSIS

The symbols $\mathbf{e}_x$, $\mathbf{e}_y$, and $\mathbf{e}_z$ denote unit vectors lying parallel to the x, y, and z axes, respectively, of the rectangular coordinate system. Similarly, $\mathbf{e}_r$, $\mathbf{e}_\phi$, and $\mathbf{e}_z$ are unit vectors for cylindrical coordinates. The unit vectors $\mathbf{e}_r$ and $\mathbf{e}_\phi$ vary in direction as the angle ϕ changes. The conversion from cylindrical to rectangular coordinates is made through the relationships

$$x = r \cos \phi \qquad y = r \sin \phi \qquad z = z$$

B-2-1 Rectangular Coordinates

$$\text{Gradient } \nabla f = \frac{\partial f}{\partial x}\mathbf{e}_x + \frac{\partial f}{\partial y}\mathbf{e}_y + \frac{\partial f}{\partial z}\mathbf{e}_z$$

$$\text{Divergence } \nabla \cdot \mathbf{A} = \frac{\partial A_x}{\partial x} + \frac{\partial A_y}{\partial y} + \frac{\partial A_z}{\partial z}$$

$$\text{Curl } \nabla \times \mathbf{A} = \begin{vmatrix} \mathbf{e}_x & \mathbf{e}_y & \mathbf{e}_z \\ \dfrac{\partial}{\partial x} & \dfrac{\partial}{\partial y} & \dfrac{\partial}{\partial z} \\ A_x & A_y & A_z \end{vmatrix}$$

$$\text{Laplacian } \nabla^2 f = \frac{\partial^2 f}{\partial x^2} + \frac{\partial^2 f}{\partial y^2} + \frac{\partial^2 f}{\partial z^2}$$

B-2-2 Cylindrical Coordinates

$$\text{Gradient } \nabla f = \frac{\partial f}{\partial r}\mathbf{e}_r + \frac{1}{r}\frac{\partial f}{\partial \phi}\mathbf{e}_\phi + \frac{\partial f}{\partial z}\mathbf{e}_z$$

$$\text{Divergence } \nabla \cdot \mathbf{A} = \frac{1}{r}\frac{\partial (rA_r)}{\partial r} + \frac{1}{r}\frac{\partial A_\phi}{\partial \phi} + \frac{\partial A_z}{\partial z}$$

$$\text{Curl } \nabla \times \mathbf{A} = \begin{vmatrix} \dfrac{1}{r}\mathbf{e}_r & \mathbf{e}_\phi & \dfrac{1}{r}\mathbf{e}_z \\ \dfrac{\partial}{\partial r} & \dfrac{\partial}{\partial \phi} & \dfrac{\partial}{\partial z} \\ A_r & rA_\phi & A_z \end{vmatrix}$$

$$\text{Laplacian } \nabla^2 f = \frac{1}{r}\frac{\partial}{\partial r}\left(r\frac{\partial f}{\partial r}\right) + \frac{1}{r^2}\frac{\partial^2 f}{\partial \phi^2} + \frac{\partial^2 f}{\partial z^2}$$

B-2-3 Vector Identities

$$\nabla \times (\nabla \times \mathbf{A}) = \nabla (\nabla \cdot \mathbf{A}) - \nabla^2 \mathbf{A}$$

$$\nabla^2 \mathbf{A} = \nabla^2 A_x \mathbf{e}_x + \nabla^2 A_y \mathbf{e}_y + \nabla^2 A_z \mathbf{e}_z$$

B-3 INTEGRALS

$$\int \sin x \, dx = -\cos x$$

$$\int \cos x \, dx = \sin x$$

$$\int \sqrt{a^2 - x^2} \, dx = \frac{1}{2}\left(x\sqrt{a^2 - x^2} + a^2 \arcsin \frac{x}{a}\right)$$

$$\int x\sqrt{a^2 - x^2} \, dx = -\tfrac{1}{3}(a^2 - x^2)^{3/2}$$

$$\int x^2 \sin^2 x \, dx = \frac{x^3}{6} - \left(\frac{x^2}{6} - \frac{1}{8}\right) \sin 2x - \frac{x \cos 2x}{4}$$

$$\int \frac{dx}{\cos^n x} = \frac{1}{n-1}\frac{\sin x}{\cos^{n-1} x} + \frac{n-2}{n-1}\int \frac{dx}{\cos^{n-2} x}$$

$$\int u \, dv = uv - \int v \, du$$

$$\int e^{ax} \, dx = \frac{1}{a} e^{ax}$$

$$\int \sin^2 x \, dx = \frac{x}{2} - \frac{1}{4} \sin 2x$$

$$\int \sin^n x \, dx = -\frac{\sin^{n-1} x \cos x}{n} + \frac{n-1}{n}\int \sin^{n-2} x \, dx$$

$$\int \cos^2 x \, dx = \frac{x}{2} + \frac{1}{4} \sin 2x$$

$$\int \cos^n x \, dx = \frac{1}{n} \cos^{n-1} x \sin x + \frac{n-1}{n} \cos^{n-2} x \, dx$$

$$\int_{-\infty}^{\infty} \frac{e^{jpx}}{(\beta + jx)^n} \, dx = \begin{cases} 0 & \text{if } p < 0 \\ \dfrac{2\pi(p)^{n-1}e^{-\beta p}}{\Gamma(n)} & p \geq 0 \end{cases} \qquad \text{where } \Gamma(n) = (n-1)!$$

$$\int_{-\infty}^{\infty} e^{-p^2 x^2 + qx} \, dx = e^{q^2/4p^2} \frac{\sqrt{\pi}}{p}$$

$$\int_{0}^{\infty} \frac{1}{1 + (x/a)^2} \, dx = \frac{\pi a}{2}$$

$$\frac{2}{\sqrt{\pi}} \int_{0}^{t} e^{-x^2} \, dx = \text{erf}(t)$$

B-4 SERIES EXPANSIONS

$$(1 + x)^n = 1 + nx + \frac{n(n-1)}{2!}x^2 + \frac{n(n-1)(n-2)}{3!}x^3 + \cdots \text{ for } |nx| < 1$$

$$e^x = 1 + x + \frac{x^2}{2!} + \frac{x^3}{3!} + \cdots$$

$$\sin x = x - \frac{x^3}{3!} + \frac{x^5}{5!} - \cdots$$

$$\cos x = 1 - \frac{x^2}{2!} + \frac{x^4}{4!} - \cdots$$

REFERENCES

1. R. S. Burington, *Handbook of Mathematical Tables and Formulas,* 5th Ed., McGraw-Hill, New York, 1973.
2. J. J. Tuma, *Engineering Mathematics Handbook,* 2d ed., McGraw-Hill, New York, 1982.
3. M. Abramowitz and I. A. Stegun, *Handbook of Mathematical Functions,* Dover, New York, 1965.
4. I. S. Gradshteyn and I. M. Ryzhik, *Table of Integrals, Series, and Products,* Academic, New York, 1980.

BESSEL FUNCTIONS

This appendix lists the definitions and some recurrence relations for integer-order Bessel functions of the first kind $J_\nu(z)$ and modified Bessel functions $K_\nu(z)$. Detailed mathematical properties of these and other Bessel functions can be found in Refs. 24 through 26 of Chap. 2. Here the parameter ν is any integer and n is a positive integer or zero. The parameter $z = x + jy$.

C-1 BESSEL FUNCTIONS OF THE FIRST KIND

C-1-1 Various Definitions

A Bessel function of the first kind of order ν and argument z, commonly denoted by $J_\nu(z)$, is defined by

$$J_\nu(z) = \frac{1}{2\pi} \int_{-\pi}^{\pi} e^{jz \sin \theta - jn\theta} \, d\theta$$

or, equivalently,

$$J_\nu(z) = \frac{1}{\pi} \int_0^{\pi} \cos (z \sin \theta - n\theta) \, d\theta$$

Just as the trigonometric functions can be expanded in power series, so can the Bessel function $J_\nu(z)$:

$$J_\nu(z) = \sum_{k=0}^{\infty} \frac{(-1)^k(\tfrac{1}{2}z)^{\nu+2k}}{k!(\nu + k)!}$$

In particular, for $\nu = 0$,

$$J_0(z) = 1 - \frac{\tfrac{1}{4}z^2}{(1!)^2} + \frac{(\tfrac{1}{4}z^2)^2}{(2!)^2} - \frac{(\tfrac{1}{4}z^2)^3}{(3!)^2} + \cdots$$

For $\nu = 1$,

$$J_1(z) = \tfrac{1}{2}z - \frac{(\tfrac{1}{2}z)^3}{2!} + \frac{(\tfrac{1}{2}z)^5}{2!3!} - \cdots$$

and so on for higher values of ν.

C-1-2 Recurrence Relations

$$J_{\nu-1}(z) + J_{\nu+1}(z) = \frac{2\nu}{z}J_\nu(z)$$

$$J_{\nu-1}(z) - J_{\nu+1}(z) = 2J_\nu'(z)$$

$$J_\nu'(z) = J_{\nu-1}(z) - \frac{\nu}{z}J_\nu(z)$$

$$J_\nu'(z) = -J_{\nu+1}(z) + \frac{\nu}{z}J_\nu(z)$$

$$J_0'(z) = -J_1(z)$$

C-2 MODIFIED BESSEL FUNCTIONS

C-2-1 Integral Representations

$$K_0(z) = \frac{-1}{\pi}\int_0^\pi e^{\pm z \cos \theta}[\gamma + \ln (2z \sin^2 \theta)] \, d\theta$$

where Euler's constant $\gamma = 0.57722$.

$$K_\nu(z) = \frac{\pi^{1/2}(\tfrac{1}{2}z)^\nu}{\Gamma(\nu + \tfrac{1}{2})}\int_0^\infty e^{-z \cosh t} \sinh^{2\nu}t \, dt$$

$$K_0(x) = \int_0^\infty \cos (x \sinh t) \, dt = \int_0^\infty \frac{\cos (xt)}{\sqrt{t^2 + 1}} \, dt \qquad (x > 0)$$

$$K_\nu(x) = \sec (\tfrac{1}{2}\nu\pi) \int_0^\infty \cos (x \sinh t) \cosh (\nu t) \, dt \qquad (x > 0)$$

C-2-2 Recurrence Relations

If $L_\nu = e^{j\pi\nu}K_\nu$, then

$$L_{\nu-1}(z) - L_{\nu+1}(z) = \frac{2\nu}{z}L_\nu(z)$$

$$L_\nu'(z) = L_{\nu-1}(z) - \frac{\nu}{z}L_\nu(z)$$

$$L_{\nu-1}(z) + L_{\nu+1}(z) = 2L_\nu'(z)$$

$$L_\nu'(z) = L_{\nu+1}(z) + \frac{\nu}{z}L_\nu(z)$$

C-3 ASYMPTOTIC EXPANSIONS

For fixed ν ($\neq -1, -2, -3, \ldots$) and $z \to 0$,

$$J_\nu(z) \simeq \frac{(\frac{1}{2}z)^\nu}{\Gamma(\nu + 1)}$$

For fixed ν and $|z| \to \infty$,

$$J_\nu(z) \simeq \left(\frac{2}{\pi z}\right)^{1/2} \cos\left(z - \frac{\nu\pi}{2} - \frac{\pi}{4}\right)$$

For fixed ν and large $|z|$,

$$K_\nu(z) \simeq \left(\frac{\pi}{2z}\right)^{1/2} e^{-z}\left[1 - \frac{\mu - 1}{8z} + \frac{(\mu - 1)(\mu - 9)}{2!(8z)^2} + \cdots\right]$$

where $\mu = 4\nu^2$.

C-4 GAMMA FUNCTION

$$\Gamma(z) = \int_0^\infty t^{z-1}e^{-t}\,dt$$

For integer value n,

$$\Gamma(n + 1) = n!$$

For fractional values,

$$\Gamma(\tfrac{1}{2}) = \pi^{1/2} = (-\tfrac{1}{2})! \simeq 1.77245$$

$$\Gamma(\tfrac{3}{2}) = \tfrac{1}{2}\pi^{1/2} = (\tfrac{1}{2})! \simeq 0.88623$$

DECIBELS

D-1 DECIBEL DEFINITION

In designing and implementing an optical fiber link, it is of interest to establish, measure, and/or interrelate the signal levels at the transmitter, at the receiver, at the cable connection and splice points, and in the cable. A convenient method for this is either to reference the signal level to some absolute value or to a noise level. This is normally done in terms of a power ratio measured in *decibels* (dB) defined as

$$\text{Power} = 10 \log \frac{P_2}{P_1} \quad \text{dB} \qquad \text{(D-1)}$$

where P_1 and P_2 are electric or optical powers.

The logarithmic nature of the decibel allows a large ratio to be expressed in a fairly simple manner. Power levels differing by many orders of magnitude can be simply compared when they are in decibel form. Some very helpful figures to remember are given in Table D-1. For example, doubling the power means a 3-dB gain (the power level increases by 3 dB), halving the power means a 3-dB loss (the

Table D-1 Examples of decibel measures of power ratios

Power ratio	10^N	10	2	1	0.5	0.1	10^{-N}
dB	$+10N$	$+10$	$+3$	0	-3	-10	$-10N$

Table D-2 Examples of dBm units (decibel measure of power relative to 1 mW)

Power (mW)	100	10	2	1	0.5	0.1	0.01	0.001
Value (dBm)	+20	+10	+3	0	−3	−10	−20	−30

power level decreases by 3 dB), and power levels differing by factors of 10^N or 10^{-N} have decibel differences of $+10N$ dB and $-10N$ dB, respectively.

D-2 THE dBm

The decibel is used to refer to ratios or relative units. For example, we can say that a certain optical fiber has a 6-dB loss (the power level gets reduced by 75 percent in going through the fiber) or that a particular connector has a 1-dB loss (the power level gets reduced by 20 percent at the connector). However, the decibel gives no indication of the absolute power level. One of the most common derived units for doing this in optical fiber communications is the *dBm*. This is the decibel power level referred to 1 mW. In this case the power in dBm is an absolute value defined by

$$\text{Power level} = 10 \log \frac{P}{1 \text{ mW}} \tag{D-2}$$

An important relationship to remember is that 0 dBm = 1 mW. Other examples are shown in Table D-2.

D-3 THE NEPER

The *neper* (N) is an alternative unit that is sometimes used instead of the decibel. If P_1 and P_2 are two power levels, with $P_2 > P_1$, then the power ratio in nepers is given as the natural (or naperian) logarithm of the power ratio

$$\text{Power} = \frac{1}{2} \ln \frac{P_2}{P_1} \quad \text{N} \tag{D-3}$$

where

$$\ln e = \ln 2.71828 = 1$$

To convert nepers to decibels, multiply the number of nepers by

$$20 \log e = 8.686$$

TOPICS FROM SEMICONDUCTOR PHYSICS

Details of semiconductor physics can be found in the literature.[1-7]

E-1 ENERGY BANDS

Semiconductor materials have conduction properties that lie somewhere between metals and insulators. As an example material, we consider silicon (Si), which is located in the fourth column (group IV) of the periodic table of elements. A Si atom has four electrons in its outer shell by which it makes covalent bonds with its neighboring atoms in a crystal.

The conduction properties can be interpreted with the aid of the *energy band diagrams* shown in Fig. E-1. In a pure crystal the *conduction band* is completely empty of electrons and the *valence band* is completely full. These two bands are separated by an *energy gap* or *forbidden band* in which no energy levels exist. If electrons are excited by some means (for example, optically or thermally) from the valence to the conduction band, a current will flow through the crystal under the influence of an applied electric field. For Si this excitation energy must be greater than 1.1 eV, which is the band gap energy. For each electron excited to the conduction band, there is one electron missing from the valence band. This vacancy is called a *hole*. Both electrons and holes contribute to current flow, as shown in Fig. E-l*a*, that is, an electron in the valence band can move into a vacant hole. This action makes the hole move in the opposite direction to the electron flow.

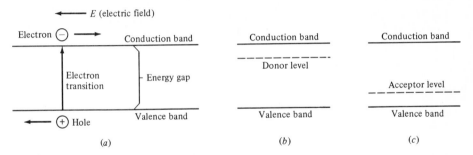

Figure E-1 (*a*) Energy level diagram showing a free electron and a free hole moving under the influence of an external electric field E; (*b*) donor level in an *n*-type material; (*c*) acceptor level in a *p*-type material.

The conduction can be greatly increased by adding traces of impurities from the group V elements (such as P, As, Sb). This process is called *doping*. These elements have five electrons in the outer shell. When they replace a Si atom, four electrons are used for covalent bonding, and the fifth loosely bound electron is available for conduction. As shown in Fig. E-1*b*, this gives rise to an occupied level just below the conduction band called the *donor level*. The impurities are called *donors* because they can give up an electron to the conduction band. Since in this type of material the current is carried by (negative) electrons, it is called *n-type* material.

The conduction can also be increased by adding group III elements, which have three electrons in the outer shell. In this case, three electrons make covalent bonds, and a hole with properties identical to that of the donor electron is created. As shown in Fig. E-1*c*, this gives rise to an unoccupied level just above the valence band. Conduction occurs when electrons are excited from the valence band to this *acceptor level* (so called since the impurity atoms have accepted electrons from the valence band). This material is called *p-type* because the conduction is a result of (positive) hole flow.

E–2 INTRINSIC AND EXTRINSIC MATERIAL

A material containing no impurities is called *intrinsic* material. Because of thermal vibrations of the crystal atoms, some electrons in the valence band gain enough energy to be excited to the conduction band. This *thermal generation* process produces free electron-hole pairs. In the opposite *recombination* process, a free electron releases its energy and drops into a free hole in the valence band. The generation and recombination rates are equal in equilibrium. If n is the electron concentration and p is the hole concentration, then, for an intrinsic material,

$$pn = p_0 n_0 = n_i^2$$

where p_0 and n_0 refer to the equilibrium hole and electron concentrations, respectively, and n_i is the carrier density of the intrinsic material.

The introduction of small quantities of chemical impurities into a crystal produces an *extrinsic* semiconductor. Two types of charge carriers are defined for this material:

1. *Majority carriers* refer either to electrons in *n*-type material or to holes in *p*-type material.
2. *Minority carriers* refer either to holes in *n*-type material or to electrons in *p*-type material.

The operation of semiconductor devices is essentially based on the *injection* and *extraction* of minority carriers.

E-3 THE *pn* JUNCTIONS

Doped *n*- or *p*-type semiconductor material by itself serves only as a conductor. To make devices out of these semiconductors, it is necessary to use both types of materials (in a single, continuous crystal structure). The junction between the two material regions, which is known as the *pn junction,* is responsible for the useful electrical characteristics of a semiconductor device.

When a *pn* junction is created, the majority carriers diffuse across it. This causes electrons to fill holes in the *p* side of the junction and causes holes to appear on the *n* side. As a result an electric field (or *barrier potential*) appears across the junction, as is shown in Fig. E-2. This field prevents further net movements of charges once equilibrium has been established. The junction area now has no mobile carriers, since its electrons and holes are locked into a covalent bond structure. This region is called either the *depletion region* or the *space charge region.*

When an external battery is connected to the *pn* junction with its positive terminal to the *n*-type material and its negative terminal to the *p*-type material, the junction is said to be *reverse-biased.* This is shown in Fig. E-3. As a result of the reverse bias, the width of the depletion region will increase on both the *n* side and the *p* side. This effectively increases the barrier potential and prevents any majority

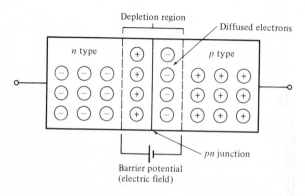

Figure E-2 Electron diffusion across a *pn* junction creates a barrier potential.

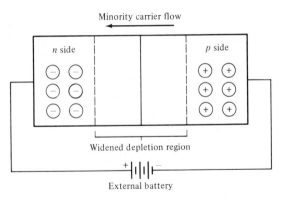

Minority carrier flow

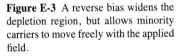

Figure E-3 A reverse bias widens the depletion region, but allows minority carriers to move freely with the applied field.

carriers from flowing across the junction. However, minority carriers can move with the field across the junction. The minority carrier flow is small at normal temperatures and operating voltages, but it can be significant when excess carriers are created as, for example, in an illuminated photodiode.

When the *pn* junction is *forward-biased,* as shown in Fig. E-4, the magnitude of the barrier potential is reduced. Conduction band electrons on the *n* side and valence band holes on the *p* side are, thereby, allowed to diffuse across the junction. Once across, they significantly increase the minority carrier concentrations, and the excess carriers then recombine with the oppositely charged majority carriers. The recombination of excess minority carriers is the mechanism by which optical radiation is generated.

E-4 DIRECT AND INDIRECT BAND GAPS

In order for electron transitions to take place to or from the conduction band with the absorption or emission of a photon, respectively, both energy and momentum must be conserved. Although a photon can have considerable energy, its momentum $h\nu/c$ is very small.

Semiconductors are classified either as *direct-band-gap* or *indirect-band-gap* materials depending on the shape of the band gap as a function of the momentum

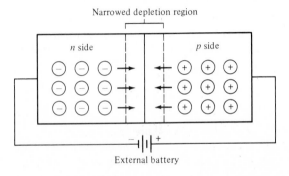

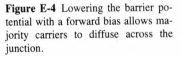

Figure E-4 Lowering the barrier potential with a forward bias allows majority carriers to diffuse across the junction.

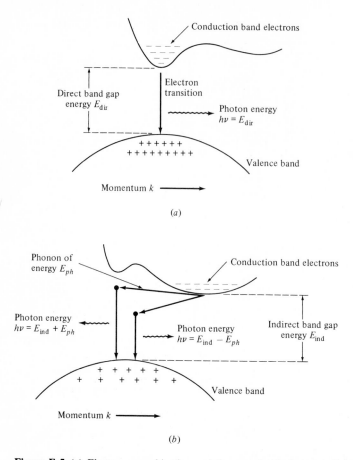

Figure E-5 (*a*) Electron recombination and the associated photon emission for a direct-band-gap material; (*b*) electron recombination for indirect-band-gap materials requires a phonon of energy E_{ph} and momentum k_{ph}.

k, as shown in Fig. E-5. Let us consider recombination of an electron and hole accompanied by the emission of a photon. The simplest and most probable recombination process will be that where the electron and hole have the same momentum value (see Fig. E-5*a*). This is a direct-band-gap material.

For indirect-band-gap materials, the conduction band minimum and the valence band maximum energy levels occur at different values of momentum, as shown in Fig. E-5*b*. Here band-to-band recombination must involve a third particle to conserve momentum, since the photon momentum is very small. *Phonons* (i.e., crystal lattice vibrations) serve this purpose.

E–5 SEMICONDUCTOR DEVICE FABRICATION

In fabricating semiconductor devices, the crystal structure of the various material regions must be carefully taken into account. In any crystal structure, single atoms (e.g., Si or Ge) or groups of atoms (e.g., NaCl or GaAs) are arranged in a repeated pattern in space. This periodic arrangement defines a *lattice,* and the spacing between the atoms or groups of atoms is called the *lattice spacing* or the *lattice constant.* Typical lattice spacings are a few angstroms.

Semiconductor devices are generally fabricated by starting with a crystalline substrate which provides mechanical strength for mounting the device and for making electric contacts. A technique of crystal growth by chemical reaction is then used to grow thin layers of semiconductor materials on the substrate. These materials must have lattice structures that are identical to those of the substrate crystal. In particular, the lattice spacings of adjacent materials should be closely matched to avoid temperature-induced stresses and strains at the material interfaces. This type of growth is called *epitaxial,* which is derived from the Greek words *epi* meaning "on" and *taxis* meaning "arrangement," that is, it is an arrangement of atoms from one material on another material. An important characteristic of epitaxial growth is that it is relatively simple to change the impurity concentration of successive material layers, so that a layered semiconductor device can be fabricated in a continuous process. Epitaxial layers can be formed by growth techniques of either vapor phase,[2] liquid phase,[8–10] or molecular beam.[11]

REFERENCES

1. S. M. Sze, *Physics of Semiconductor Devices,* 2d ed., Wiley, New York, 1981.
2. A. G. Grove, *Physics and Technology of Semiconductor Devices,* Wiley, New York, 1967.
3. J. L. Moll, *Physics of Semiconductors,* McGraw-Hill, New York, 1964.
4. R. A. Smith, *Semiconductors,* 2d ed., Cambridge University Press, London, 1979.
5. I. Tepper, *Solid State Devices,* Addison-Wesley, Reading, Mass., 1972.
6. O. Madelung, *Physics of III-V Compounds,* Wiley, New York, 1964.
7. C. Kittel, *Introduction to Solid State Physics,* Wiley, New York, 1976.
8. H. Kressel and J. K. Butler, *Semiconductor Lasers and Heterojunction LEDs,* Academic, New York, 1977.
9. H. C. Casey, Jr., and M. B. Panish, *Heterostructure Lasers: Part A—Fundamental Principles; Part B—Materials and Operating Characteristics,* Academic, New York, 1978.
10. G. H. B. Thompson, *Physics of Semiconductor Laser Devices,* Wiley, New York, 1980.
11. A. Y. Cho, "Recent developments in molecular beam epitaxy," *J. Vac. Sci. Tech.,* **16,** 275, 1979.

TOPICS FROM COMMUNICATION THEORY

F-1 CORRELATION FUNCTIONS

A spectral density function is often used in communication theory to describe signals in the frequency domain. To define this, we first introduce the *auto-correlation function $R_v(\tau)$*, defined as

$$R_v(\tau) = \frac{1}{T_0} \int_{-T_0/2}^{T_0/2} v(t)v(t - \tau) \, dt \qquad \text{(F-1)}$$

where $v(t)$ is a periodic signal of period T_0. The autocorrelation function measures the dependence with time τ of $v(t)$ and $v(t - \tau)$. An important property of $R_v(\tau)$ is that, if $v(t)$ is an energy or power signal, then

$$R_v(0) = \|v\|^2 = \frac{1}{T_0} \int_{-T_0/2}^{T_0/2} v^2(t) \, dt \qquad \text{(F-2)}$$

where $\|v\|^2$ represents the signal energy or power, respectively.

F-2 SPECTRAL DENSITY

Since $R_v(\tau)$ gives information about the time domain behavior of $v(t)$, the spectral behavior of $v(t)$ in the frequency domain can be found from the Fourier transform of $R_v(\tau)$. We thus define the *spectral density $G_v(f)$* by

$$G_v(f) = F[R_v(\tau)] = \int_{-\infty}^{\infty} R_v(\tau)e^{-j2\pi f\tau} \, d\tau \qquad \text{(F-3)}$$

The fundamental property of $G_v(f)$ is that by integrating over all frequencies we obtain $\|v\|^2$:

$$\int_{-\infty}^{\infty} G_v(f) \, df = R_v(0) = \|v\|^2 \qquad \text{(F-4)}$$

The spectral density thus tells how energy or power is distributed in the frequency domain. When $v(t)$ is a current or voltage waveform feeding a 1-Ω load resistor, then $G_v(f)$ is measured in units of watts/hertz. For loads other than 1 Ω, $G_v(f)$ is expressed either in volts²/hertz or amperes²/hertz. The symbol S is often used in electronics books to denote the spectral density.

Consider a signal $x(t)$ that is passed through a linear system having a transfer function $h(t)$. The output signal is given by

$$y(t) = h(t)x(t)$$

If $x(t)$ is modeled in the frequency domain by a spectral density $G_x(f)$ and the linear system has a transfer function $H(f)$, then the spectral density $G_y(f)$ of the output signal $y(t)$ is

$$G_y(f) = |H(f)|^2 \, G_x(f) \qquad \text{(F-5)}$$

F-3 NOISE EQUIVALENT BANDWIDTH

A *white-noise* signal $n(t)$ is characterized by a spectral density that is flat or constant over all frequencies, that is,

$$G_n(f) = \frac{\eta}{2} = \text{constant} \qquad \text{(F-6)}$$

The factor $\frac{1}{2}$ indicates that half the power is associated with positive frequencies and half with negative frequencies.

If white noise is applied at the input of a linear system having a transfer function $H(f)$, the spectral density $G_0(f)$ of the output noise is

$$G_0(f) = |H(f)|^2 \, G_n(f) = \frac{\eta}{2} |H(f)|^2 \qquad \text{(F-7)}$$

Thus the output noise power N_0 is given by

$$N_0 = \int_{-\infty}^{\infty} G_0(f) \, df = \eta \int_0^{\infty} |H(f)|^2 \, df \qquad \text{(F-8)}$$

If the same noise comes from an ideal low-pass filter of bandwidth B and amplitude $H(0)$, that is, the magnitude of an arbitrary filter transfer function at zero frequency, then

$$N_0 = \eta |H(0)|^2 \, B \qquad \text{(F-9)}$$

By combining Eqs. (F-8) and (F-9), we define a *noise equivalent bandwidth B* as

$$B = \frac{1}{|H(0)|^2} \int_0^\infty |H(f)|^2 \, df \qquad \text{(F-10)}$$

F-4 CONVOLUTION

Convolution is an important mathematical operation used by communication engineers. The convolution of two real-valued functions $p(t)$ and $q(t)$ of the same variable is defined as

$$p(t) * q(t) = \int_{-\infty}^\infty p(x)q(t-x)\, dx$$

$$= \int_{-\infty}^\infty q(x)p(t-x)\, dx$$

$$= q(t) * p(t) \qquad \text{(F-11)}$$

where the symbol $*$ denotes convolution. Note that convolution is commutative. Two important properties of convolutions are

$$F[p(t) * q(t)] = P(f)Q(f) \qquad \text{(F-12)}$$

that is, the convolution of two signals in the time domain corresponds to the multiplication of their Fourier transforms in the frequency domain, and

$$F[p(t)q(t)] = P(f) * Q(f) \qquad \text{(F-13)}$$

that is, the multiplication of two functions in the time domain corresponds to their convolution in the frequency domain.

INDEX